FUNDAMENTALS OF MICROANALYTICAL ENTOMOLOGY

A Practical Guide to Detecting and Identifying Filth in Foods

FUNDAMENTALS OF MICROANALYTICAL ENTOMOLOGY

A Practical Guide to Detecting and Identifying Filth in Foods

Edited by
ALAN R. OLSEN
Director and Chief Entomologist
San Francisco District Laboratory
U.S. Food and Drug Administration
Alameda, California

THOMAS H. SIDEBOTTOM
Associate Editor

SHERRY A. KNIGHT
Editorial Assistant

CRC Press
Boca Raton New York London Tokyo

Library of Congress Cataloging-in-Publication Data

Fundamentals of microanalytical entomology : a practical guide to detecting and identifying filth in foods /
Alan R. Olsen, editor, Thomas H. Sidebottom, associate editor, Sherry A. Knight, editorial assistant.
p. cm.
Includes bibliographical references and index.
ISBN 0-8493-8925-9 (alk. paper)
1. Food adulteration and inspection. 2. Microscopy. 3. Entomology. I. Olsen, Alan R.
TX563.F86 1995
664'.07--dc20
DNLM/DLC
for Library of Congress 95-6737
CIP

Direct all inquiries to CRC Press, Inc., 2000 Corporate Blvd., N.W., Boca Raton, Florida 33431.

International Standard Book Number 0-8493-8925-9
Library of Congress Card Number 95-6737
Printed in the United States of America 1 2 3 4 5 6 7 8 9 0
Printed on acid-free paper

DEDICATION

In memory of my father
Raymond F. Olsen
Who always encouraged my pursuit of Entomology

ACKNOWLEDGMENTS

First and foremost, I acknowledge the invaluable assistance and support of the love of my life and partner in this venture, Angelina Smaserui Olsen, who provided artistic advice, helped prepare manuscripts, uplifted my spirits and supported me every day in every way.

Special thanks to Thomas Cairns of the U.S. Food and Drug Administration (FDA) Division of Field Science, without whose advice and support this book would not be possible. Professor Cairns also reviewed the final manuscript in its entirety.

For all their help, consultations and advice, we thank F.G. Andrews, D.S. Chandler, C.E. Griswold, D.G.H. Halstead, C.D. Johnson, J.M. Kingsolver, S. Larcher, J.F. Lawrence, G.S. Pfaffenberger, S.A. Slipinski, R.L. Smiley, R. Snelling, M. Sorenson, M.C. Thomas and S.L. Wood.

Thanks also to Paul Arnaud, Vincent Lee and the Entomology Department of the California Academy of Sciences; Jon Coddington and the Entomology Department of the U.S. National Museum of Natural History; Vincent Roth of the Southwestern Research Station; and Nancy Axelrod of the Entomology Library, Wellman Hall, University of California–Berkeley.

From the FDA: Steven Angold, Rosita Camarillo, David Chesney, Marshalette Edwards, Kris Foster, Sharon Friedman, Dan Gross, Ronald Johnson, Lee Kyker, Pete Santamaria and Luis Solorzano. Our FDA Science Advisors, Barry Rothman and Richard Bernstein (San Francisco) and Ric Greenwood (Los Angeles) provided valuable advice and reviews of manuscripts.

Finally, two departed colleagues who helped in many ways, John Redd VI of the FDA and Charles Hogue of the Los Angeles County Museum of Natural History, will be in our thoughts and prayers always.

FOREWORD

Publication of *Fundamentals of Microanalytical Entomology* was stimulated by an FDA training course sponsored by the Division of Field Science in the Office of Regulatory Affairs. It was evident during the planning sessions that a wealth of diverse information was being gathered for presentation by an experienced group of distinguished scientists. As part of the planning group, I had the task of playing the role of Editor in combining the various chapters into a cohesive volume for the students. It occurred to me during that process that a textbook on the subject was being compiled. I thus urged the entomologists to consider transforming their various chapters into a textbook to share the practical experience of microanalytical entomology with a much wider audience than the training course provided.

Although entomology as a scientific discipline is well represented by several publications and a number of professional societies, the application to consumer protection is not well documented in the literature. This textbook, however, provides concise reports on the multifaceted components of microanalytical entomology and is augmented with a pertinent bibliography.

The distinguished authors who have contributed to this publication all agreed that a platform existed for a premier textbook with high standards to cover comprehensively all aspects of the topic; hence the selected subtitle—*A Practical Guide to Detecting and Identifying Filth in Foods*.

I firmly believe that my colleagues at the FDA who have worked so diligently to bring this unique treatise to life, have exceeded their expectations. This book stands as a testimony to the efforts and scientific skills of regulatory entomologists to provide a full measure of consumer protection.

Thomas Cairns, Ph.D., D.Sc.
Senior Research Scientist
Office of Regulatory Affairs

PREFACE

The chapters in this volume were compiled from a unique perspective representing the combined collective intelligence and practical experience of scientists intimately involved in microanalytical entomology on a daily basis.

Although many scientists have studied various aspects of stored-product entomology, relatively few have been involved in the forensic aspect of trying to piece together the etiological puzzles of how insect filth gets into processed food products. Some of these talented scientists have combined their knowledge to compile this treatise gained from working in the active arena of food sanitation regulatory laboratories. These perspectives, I believe, are unparalleled in the scientific literature. Such an approach should prove valuable and informative to readers interested in learning how microanalytical entomology works.

The various authors have assumed the reader has a basic knowledge of the principles of entomology, including basic insect taxonomy and morphology. Discussions and identification keys, therefore, have been presented with regard to that level of understanding. Many excellent reference texts are available that explain the basics and numerous citations have been listed in the reference sections of the various chapters. In addition, I stress the absolute necessity of a reference collection of pinned specimens, microscope slide mounts of insect fragments and other authenticated reference specimens.

In addition to describing and discussing the insects and other pests that are forensically significant for food sanitation control investigations, the authors have endeavored to give insights into the operational details of practicing microanalytical entomology. Furthermore, attention has been paid to the legal ramifications and regulatory significance of filth in foods. The chapters on performing acceptable identifications of insect and hair filth, laboratory extractions of filth from product substrates and maintaining a reference collection have been amplified with the details of operating a microanalytical entomology laboratory. The chapters on regulatory procedures and quality assurance offer an insider's view regulating food sanitation.

Every chapter in this book is constructed to emphasize the fundamental purpose of microanalytical entomology: to protect the consumer from food and other products that have become filthy through negligence of good sanitation practices during periods of manufacture or storage. Unsanitary conditions surrounding the production and storage of our food supply are conducive to the spread of disease. Many of the pests discussed in this book are known or suspected carriers of food-borne disease organisms. Regardless of whether a given specimen of a food-contaminating pest is actually infected with a pathogen, the conditions that allowed that potential disease vector access to a food product are an unacceptable risk that must be reduced or eliminated. The science of microanalytical entomology helps us understand the etiology of the contamination in order to accurately assess the gravity of the risk and devise a solution to remove unnecessary risk.

It is the job of the microanalytical entomologist to detect and identify the filth from food-contaminating invertebrate and vertebrate pests. The methods, technical descriptions and identification keys in this book have been deliberately designed to assist in those tasks.

Alan R. Olsen
President
Kerallang Natural History Institute
Palau

THE EDITOR

Alan R. Olsen is the Director and Chief Entomologist of the San Francisco District Laboratory of the U.S. Food and Drug Administration (FDA) in Alameda, California.

Over the past 22 years, he has served as an entomologist in FDA laboratories. His experience includes thousands of laboratory analyses of foods for filth contamination, more than 100 field investigations of food-handling establishments and a dozen scientific research projects. He has lectured at national and international symposia and chaired regional and national workshops in microanalytical entomology.

Chief Entomologist Olsen served as a Peace Corps volunteer from 1966 to 1969 in the Palau Islands in Micronesia, where he taught school and studied island insects and other fauna. In 1971 he received a B.S. in Entomology from the University of Arizona.

He first worked for the FDA in Seattle, Washington, from 1972 to 1978. From 1978 to 1992 he worked in the Los Angeles FDA laboratory, where he supervised the Entomology Section since 1982. In 1992, he assumed directorship of the San Francisco FDA laboratory. He is recipient of the FDA Award of Merit, the agency's highest honor, for his individual scientific achievements in entomology.

Director Olsen is a member of the Pacific Coast Entomological Society, the Entomological Society of Washington, the Society for Vector Ecology, the American Arachnological Society and the Northern California Spider Society. He is recognized in *A Decade of Acarology* (1987). He is the author of chapters in two FDA technical manuals and of 15 articles on a wide variety of topics in entomology including insect fragment identification, animal hair identification, mite taxonomy, field studies and review articles. He is the scientific review authority for the FDA laboratory in Alameda and a recognized expert in microanalytical entomology.

His current research interests are the taxonomy of stored-product mites, ethnoentomology and the natural history of the terrestrial arthropods of the Palau Islands. He is President of the Kerallang Natural History Institute in Palau and a fluent speaker of the Palauan language.

This book is in part a tribute to Paris M. Brickey, Jr. who passed away on July 24, 1995. All of the contributing authors wish to acknowledge his accomplishments and recognize his dedication and contributions to the field of microanalytical entomology. He will be deeply missed, but his legacy will be carried on as witnessed by this publication.

Alan R. Olsen
Michael L. Zimmerman

CONTRIBUTORS

Vernon K. Avaritt III
U.S. Food and Drug Administration
Los Angeles, California

Richard A. Baldwin
U.S. Food and Drug Administration
Rockville, Maryland

Paris M. Brickey, Jr.
U.S. Food and Drug Administration
Washington, D.C.

John R. Bryce
U.S. Food and Drug Administration
Washington, D.C.

J. Richard Gorham
U.S. Food and Drug Administration
Washington, D.C.

Kenneth R. Halcrow
U.S. Food and Drug Administration
Los Angeles, California

Dana Hayakawa Ludwig
U.S. Food and Drug Administration
Alameda, California

James J. Madenjian
U.S. Food and Drug Administration
Los Angeles, California

Alan R. Olsen
Kerallang Natural History Institute
Ngaremlengui, Republic of Belau

Jeremy M. Olsen
Kerallang Natural History Institute
Ngaremlengui, Republic of Belau

Richard W. Potter
U.S. Food and Drug Administration
Los Angeles, California

Nicholas J. Richter
U.S. Food and Drug Administration
Los Angeles, California

Warren E. Savary
U.S. Food and Drug Administration
Alameda, California

Howard G. Semey
U.S. Food and Drug Administration
Alameda, California

Thomas H. Sidebottom
U.S. Food and Drug Administration
Alameda, California

Darrell Ubick
California Academy of Sciences
Golden Gate Park
San Francisco, California

Edwin C. Washbon
U.S. Food and Drug Administration
Los Angeles, California

Michael L. Zimmerman
U.S. Food and Drug Administration
Baltimore, Maryland

TABLE OF CONTENTS

Chapter 1
Introduction ... 1
Alan R. Olsen

Chapter 2
Quality Assurance for Regulatory Science ... 11
Richard A. Baldwin

Chapter 3
Identification of Filth ... 21
Alan R. Olsen

Chapter 4
Ametabolous and Hemimetabolous Insects ... 29
Thomas H. Sidebottom

Chapter 5
Beetles (Insecta: Coleoptera) ... 39
James J. Madenjian

Chapter 6
Flies (Insecta: Diptera) ... 71
Alan R. Olsen

Chapter 7
Moths (Insecta: Lepidoptera) ... 93
Alan R. Olsen

Chapter 8
Ants and Parasitoids (Insecta: Hymenoptera) ... 105
Vernon K. Avaritt III and Nicholas J. Richter

Chapter 9
Mites (Arachnida: Acarina) ... 123
Alan R. Olsen and Richard W. Potter

Chapter 10
Spiders (Arachnida: Araneae) ... 135
Alan R. Olsen, Warren E. Savary and Darrell Ubick

Chapter 11
Other Arthropods ... 151
Alan R. Olsen

Chapter 12
Hairs and Feathers ... 157
Dana Hayakawa Ludwig and John R. Bryce

Chapter 13
Ethnoentomology: Insects for Human Consumption 217
Alan R. Olsen and Jeremy M. Olsen

Chapter 14
Reference Collections and Materials 227
Howard G. Semey

Chapter 15
Laboratory Procedures 241
Edwin C. Washbon and Kenneth A. Halcrow

Chapter 16
Forensic Evidence Development 261
Michael L. Zimmerman and Paris M. Brickey, Jr.

Chapter 17
Reflections on Food-Borne Filth in Relation to Human Disease 269
J. Richard Gorham

Subject Index 277

Taxonomic Index 289

Chapter 1

INTRODUCTION

Alan R. Olsen

CONTENTS

I. Scope of Microanalytical Entomology 1

II. Definitions 2

A. Adulteration 2
B. Filth 2
C. Microanalytical Entomology 3
D. Stored-Product 3

III. Historical Background 3

IV. Insects in Processed Foods 4
A. Categories of Food-Contaminating Insects 5
1. Stored-Product 5
2. Habitat Categories 5
3. Trophic Level Categories 5
4. Feeding Damage Categories 5
5. Public Health Categories 6
6. Microanalytical Entomology Categories 6
B. Risk Categories of Filth in Foods 6
1. Filth from Vectors of Food-Borne Disease 7
a. Vertebrate Vector Filth 7
b. Invertebrate Vector Filth 7
2. Filth That Poses a Nonpathogenic Potential Hazard 7
3. Live Infestations and Other Outrageous Adulterants 7
4. Filth from Stored-Product Insects 7
5. Incidental Filth 8
6. Risk Assessment 8

References 9

I. SCOPE OF MICROANALYTICAL ENTOMOLOGY

Microanalytical entomology is the scientific discipline concerned with the detection, identification and interpretation of insect filth in foods. Under the federal statutes and regulations that are administered by the U.S. Food and Drug Administration (FDA), insects are recognized as adulterants, or contaminants, that do not belong in the food we eat. Similar laws exist in other countries, and in states and local municipalities. Adulteration by insects is often the result of insanitary conditions and practices at factories or warehouses where the food products are handled prior to entering the marketplace. Applied microanalytical entomology is used to detect insect adulteration, to determine the etiology of the adulteration and to prevent adulterated products from being distributed to consumers.

Microanalytical entomology is a branch of forensic entomology. There are three generally recognized categories of forensic entomology: urban forensic entomology, stored-product forensic entomology and medicolegal forensic entomology.[1] Stored-product forensic entomology deals with the infestation or contamination of commercial products, including foods, by insects or other pests.[2] Microanalytical entomology is the subdiscipline of stored-product forensic entomology that is concerned with the adulteration of commercial food, drug, cosmetic and medical device products. The term "adulteration" has a specific legal meaning associated with equally specific illegal acts (see chapter 2). As a forensic scientist, the microanalytical entomologist helps determine whether the "crime" of adulteration has been committed and, if so, develops entomological evidence that helps establish the guilt or innocence of suspects. What sets microanalytical entomology apart from other disciplines of forensic entomology is its emphases on (1) the identification of insect fragments, (2) the interpretation of the sanitary significance of insect filth in food and (3) the detection of postharvest adulteration that may occur during the manufacture, packing or warehousing of food products.

From the governmental viewpoint, the applications of microanalytical entomology are definitely regulatory and forensic because the objectives are to establish compliance or noncompliance of the food product with the law and to establish guilt or innocence of a particular individual or corporation that is suspected or accused of causing the adulteration of a food product. Outside the regulatory arena, microanalytical entomology is applied to prevent the adulteration of foods by insects. Public service organizations, consultants and food manufacturers can use microanalytical entomology to achieve a level of quality control in food-handling venues that will ensure a clean and wholesome food product and, it is hoped, to avoid a day in court as a result of insanitation.

II. DEFINITIONS

The technical and scientific terms in this book conform to standard usage in the field of general entomology. For a full understanding of entomological terminology, refer to one of the general textbooks on entomology.[3,4] For more detailed information refer to Torre-Bueno's general glossary[5] or Tuxen's glossary for genitalia.[6] A few terms are explained below because of their unique importance to the practice of microanalytical entomology.

A. ADULTERATION

Within the context of the U.S. Food, Drug and Cosmetic Act, "adulteration" has two special meanings for microanalytical entomology:

1. Contamination that renders an item filthy.
2. Insanitary conditions whereby an item may have been rendered filthy.

This dual application of the term "adulteration" broadly covers not only filth and any other potentially harmful or deleterious substance but also the insanitary conditions that allow pest infestations and contamination to occur.[7] It is important to note that insanitary conditions are considered to be adulteration regardless of whether actual contamination of a product can be demonstrated. The legalistic details of the definition of this term are further discussed in chapter 2.

B. FILTH

In this book, as in food and drug law, the terms "filth" and "filthy" are construed in their usual and ordinary meanings without being confined to any special scientific or medical definition. Insect filth, for example, is any extraneous material from insects that an ordinary person would find offensive, objectionable or disgusting if they knew it was in their food, drugs or cosmetics. This includes not only the insects themselves but also their body parts, internal organs or fluids, excrement, webbing, cast skins, frass, detritus and other extraneous materials associated with insects. Filthy conditions in a food manufacturing plant, as another example, are unclean conditions involving insects or any other insanitary practices or situations.

C. MICROANALYTICAL ENTOMOLOGY

The scientific discipline of microanalytical entomology is the application of entomological knowledge to issues involved in the civil and criminal law concerned with the cleanliness and safety of foods, drugs and cosmetics. In practice, it is the forensic scientific examination of these products for contamination with filth from insects or other pests and the detection of filthy conditions involving these pests in factories, warehouses and other places where foods, drugs or cosmetics are manufactured, processed, packed or stored.

The term "microanalysis" connotes the analysis of minute amounts of materials, usually requiring the use of microscopes to observe the analytical results. In an entomological context, the term refers to the small sizes of the insects involved and to the minute sizes of the insect fragments that are isolated and identified from highly processed foods. Because each insect or fragment is individually studied under the microscope by the microanalytical entomologist, it is the minute size of each item, or filth element, that is designated by the term rather than the quantities of insects or insect fragments.

D. STORED-PRODUCT

Throughout this book, the term "stored-product" is used in a restricted sense to mean stored food, drugs and cosmetics. Products excluded by this narrow definition (e.g., textiles, wool, hides, lumber and wood products, furniture, etc.) are beyond the scope of this book.

III. HISTORICAL BACKGROUND

Because the FDA is primarily responsible for enforcing the statutes that generate the need for microanalytical entomology applications, the history, development and scope of this scientific discipline are inextricably entwined with FDA activities. This brief historical review provides a reference to assist all parties, inside the FDA and outside, in understanding both the science and the organization that produced it.

In the United States, the federal Food and Drug Act of 1906 marks the beginning of the development of microanalytical entomology. During this era, food adulterants were studied by the same microscopists who performed the microchemical testing of drugs and other analyses that required the use of a microscope. These other analyses often included examinations for mold and rot, for falsification of ingredients, and for adulteration involving insects and other pests. Microanalytical entomology grew out of the need for some microscopists to specialize in examinations for insect adulteration. Very little information reached the entomological community from these scientists because they seem to have communicated through internal correspondence between their offices and through occasional notices to the trades whose products were being tested. It was during this period that the framework for microanalytical entomology evolved to include the basic elements of identification of insect fragments using microscopic characteristics, interpretation of the sanitation significance of different kinds of insect filth and evaluation of postharvest conditions in processing and storage facilities.

In 1938, the U.S. Food, Drug and Cosmetic Act replaced earlier federal statutes in the United States. This landmark legislation defined two types of adulteration that relate to microanalytical entomology. They are Sections 402(a)(3) and 402(a)(4) of the Food, Drug and Cosmetic Act [21 U.S.C. 342(a) and (a)(4)]. These provisions, discussed more fully later, require food products to be clean and unadulterated by filth and require the places and people that produce food products to be clean and sanitary. This legislation increased the emphasis on detection and prevention of insect-related insanitation by encouraging inspections of food processing, packing and storage establishments for insect infestations and by increasing the need for more sophisticated levels of entomological identifications. Microanalytical entomology emerged as a distinct scientific discipline shortly after the enactment of this legislation.

By 1944, the FDA had issued three manuals that mark the full emergence of microanalytical entomology from the confines of internal procedures of governmental and trade organizations into the scientific community. The first manual was circulated within the agency and was entitled *Manual of Microanalytical Methods*. It was a compendium of analytical techniques for the detection and quantification of filth from food products. Methods for isolating insect filth were contained in this manual along with methods for detecting mold, rot and other extraneous materials. The methods for insect filth also detect noninsect contaminants such as hairs, excreta, feathers, textile fibers and mites. Collectively, insect filth and these other contaminants that are concurrently recovered by the same analytical technique became known as "light filth." The manual experienced several changes over the years, as methodologies improved, and a major revision was done in 1959.[(8)] It was later reissued as the *Macroanalytical Procedures Manual*.

The second FDA manual was the *Manual of Microanalytical Photomicrographs*, also circulated internally. It was a compendium of photomicrograph examples of light filth, including insect fragments, to be used to identify and interpret the significance of the microscopic adulterants that may be found through analysis by the methods in the first manual described above. FDA laboratories also received a set of authentic reference specimens for use with the book. This was the first attempt to organize the identification of insect fragments and other microscopic filth. Finally, in 1944, a publication entitled *Microanalysis of Food and Drug Products*[(9)] was distributed to the general scientific community. It contained a summary of the knowledge concerning microanalysis and included a lengthy chapter on insect filth. Both of these manuals are long out of print. Complete reviews of FDA internal documents and scientific literature are available to provide background on the period from 1906 through the publication of the 1944 manuals.[(10,11)]

At the same time, the Association of Official Agricultural Chemists, later to become the Association of Official Analytical Chemists (AOAC), prepared a chapter on microanalytical methods for light filth to be included in their compendium of methods.[(12)] The emphasis here was on chemistry and statistics. Insects and insect fragment identification received only generalized treatment. The association had been concerned with insect adulterants since 1907, as indicated by their annual reports over the years, but this new chapter in the methods compendium gave increased emphasis that coincided with the emergence of microanalytical entomology as a recognized discipline.

Microanalytical entomology grew rapidly after 1944. The AOAC methods chapter expanded with each successive edition of the methods compendium to the present.[(13)] Beginning in 1950, reports of research on the identification of insect fragments began appearing in the literature.[(14,15)] Textbooks and manuals appeared that emphasized insect filth or were totally devoted to the topic of microanalytical entomology.[(16,17)] More recently, new applications have appeared for scanning electron microscopy, and fragment identification has reached species-level precision.[(18-20)]

IV. INSECTS IN PROCESSED FOODS

Since the time of the pharaohs, human beings have stockpiled foodstuffs in times of plenty for use during leaner periods or for sale at a later date.[(21)] Equally ancient are the stored-product insects that infested the pharaohs' food storage facilities. Some of the same pests that infest foods in the late 20th century have been found in the tombs of the pharaohs.[22] Although the problem is age-old, our scientific knowledge of stored-product insects has increased more in the latter half of this century than at any other time. As a result, much thought has recently been devoted to understanding the origins and interrelationships of the insects that infest and contaminate our reserves of stored food. The microanalytical entomologist must be aware of the competing approaches for organizing stored-product insects into groups or categories that reflect various modern schools of thought on the subject.

A. CATEGORIES OF FOOD-CONTAMINATING INSECTS

There are many species of insects that are attracted to stored food products and many methods of arranging them into manageable groups. Estimates of the number of species of stored-product insects range up to 600 or more.[23] Methods of grouping this array of species vary depending on the scientific interests of the researcher. The following are among the most popular approaches.

1. Stored-Product

This is a broad category of insects that share biological characteristics that are adaptive for survival under conditions of neglect in food handling and storage areas. These characteristics are (1) a wide range of tolerance for physical factors of the environment; (2) a greater range of food habits than most other insects; (3) more or less continuous reproductive activity spanning periods of up to three years; (4) an ability to survive long periods without food; and (5) a capacity to build up large populations undetected, due to their relatively small size.[24] These biological characteristics are perfectly adapted to the typical cycles of insanitation. Stored-product insect populations build up in static material that accumulates in uncleaned machinery or work areas, having arrived in raw materials that were not inspected. After a considerable period of neglect, the populations grow to the point of overflowing into production lines and storage areas. At this point, the insects may become noticeable, provoking remedial sanitary measures, which, if not maintained, gradually decline and allow the cycle to begin anew. The sort of neglect of sanitation that perpetuates the natural cycle of stored-product insects is indicative of a general, long-term neglect of basic cleanliness that is a genuine, if indirect, public health hazard.

2. Habitat Categories

Some ecologists group stored-product insects according to their apparent or presumed habitats in nature. The "original habitat" categories for grouping stored-product insects are (1) seed-infesting species; (2) fungus-feeding species; (3) scavengers on plant detritus; (4) scavengers on dead animals; (5) species that live under bark, (6) wood borers and wood scavengers; and (7) inquilines in nests of other animals.[24,25] This approach helps us understand the evolution and bionomics of stored-product pests.

3. Trophic Level Categories

Another approach is to group stored-product insects by their position in the food chain of a storage environment. This system classifies pest species as primary, secondary or tertiary consumers according to the respective trophic levels that they occupy in, say, a grain elevator.[26] The concept of a stored-product food-chain pyramid helps us understand and prevent economic losses due to the feeding activities of stored-product insects.

4. Feeding Damage Categories

A related approach is to consider the kinds of economic damage caused by stored-product insects, including: (1) direct weight loss from feeding; (2) quality loss from contamination of various kinds; and (3) public health risks.[27] This damage-assessment approach helps determine economic thresholds and provides a basis for various integrated pest management programs.

5. Public Health Categories

Public health entomologists interpret stored-product insects as they do other insects, in terms of their effects on human health. Among these are (1) direct effects of biting, stinging and infesting human flesh; (2) mechanical transmission of disease; and (3) biological transmission of disease.[28] This interpretation draws into the picture groups of insects other than the traditional stored-product pests. For example, filth-inhabiting insects, parasitic insects and disease vectors. This interpretation generally excludes traditional stored-product pests except in cases of canthariasis, myiasis, parasite life cycles and other direct clinical manifestations.[29,30] The public health significance of insects as direct food-borne pathogenic organisms and as vectors of food-borne pathogens is recognized by both medical and public health professionals.

6. Microanalytical Entomology Categories

For microanalytical entomology, the above hierarchies do not provide a fully satisfactory means for interpreting the types of adulteration that are found in processed foods. This is because the forensic objectives of microanalytical entomology—to know what insect is involved as an adulterant and then to apply the biology of the insect to effective law enforcement—are different from the objectives of the scientists who developed the above hierarchies of categorization.[17] Certainly elements of ecology, economic entomology, stored-product entomology, public health entomology and medical entomology are applicable to microanalytical entomology, but a key conceptual factor that is missing from the categorical hierarchies of these other fields is temporal sequencing. For microanalytical entomology, it is necessary to interpret adulterants in terms of when the contamination is likely to have occurred relative to the technology and production schedules of a food handling process. In this respect, microanalytical entomology is more closely aligned with forensic entomology, where insect biology is used to establish event sequences relative to a specific criminal act.[31] In food sanitation, it is important to know, for instance, whether contamination occurred before or after harvest, milling, biocidal processing steps, packing or distribution in commerce. Addition of temporal sequencing to the appropriate concerns of the other entomological hierarchies discussed above results in the establishment of a hierarchic grouping of insect pests that is uniquely suited for microanalytical entomology.

The categorization of insect contaminants that is used by microanalytical entomologists is (1) contaminants from insects in fields or agricultural areas where food crops are grown and harvested; (2) contaminants from insects whose activity may include both field and storage habitats; (3) contaminants from insects that infest foods in storage facilities, processing plants and warehouses; and (4) insect contaminants that have other specific etiological interpretations.[32] This sequencing in time, from harvest to consumer's table, recognizes the interests of microanalytical entomology in differentiating when contamination occurs and developing evidence that will pinpoint the insanitary conditions and responsible parties.

B. RISK CATEGORIES OF FILTH IN FOODS

Although the adulteration of a product is illegal *per se*, the outcome of a legal or regulatory procedure involving an adulterated product may proceed along different lines depending on many factors, including among other things, degree of negligence, premeditation, motive, recidivism and relative risk to the consumer of a product. This latter factor (risk) is a special concern of microanalytical entomologists because of the public health protection underpinnings of most food sanitation laws. The dual intent of these laws is to prevent injury as well as to establish culpability when injury has occurred. Assessment of the relative risks associated with different types of filth adulterants is required in either case but more so in the prevention of injury.

Filth adulterants can be partitioned into five broad categories that pose different types of potential risk to the consumer of a product. These are

1. Filth from vectors of food-borne disease.
2. Filth that poses a nonpathogenic potential hazard.
3. Live infestations and other outrageous adulterants.
4. Filth from stored-product insects not included in the above categories.
5. Incidental filth.

1. Filth from Vectors of Food-Borne Disease

The filth in this category is subdivided into two groups: filth from vertebrates and from invertebrates.

a. Vertebrate Vector Filth

The filth from vertebrate vectors includes hairs, feathers, excrement and other filth from potential carriers of food-borne diseases. The commensal murine rodents are the most frequent contributors of this type of filth. Birds that contaminate food with feather filth and excrement include pigeons, sparrows and other commensal species. Filth from bats and the Asian musk shrew is found in a number of products. Other types of hairs found in samples include pet hairs, human hairs, other miscellaneous mammalian hairs and the occasional hair from apparel (wool or fur-lined clothing).

b. Invertebrate Vector Filth

Invertebrate vectors of food-borne diseases include the filth-inhabiting flies, roaches and ants. Others in this category are beetle families whose filthy habits enhance their potential for mechanical transmission of pathogens (e.g., Cleridae, Histeridae and Staphylinidae). The stored-product mite, *Tyrophagus putrescentiae* (Schrank) is included on the basis of labortory observations of mechanical pathogen transport.

2. Filth That Poses a Nonpathogenic Potential Hazard

This category includes the dermestid beetle larvae (family Dermestidae), whose spicasetae and hastisetae are an ingestant health hazard, and the allergenic house dust mites (family Pyroglyphidae). Also included are any large, sharp or pointed fragments of arthropods, which may pose a physical hazard if ingested.

3. Live Infestations and Other Outrageous Adulterants

Certain types of contaminants carry a high "outrage" factor due to the repugnance of the type of filth. Live infestations, ectoparasites and arthropods that are repugnant in appearance to the average consumer (earwigs, spiders, etc.) are in this category. Quite often the very knowledge that significant amounts of insect fragments, feces or internal organs are present, albeit unseen, in a food product is sufficiently repugnant to the average consumer to warrant corrective measures against foods that were manufactured with infested raw materials or ingredients.

4. Filth from Stored-Product Insects

This category includes the stored-product pests not included in the above categories. The occurrence of these pests is an indication of insanitary conditions that may be conducive to the spread of food-borne disease, but the known direct associations of specific pathogens with members of this category are a relatively low-risk proposition in most cases.

TABLE 1
Filth Adulterants Detected in 2906 Food Samples Examined in Los Angeles (4/87–4/91)

Category of Filth	No. of Incidents[a]
Vectors of disease	
Vertebrate	
rodent hair	1376
feather filth	1044
Asian musk shrew hair	106
bat hair	93
other hair	916
rodent excrement	45
bird excrement	6
dung	4
bat excrement	1
Invertebrate	
ant	848
filth fly	301
Tyrophagus mite	258
cockroach	78
rove beetle	46
clerid beetle	27
Nonpathogen hazards	
house dust mite	44
dermestid beetle larva	25
Outrageous filth	
booklouse	306
live infestation	132
ectoparasite	19
earwig, etc.	12
Stored-product pests (not included above)	
stored-product insect	1381
stored-product mite	618

[a] Many incidents involved multiple filth contaminants, which are listed separately under each category, causing the total in this column to exceed the total number of examinations performed.

5. Incidental Filth

The remaining types of filth are those types that pose little risk to the consumer but are contaminants nonetheless because they do not belong in the product. In this category, levels of filth that are nonhazardous and unavoidable under the prevailing conditions of the industry are usually not considered actionable by agencies such as the FDA, although this is not always the case if other factors indicate that their presence in a product is actually a result of insanitation or some other prohibited act.

6. Risk Assessment

Risk assessment of filth in foods is a complicated process that is discussed more fully in chapter 2, chapter 18, and in the literature.[33] An overview of the impact of filth from different risk categories can be obtained from the results of sample analyses performed by the FDA Entomology Section in Los Angeles. Over a four-year period, 2906 samples of food products were found to contain avoidable types of filth adulteration. An approximately equal number of samples examined during the same period contained no filth or insignificant levels of unavoidable incidental filth. Table 1 lists the results of these examinations of food products, organized into risk categories. Within each risk category, the types of contaminants are listed in order of prevalence. Although Table 1 contains the relative distribution of types of filth from only a small, selective sample base, it illustrates the importance of evaluating the degree of potential risk that can be imposed on unsuspecting consumers from products that are adulterated with filth.

REFERENCES

1. Lord, W.D. and Stevenson, J.R., *Directory of Forensic Entomologists*, 2nd ed., Am. Registry Prof. Entomol., Washington, DC, 1-42, 1986.
2. Catts, E.P. and Goff, M.L., Forensic entomology in criminal investigations, *Annu. Rev. Entomol.*, 37, 253-272, 1992.
3. Borror, D.J., Triplehorn, C.A., and Johnson, N.F., *An Introduction to the Study of Insects*, 6th ed., Saunders Coll. Publ., Philadelphia, 1-875, 1989.
4. Snodgrass, R.E., *Principles of Insect Morphology*, McGraw-Hill, NY, 1-667, 1935.
5. Schuh, R.T., Ed., *The Torre-Bueno Glossary of Entomology*, revised, New York Entomol. Soc., NY, 1-840, 1989.
6. Tuxen, S.L., Ed., *Taxonomist's Glossary of Genitalia in Insects*, 2nd ed., Munksgaard, Copenhagen, 1-359, 1970.
7. Food and Drug Administration (FDA), *Requirements of Laws and Regulations Enforced by the U.S. Food and Drug Administration*, U.S. Dept. Health, Education and Welfare Publ. No. (FDA)79-1042, 1-72, 1979.
8. FDA, *Microanalytical Methods* (reissued as *Microanalytical Procedures Manual*), U.S. Dept. Health, Education and Welfare, Washington, DC, 1-180, 1959.
9. Welch, H., Ed., *Microanalysis of Food and Drug Products*, Food & Drug Circ. No. 1, 1-171, 1944.
10. Harris, K.L., Foreign matter in corn meal, *Cereal Chem.*, 18, 655–661, 1941.
11. Harris, K.L., Annotated bibliography of methods for the examination of foods for filth, *J. Assoc. Off. Agric. Chem.*, 29, 420-439, 1946.
12. Lepper, H.A., Ed., *Official and Tentative Methods of Analysis of the Association of Official Agricultural Chemists*, 6th ed., Assoc. Off. Agric. Chem., Washington, DC, 1-932, 1945.
13. Boese, J.L. and Bandler, R., Extraneous Materials: Isolation, in *Official Methods of Analysis of the Association of Official Analytical Chemists*, 15th ed., Vol. 1, Helrich, K., Ed., Assoc. Off. Anal. Chem., Arlington, VA, 369-406, 1990.
14. Harris, K.L., Identification of stored product insects by the micromorphology of the exoskeleton, *J. Assoc. Off. Agric. Chem.*, 33, 878-933, 1950.
15. Kurtz, O.L., An approach to insect larval identification by structures isolated as fragments from cereal products, *J. Assoc. Off. Agric. Chem.*, 37, 167-169, 1954.
16. Harris, K.L. and Reynolds, H.L., Eds. *Microscopic-Analytical Methods in Food and Drug Control*, FDA Tech. Bull. No. 1, 1-270, 1960.
17. Peace, D.M., *Key for Identification of Mandibles of Stored-Food Insects*, Assoc. Off. Anal. Chem., Arlington, VA, 1-169, 1985.
18. Brickey, P.M., Jr. and Gecan, J.S., Introductory study of scanning electron microscopy of elytral fragments of stored product beetles, *J. Assoc. Off. Anal. Chem.*, 57, 1235-1247, 1974.
19. Kvenberg, J.E., Scanning electron microscope study of adult stored product beetle elytra, *J. Assoc. Off. Anal. Chem.*, 61, 76-87, 1978.
20. Zimmerman, M.L., Identification of *Ahasverus advena* (Waltl) (foreign grain beetle) and *Ahasverus rectus* (LeConte) (Coleoptera: Cucujidae) by micromorphology of adult fragments including genitalia, *J. Assoc. Off. Anal. Chem.*, 70, 484-495, 1987.
21. Genesis 41:53-57.
22. Levinson, H.Z. and Levinson, A.R., Storage and insect species of stored grain and tombs in ancient Egypt, *Z. Angst. Entomol.*, 100, 321-339, 1985.
23. Sinha, R.N. and Watters, F.L., *Insect Pests of Flour Mills, Grain Elevators, and Feed Mills and Their Control*, Agric. Can. Publ. 1776, Ottawa, 1-290, 1985.
24. Linsley, E.G., Natural sources, habitats and reservoirs of insects associated with stored food products, *Hilgardia*, 16, 187-224, 1944.
25. Freeman, J.A., Methods of spread of stored products insects and origin of infestation in stored products, *Proc. Int. Congr. Entomol.*, 8, 815-825, 1948.

26. Sinha, R.N., Ecology of storage, *Ann. Trop. Agric.*, 22, 351-369, 1973.
27. Howe, R.W., Losses caused by insects and mites in stored foods and feedingstuffs, *Nutr. Abstr. Rev.*, 35, 285-295, 1965.
28. Scott, H.G., Human myiasis in North America (1952-1962 inclusive), *Florida Entomol.*, 47, 255-261, 1964.
29. Harwood, R.F. and James, M.T., *Entomology in Human and Animal Health*, 7th ed., Macmillan, NY, 1-548, 1979.

Chapter 2

QUALITY ASSURANCE FOR REGULATORY SCIENCE

Richard A. Baldwin

CONTENTS

I. Regulatory Role of Quality Assurance 11
- A. Establishing Credibility 12
- B. Importance of Training 13
- C. Elements of Quality 13
  - 1. Definitions 13
  - 2. Quality Elements 14

II. Credibility (Quality) Assurance Issues 14
- A. Credibility (Quality) of Equipment 14
  - 1. General Concerns 14
  - 2. Equipment Capabilities 15
- B. Credibility (Quality) of Standards, Reagents and Environmental Controls 15
  - 1. Standards 15
  - 2. Reagents and Environment 16
- C. Credibility (Quality) of Analytical Work 16
  - 1. General Concerns 16
  - 2. Managerial Responsibilities 16
- D. Credibility (Quality) of Analytical Documentation 17
  - 1. General Concerns 17
  - 2. Documentation Issues 17
- E. Credibility (Quality) of Sample Accountability 17
- F. Credibility (Quality) of Laboratory Administrative Systems 18

III. Quality Assurance Activities 18
- A. Check Sample Program 18
- B. Oral and On-Site Credibility (Quality) Reviews 19
- C. Credibility (Quality) Assurance Assessment Reviews 19

IV. Conclusions 20

References 20

I. REGULATORY ROLE OF QUALITY ASSURANCE

The concept of *quality assurance* can convey different messages and meanings to people. For some, it connotes the idea of a high level of reliability and dependability, built into a product that might be something we use everyday such as a refrigerator, toothbrush or automobile. The message conveyed is that the "finished product" is always there to provide consistent, dependable and effective service. The "product" may provide day to day basic needs in plain or fancy packaging or absolutely vital, intermittent needs, such as a parachute. If one's breakfast toast is delivered as a charcoal briquette because a timer was improperly installed or faulty, then we might not buy that manufacturer's product again. On the other hand, if a parachute does not properly open, as we hurtle toward the ground, then we probably will not even get the opportunity to express our disdain to the responsible person (unless we packed our own parachute). In another person's experience, the concept of quality assurance may convey the sense of "paying attention to the details," in a product that might be noted for its style and appeal to the senses, such as the handling, comfort, amenities and styling of a luxury car. The extent to which quality might be reflected in these characteristics would depend on how consistently and effectively tight tolerances and specifications are met in the raw materials and finished product, time after time. Quality assurance might also be reflected in how clothing amenities flatter the human form, how the material is sized and the durability of the garment. All of these expectations are valid for their intended product and the environment within which the product will be used.

Other situations where quality assurance manifests itself in different ways would be the "service areas." For example, was a problem with the auto engine properly diagnosed and corrected? Did your hotel reservations get properly conveyed to the right location and did you get what you were led to expect from room service, the valet and the front desk during your entire stay? If you sent some clothes to the dry cleaners, did they come back "really clean" and with the buttons present and intact? Is it obvious that someone really paid attention to your food order and it really was served medium rare? How about the next time you use your dry cleaners or visit your favorite restaurant; are they consistent?

A. ESTABLISHING CREDIBILITY

Regardless of your experience or expectations about quality in other situations, *regulatory science* expands the meaning of quality assurance to include the added dimensions of such concepts as accuracy, precision, reliability, ruggedness, sensitivity of the method and other scientific criteria. The essence of this value-added dimension is what I will call *credibility*. Credibility connotes trust, believability, reliability, plausibility, likelihood, convincibility and truth. This credibility becomes the basis for regulatory decisions that effect the protection of the public health. Are consumer products such as foods, drugs, cosmetics, devices and biologics suitable for their intended purpose? In other words, are they safe and effective within the context of the law [Food, Drug and Cosmetic Act (FD&C Act) and related laws and regulations], or do they constitute articles that are misbranded, adulterated, unsafe, ineffective, fraudulent, counterfeit or bogus? Furthermore, one may ask the following questions: How is the performance of the product suitably demonstrated as measured? What characteristics must be assessed and how can the product be documented as acceptable or unacceptable? What role does science play and how does one have confidence in the legitimacy of the science? On what empirical and experimental foundation is the science based? Is/are the test(s) appropriate for the intended regulatory decision? Are the tests reliable, reproducible, accurate and made on a representative sample? Ultimately, will the truth prevail and be substantiated?

The role that regulatory quality assurance plays in the area of public health and consumer protection is to assure that the science on which regulatory agencies make decisions is based on correct and sufficiently convincing data. That scientific data must be able to withstand the careful scrutiny of all interested parties, be they consumers, manufacturers, brokers, scientists, lawyers, bureaucrats, judges, the media, congress and any others touched by the event.

Although this chapter will focus primarily on the regulatory analyses of samples to determine compliance with the FD&C Act, one should also understand that this concept of credibility control/assurance extends to all other activities performed in the U.S. Food and Drug Administration's (FDA) field labs. For example, analytical methods development and research undergo similar kinds of scrutiny in the form of controls and reviews to demonstrate the scientific credibility of the new or modified method. Even the preparation of a publication or management study is expected to have been checked by knowledgeable peers and management for completeness, pertinence, accuracy of data and appropriateness of conclusions. The goal and intent of management is to establish and maintain an organizational culture that assimilates in each person a work ethic that lives and breathes credibility into everything that is done.

In the regulatory analysis of foods for filth (i.e., extraneous materials/sanitation) how does one build and maintain a credibility (quality) assurance program [C(Q)AP]? The life's blood of any C(Q)AP is the attention that individuals pay to the details of their scientific tasks and the prominence that management places on the continual importance of checking one's work, having others check the work and reassuring that there is accountability and follow-through as quality issues are uncovered—be it good or bad news.

B. IMPORTANCE OF TRAINING

An effective C(Q)AP must start with proper training, both in the scientific forum, as well as in the documentation of regulatory accountability. Proper training grounds a scientist or analyst in the fundamentals of their profession. It is the basis of credibility (quality) control. Training should be a blending of on-the-job and formal experiences. Due to the uniqueness of the FDA's regulatory analyses, it is critical that new employees be trained in what constitutes appropriate analytical documentation, use of methodology, sample accountability and other areas related to their role. That initial training must be reinforced as they progress in their careers. Without a fundamental grounding in regulatory science, the public health significance of the analytical work that is produced can suffer drastically. Periodic technical update training is especially important in areas that rely on sensory or visual skills, such as organoleptic and microanalytical entomology analyses. However, updates should not be ignored in the more traditional areas either. These kinds of training allow the scientist/analyst to "calibrate" themselves against properly prepared samples and their peers. Training is also a viable way to introduce new technologies for routine use, such as the use of CD-ROM data retrieval systems to obtain and record reference criteria and documentation that correlate with incurred extraneous materials.

C. ELEMENTS OF QUALITY

1. Definitions

Credibility (quality) control activities: Actions that a scientist/analyst takes to demonstrate, with certainty and confidence, that the workproduct of their science is performed in a reliable, reproducible, accurate and truthful manner.

Credibility (quality) assurance program: Includes those activities that management undertakes to demonstrate that the credibility (quality) controls are performed in an appropriate, timely and proper manner.

There must be a written plan that defines who does what, when they do it, where the activities take place, how the activities are to occur and the rationale for why all of these things are the way they are. The written plan needs to be communicated in such a way that there is no ambiguity about what is expected. Such documents exist in the FDA in the form of the Office of Regulatory Affairs (ORA) *Quality Assurance Program*[(1)] and other related material, including the ORA *Regulatory Procedures Manual,*[(2)] *Laboratory Procedures Manual*[(3)] and others. These documents articulate the credibility (quality) expectations of the organization. It is important that the program mission of the organization be considered in developing or modifying a C(Q)AP, but it should include some of the basics as follows.

2. Quality Elements

Credibility (quality) of equipment
Credibility (quality) of standards, reagents and environmental controls
Credibility (quality) of analytical work
Credibility (quality) of analytical documentation
Credibility (quality) of sample accountability
Credibility (quality) of laboratory administrative systems
Check sample programs
Oral and on-site credibility (quality) reviews
Credibility (quality) assurance assessment reviews

Within each of the areas noted above, it is essential that the written plan define the scope of the review, the objective(s) to be achieved and what other resources or references should be used or consulted. Other issues that should be covered include how often audits should be performed, who should conduct the audits, what the audits should cover, what should be done to deal with and resolve deficiencies and discrepancies that are uncovered and how to prevent future problems, while sustaining excellence of credibility.

It is important to look at some examples of discrete issues that better focus on the reasons for certain control and assurance steps. It is not my intent to be all inclusive, but rather to provide various examples that give a fuller understanding of regulatory credibility.

II. CREDIBILITY (QUALITY) ASSURANCE ISSUES

A. CREDIBILITY (QUALITY) OF EQUIPMENT

With microanalytical determinations one might be prone to diminish the importance of this area, but that kind of lax attitude will quickly lead to inaccurate or inconclusive results. The regulatory expectation of the FDA is that the scientist/analyst knows the significance of and takes appropriate steps to ensure that items such as balances are routinely calibrated and maintenance performed, so that they can be used properly. This, then, gets to such areas of significance as accuracy of measurements, repeatability from sample to sample and that representative portions of the sample can be consistently evaluated.

1. General Concerns

Are items such as beakers, flasks, stirrers, sieves, separators, etc., properly cleaned and maintained so that there is no cross-contamination among samples or so that recoveries of fragments are not compromised because of the build-up of residues of debris, detergent, fibers or other materials? Should anyone even think of using glass beakers or stirring rods in the analyses of a product for glass contamination? Are the containers used during analysis properly labeled to avoid confusion or inappropriate transfers/discards? Are microscopes or other visualization devices properly maintained and in good working order so that the appropriate quantity and physical details of the recovered fragment(s) can be fully assessed and characterized? Are solutions or reagents properly and freshly prepared so that they yield the desired effect on the analysis? Who maintains the equipment and is responsible for its calibration, cleanliness, etc., and how is that information appropriately documented for future reference? (For example: Is the thermometer accurate? How do you know it is? What are the performance criteria for this and other equipment?) How often are various items of equipment calibrated/maintained? How do others know if a piece of equipment is acceptable or "out of order"? Is there a track record of equipment performance and maintenance such as a log book or computer-based record? Is the record up to date and complete? Did the scientist/analyst employ proper safety precautions to protect oneself and others and maintain the integrity of the sample? How are these things evidenced and who is/was involved? Many of these concerns are fully discussed in chapter 15.

2. Equipment Capabilities

Another aspect to assuring credible quality in the area of equipment and instrumentation is the need to have access to current, generally recognized, state-of-the-art apparatus. Furthermore, when it is necessary, equipment and instrumentation that are on the cutting edge of scientific capabilities should also be used. Expense and complexity of use may dictate the need to limit the number of these high-tech devices and locate them in labs where they can be most effectively and efficiently applied to the organizational mission. In this context it is also essential that staff and management understand the limitations of an equipment/instrumentation-based technique. Analytical results and interpretations can be inappropriately biased if they are not properly judged and assessed in the context of actual instrumentation and equipment capabilities. Specific creditibility (quality) assurance issues involving the equipment used by microanalytical entomologists are discussed in other chapters. See chapter 3 for information about specialized equipment such as microscopes and chapter 15 for information about general equipment.

B. CREDIBILITY (QUALITY) OF STANDARDS, REAGENTS AND ENVIRONMENTAL CONTROLS

Analytical testing requires the use of not only what we traditionally consider equipment, but also various solutions, reagents, reference materials, expendables and other items, which must be employed in an environment that is designed to protect the integrity of the analysis. Environmental controls that ensure access to reliable and appropriate humidity, temperature, refrigeration, compressed air, vacuum source, constant temperature baths and so forth, are essential to the credibility of regulatory analyses.

1. Standards

How does the scientist/analyst know the origin and/or identity of extraneous material they have found? Was the identification based on proper training, experience and ongoing comparison with authentic samples? If not, then the regulatory and public health significance of the case is seriously compromised and diminished. Where did the authentics come from? How are/were they maintained, identified and characterized? What is the experience of those doing the characterizations? Does the scientist/analyst account for the expected and unexpected anomalies that can be found in dealing with the attributes of living organisms? Authentic specimens and reference collections (see chapter 14) are benchmarks to the entomologist (or other scientist) analyzing food for adulteration. This is analogous to the way a reference standard, such as a pharmaceutical or pesticide, is used by a chemist who is testing the purity/potency of a human drug dosage form or the presence of a pesticide residue in a raw agricultural commodity. If the benchmark (i.e., the authentic specimen) has no credibility or accountability, then neither can the results of the analysis. Definitive identification of extraneous materials and fragments is of regulatory importance, because it helps to focus on who is responsible for the contamination and the level of public health risk involved. There are clear-cut legal significances to the presence of field versus storage insects and the quantity of contamination. There is likewise public health significance to the kinds of extraneous matter found, in that certain insects tend to be more susceptible to carrying and transmitting human pathogens than others. This rationale is consistent with not only insects but also glass, metal or other debris. What kind of glass is it and where could it have been introduced into the product? What are the sizes/shapes/quantities of the extraneous material?

2. Reagents and Environment

The manner in which the work site is maintained is also an important consideration for credibility and safety. Are the water, solvents, reagents and other expendable items free of interfering contaminants? For example, have the scientists/analysts and management ensured that the water system does not contain particulate contamination that would compromise analysis? Are there periodic checks to determine that the site is free of the kind of adulteration being tested (e.g., insects)? Is there a plan for continuously ensuring the prevention and elimination of possible contamination of the building and/or work site? If animals are used by other units for scientific purposes, are there procedures to ensure that debris, hairs, etc., from those animals do not contaminate the environment and subsequent samples? Who is responsible? What is to be done? When is it to be done? Are proper temperature and humidity maintained in the work site and for certain analytical steps? What are the who, how, why, when, what, where questions and answers?

C. CREDIBILITY (QUALITY) OF ANALYTICAL WORK

This area of credibility (quality) control and assurance gets to the heart of whether the analyst used appropriate analytical methodology to obtain the regulatory data on which consumer protection decisions can be properly based.

1. General Concerns

What method was used? What are the known performance characteristics of the method, i.e., accuracy, precision, sensitivity, selectivity, recovery, etc.? How does the analyst document that the method achieved the expected levels of performance? What kinds of controls were used? Did the analyst do recovery studies? How were they done and how does that speak to the validity of the method in the context of spike recoveries versus incurred extraneous materials recoveries? Are they analogous or dissimilar? Does it affect the regulatory outcome if it is grounded in quantitative risk assessment? Were quantitative transfers accomplished (see chapter 15)? Was a check analysis done on the sample to verify the original results? What analytical method was employed for the check analysis? What is the experience of the check analyst? On what portion of the sample was the check analysis performed? If the nature of the product and the contamination are such that the homogeneity or stability of the sample is not reasonably expected, then is a check analysis really necessary? If not, did someone more experienced or with equivalent experience verify the original findings, i.e., view the extraneous materials and confirm the quantities and identifications?

Was the analyst worksheet checked by a scientist/analyst peer? To what extent did the worksheet check by the peer delve into the appropriateness of the method, recoveries (if run), condition of the sample, calculations, summary and presentation of the data?

Did the supervisor review and endorse the analytical work? Did the supervisor make an appropriate judgment concerning the regulatory significance of the analytical results? Are those supervisory judgments consistent with the law, regulations and administrative guidelines and based on the analytical evidence?

2. Managerial Responsibilities

Does the next level manager concur in the findings? If an inappropriate method is used, it can yield misleading and/or erroneous analytical results that would mitigate the ability of management to effectively rely on those data. On the one hand, the significance of a violation could be diminished, whereas in other situations actual violations may not be uncovered because the method was misapplied or ineffectively validated. If it appears that management has some analytical data that indicates a regulatory problem, then they can have a dialogue on the relative significance of the data and perhaps pursue the "true" numbers more vigorously. However, if one does not know there is a problem because of a faulty method or poor application of what should be a proper method, then the opportunity for consumer protection is lost.

D. CREDIBILITY (QUALITY) OF ANALYTICAL DOCUMENTATION

1. General Concerns

Was the method properly cited and modifications noted/validated? Were deviations from the method explained? Were errors or discarded results stricken, explained and so identified by the initials/date of the scientist/analyst? Were appropriate standards, reagents and/or special equipment used and described on the worksheet? Were the calculations accurate, scientifically sound, easy to follow and checked by a peer? Is the original data on the back of the worksheet consistent with the summary on the front and/or other parts of the worksheet? If more than one person participated in an analysis, does the worksheet accurately reflect who did what, when? Is the sample reserve or disposition properly reported? Was the sample analyzed in such a manner as to permit the application of appropriate regulatory guidelines? Were analytical timeframes met? Does management agree with the level of quality of the work and does feedback documentation exist when necessary or appropriate? How does management go about sustaining excellent quality work? Do they recognize excellence as well as defects/deficiencies and document it in writing? Are the cause and significance of defects articulated in writing, if necessary? Are the results of corrective actions articulated in writing, if necessary? Was the follow-up ultimately successful?

2. Documentation Issues

Generally, these issues focus on the scientist/analyst's written record of the analytical work performed on a sample. The documentation of the analysis of a regulatory sample in the FDA is made using a worksheet and that worksheet generally comprises the entire original record of the testing of the product. The worksheet may be a written record prepared as the analysis is conducted or an electronic computer-based record incorporated into an appropriately secure system that ensures the integrity of the data. Regardless, the worksheet, in whatever form, must represent what the analysts did as they accomplished their various tasks. As such, it must be a complete and unambiguous description of what was done to the sample, who did what when, why results were obtained and how the sample was stored. The importance of this original record cannot be overstated, because the worksheet may be used as evidence in an enforcement action. Because of their public safety importance, regulatory actions may evolve quickly or because of their complexity require additional resources or time. For these and other reasons, the completeness of the record and the ability of the various reviewers, including the scientist, to be able to go back and conclusively determine exactly what was done in the past, is critical to sustaining a case. For appropriate regulatory consideration, the worksheet must detail chain of custody of the sample and accountability of the sample while under the scientist/analyst's control (and anyone else involved in the process) to distinguish one sample from another and to tie the chain of custody of the sample to each person responsible and involved in the handling of the sample/product. This also helps ensure that our handling of the sample was neutral to any impact the sample and analytical results would have on regulatory considerations. Many of the issues discussed in previous sections resurface during the conduct of the analysis and audit of the documentation of the analysis.

E. CREDIBILITY (QUALITY) OF SAMPLE ACCOUNTABILITY

The intent here is to ensure the proper handling, storage, accountability and disposition of samples. The checks and audits determine if the sample was properly stored from the time of receipt into the lab to its ultimate disposition out of the lab. It is critical to be able to demonstrate that the sample was handled in a manner that did not compromise sample integrity and thereby the ability of the agency to sustain an appropriate regulatory action.

If the sample required refrigeration, then the various documents (which may be computer-based), such as the sample collection report, sample accountability card/record, worksheet and other related records, should reflect that the sample was indeed held at refrigerated temperatures. Exceptions are for only those appropriate instances where the sample or a portion was prepared for analysis or actually under testing. FDA sample handling protocol generally, but not always, uses markings such as official seals, initials/dates of sample handlers and similar identification procedures to document who had custody of the sample, when and for what purposes. In certain circumstances, when it is necessary for the agency to share a portion of the reserve sample with a responsible party, such as the manufacturer, broker, etc., the sample accountability records must accurately document the transfer. The sample accountability records are used to determine the amount of the reserve available and it's location. In the context of enforcing the FD&C Act, analysts of the agency may find themselves testifying in court about sample accountability. The sample itself can become an exhibit and constitute evidence. Therein is the essence of the rationale for the extent of the efforts described in sample accountability.

F. CREDIBILITY (QUALITY) OF LABORATORY ADMINISTRATIVE SYSTEMS

This area may seem to be out of place if one considers just the title; however, the checks and audits described here play a vital role in the overall management of the laboratory operations. The kinds of issues addressed here include ensuring that there are appropriate procedures for handling sensitive solvents/substances such as alcohol (ethanol) and other controlled substances.

Are there procedures to account for not only the quality of the substance, but also the quantity used and/or available, who used it, when and for what purposes? Abuses in these areas would not only be illegal, but would also have some serious practical consequences. Other aspects to be considered would be waste management and recycling of solvents. Who is in charge, how are recycled products assessed for purity, etc.? Who does it? When? Are manuals, reference materials, journals, libraries, on-line computer-based reference systems reliable, up to date and secure? How do you know? Who knows? Who checks?

III. QUALITY ASSURANCE ACTIVITIES

A. CHECK SAMPLE PROGRAM

A check sample program can be used in one lab or in many labs. For the FDA's field organization it was designed to demonstrate that analytical results from various field science units are equivalent within both a scientific and regulatory context.

A valid check sample program is predicated on providing each laboratory with identical samples and having the laboratories use the same analytical methodology to test for designated analytes or attributes. The methodology that is cited for use will have a known set of performance characteristics that make it acceptable for regulatory work. The intent is to ensure a level of confidence that samples of a particular product, when analyzed for the same attribute in one lab, will yield the same regulatory results in another lab with the same program responsibilities.

The program is strictly for in-house use and provides management with another valuable piece to the bigger picture of the overall scientific credibility of the organization.

Attempts at modifying the process to gain other information must be carefully considered so that the ultimate purpose of the check sample program is not diluted or rendered ineffective. For example, the program should not be used to conduct collaborative studies on emerging methodologies or technologies or those that have not been fully characterized. If an organization has a need to do that, it is best kept as a separate exercise. The uniformity of the sample being analyzed and its ability to withstand the rigors of shipment and storage are critical to ensuring that each lab is on a level playing field. Careful pretesting of the check sample is an absolute necessity. Whatever reports that are written should reflect the significance of the analytical results not only from a scientific and/or statistical perspective, but also from the impact on regulatory outcomes. Hindsight would probably reveal that conducting a check sample of pesticides (or insect filth), in a raw agricultural product, where the residues are spiked at *exactly* the tolerance or administrative guideline, would probably yield good overall scientific results, but poor regulatory correlation. Although one attempts to achieve the *best* analytical precision and accuracy, the practical nature of science is that there is a range of precision or repeatability that is measurable and is acceptable within a defined latitude. The practical realities of science may have seeming adverse effects on the regulatory consequences. As with all of our personal and professional efforts, thinking ahead and understanding the consequences of decisions is a valued and time saving course of action.

B. ORAL AND ON-SITE CREDIBILITY (QUALITY) REVIEWS

These reviews, rather than solely audits of written documentation, include the added dimension of having a more experienced scientist and/or management person observe the actual conduct of an analysis and/or conduct an interview of the analyst based on analytical work previously performed. The advantage offered by these audit techniques is that they allow management to enter into dialogues with staff about what they are doing, why they are doing it and the consequences, in a real-time mode. This is an opportunity to observe and deal with issues that cannot be readily ascertained by the written documents. For example, observing a quantitative transfer is very different than merely making assumptions about how it was accomplished by the analyst. Therein lies a valuable opportunity to make positive adjustments in how procedures are handled and carried out. The verbal dialogue also simulates what may occur should the analyst have to testify in court. It gives the analyst and management a mechanism whereby potential issues can be addressed in a less threatening and formal environment. Often times, verbal explanations can clarify and resolve misunderstandings and at the same time sensitize the scientist/analyst to the need for clear and unambiguous worksheet documentation. The oral review of a worksheet that is 6 or 12 months old can shed considerable light on the ability of the document to yield *all necessary scientific data* that affects the credibility of the regulatory decision.

C. CREDIBILITY (QUALITY) ASSURANCE ASSESSMENT REVIEWS

These reviews are actually hybrids of the oral and on-site reviews, except that they are normally done by a team that includes scientists, analysts and management from other parts of the organization. The purpose and intent of these audits are to periodically assess the overall branch and/or district effectiveness in achieving high levels of credibility (quality) and determine if they are consistently maintained. The assessment review team should be comprised of members who are thoroughly familiar with the programs and science being assessed. The team should determine the effectiveness of the procedures used by local management to ensure credibility (quality). Work products such as sample worksheets, research reports, environmental control audits, etc., should be examined and evaluated for adherence to good science and organizational policy and procedures. Timeliness of sample analyses, number of analyses that require rework or are rejected due to quality defects should also be reviewed, evaluated and discussed with local management. An exit interview should be conducted with local management and a written report should be prepared that summarizes the assessment procedure, articulates the results of the independent evaluation of work (including specific and overall strengths and weaknesses) and recommends improvements to the effectiveness of credibility (quality) efforts. As appropriate, situations or events that may have broader impact should be brought to the attention of those who have the authority and responsibility to effect necessary changes.

IV. CONCLUSIONS

A properly run credibility (quality) assurance system strikes a reasonable balance between the resources needed to achieve high-quality work and the need to have high levels of productivity. It is extremely important to recognize the nuances and consequences of various scientific and analytical actions, so that effective analyses are conducted. It is just as important to avoid becoming totally immersed in the inefficiencies that spring from extraordinary minutia.

Depending on where you are in the process of developing a new or nurturing a maturing credibility (quality) assurance system, you should consider the need to determine levels of "trust" of the system that you find acceptable and can justify. In this context, trust is associated with both the level of frequency and depth of credibility (quality) assurance system activities. For example, the level of frequency of "credibility" activities could be expected to vary in the context of the numbers of similar samples that are analyzed over a period of time. Although the frequency and level of quality control activities may stay constant or be batched with the samples, the frequency and level of quality assurance audits may be spread out over greater numbers of samples. The basic rationale for this is that when a person does a task less frequently, there is a potential tendency for the person to become desensitized to the nuances of the methodology, technology, sample preparation or other activities. This is often rooted in the lack of familiarity that repetition brings. Consequently, the relative frequency of audits of less often performed work may be greater than when the analyst does many samples on a frequent basis. One should also fine tune the depth of quality assurance audits based on the kinds of deficiencies that tend to occur when frequency of experiences interact with varying degrees of technological difficulty.

There is no magic formula to quantify how often credibility (quality) audits should be performed. Rather, experience, understanding of technological difficulties, methodology challenges and frequency of use all play a role in helping to determine the frequency of audits and consequent trust of the process. If peers are used in the credibility (quality) assurance system, a culture needs to be established that ensures they are collegial, yet honest brokers of credibility.

Too little C(Q)AP leads to sloppy and noncredible work; too much leads to an elegant system of gridlock.

In closing, it is important to keep in perspective that the ability to produce high-quality work products is at the same time both an individual and collective responsibility. The willingness of each person to openly communicate and be accountable for their own and the organization's work makes credibility an attainable goal. An organization must determine the unique needs for which it creates and maintains its own credibility assurance program. That program must take into account the resources it takes to maintain a viable program, while keeping productivity at appropriately high levels. Develop your plan as an inclusive and vibrant guideline and then work the plan.

REFERENCES

1. DHHS/PHS/Food and Drug Administration, Office of Regulatory Affairs, *Quality Assurance Program*, Rockville, MD, June 27, 1988.
2. DHHS/PHS/Food and Drug Administration, *Regulatory Procedures Manual*, Rockville, MD.
3. DHHS/PHS/Food and Drug Administration, Office of Regulatory Affairs, *Laboratory Procedures Manual*, Rockville, MD.

Chapter 3

IDENTIFICATION OF FILTH

Alan R. Olsen

CONTENTS

I. Introduction 21

II. Identification Procedures 22
- A. Identification vs. Classification 22
  - 1. Taxa 22
  - 2. Classification 23
  - 3. Identification 23
- B. Methods of Identification 23
  - 1. Direct Comparison 24
  - 2. Identification Keys 24
  - 3. Other Resources 24

III. Fragment Identification 25
- A. Whole or Equivalent Insects 25
- B. Insect Fragments 26
  - 1. General Recognition Criteria 26
  - 2. Diagnostic Insect Fragments 26

References 27

I. INTRODUCTION

"Objects without names cannot be talked about or written about; without descriptions they cannot be identified and such knowledge as may have accumulated regarding them is sealed; their relations are unknown and the possibilities of deduction are destroyed."[1] No matter what our particular interest, we must know what insect or other animal we are dealing with. More specifically, the analysis of a food product for filth is incomplete if the analyst does not distinguish between kinds of filth contaminants by determining their scientific names, or identities. An incomplete analysis may result in error because subsequent administrative and legal reviews of the analytical results depend on a full understanding of the nature of the contaminants in order to produce correct decisions.[2]

The actual work of a microanalytical entomologist can be divided into two distinct phases: extraction of filth from a product matrix and examination of the extracted filth to quantify, identify and document it. The methods and techniques for extracting filth from a product are discussed in chapter 15. The following discussions deal with the procedures for examining filth once it has been separated from the product or collected during the course of an investigation of a filthy building or facility. As with other forensic evidence, entomological items must be identified because identifications are important clues to the etiology or possible origin of the evidence.

Taxonomy is the study of the principles and practices of naming organisms. In zoology, taxonomists establish and validate the scientific names of groups of animals so that other scientists can use those names to identify animals with confidence that there is minimal confusion about which animal they mean. Without this confidence, scientists could not compare results of experiments, test each other's hypotheses or even intelligently discuss a group of animals over coffee. Taxonomists also rank groups of animals into systematic hierarchies to study the evolutionary and biological interrelationships between groups. These studies, systematics, are fundamental to understanding the scope of the group of animals to which a particular scientific name refers.

II. IDENTIFICATION PROCEDURES

A. IDENTIFICATION VS. CLASSIFICATION

Common or vernacular names of insects and other animals are often imprecise or ambiguous, so scientists use a system of scientific nomenclature that is precise and clear. To reliably identify an animal, one must first be able to evaluate the scientific validity of the taxonomic names that are used for identifications. This section explains some basic principles of taxonomy that must be understood and followed for identification work to be scientifically valid.

1. Taxa

The term *taxon* (pl. taxa) is a modern derivative of the term *taxonomy*, which has a special meaning in zoology. A zoological taxon is a group of animals that is sufficiently distinct to be worthy of being named and being ranked in a definite category.(3,4) In zoology, for a taxon to be scientifically valid it must meet three criteria:

1. It must represent a group of actual animals.
2. It must be formally recognized by taxonomists.
3. It must be ranked in an established category of a classification system.(3)

For example, the species *Apis mellifera* L. is a taxon that encompasses all the "honeybees" in the world, in the past, present and future of the species. The taxon's name, *Apis mellifera* L., refers to the honeybees themselves (all of them) according to the first criterion. The individual honeybee that stung me last spring in the garden is merely a late and unlamented member of the taxon. It is the group, not the individual, that comprises a taxon. This is an important distinction if one is to understand identification, discussed in section 3 below.

The second criterion, formal recognition by taxonomists, appears to be a tautology—a taxon is what taxonomists say it is. Recognition involves more than merely asserting that a taxon is valid. A taxon is formally recognized by the publication of the taxon's name, description and rank in a scientific book or journal. This procedure subjects the taxon to review by other scientists to confirm its objectivity and validity. Like other scientists, microanalytical entomologists must be sufficiently familiar with this taxonomic literature to assess the validity of the scientific names they use to identify the animals they work with.

Finally, a taxon must be ranked in an established category of a classification system. Most classification systems use a form of the ranking hierarchy established by Linnaeus. The modern form of this hierarchy(3-6) that is accepted by most zoologists is, in descending order of inclusivity:

Kingdom
Phylum
Class
Order
Family
Genus
Species

2. Classification

This is the theoretical side of taxonomy. In zoology, classifiers scientifically arrange populations of animals and groups of populations into a system that reflects relationships between the groups. A classifier must interpret large amounts of information about groups of animals to establish and name taxa and then rank the taxa in a classification hierarchy.[4] Use of all the categories in a hierarchy is mandatory; a classifier cannot name and describe a taxon without properly arranging it in an established hierarchy.[6] A species, such as *A. mellifera*, must be placed in the appropriate genus (*Apis*). A genus, which is a group of species, must be placed in an appropriate family or group of genera (pl.). So the process of ranking proceeds to the pinnacle of the hierarchy.

3. Identification

This is the practical side of taxonomy, and the starting point of practically all applied zoology. It has been described as the recovery side of taxonomy because identifiers retrieve facts about taxa, facts that were stored in the scientific literature by classifiers, and associate that data with some specimen at hand.[4] In contrast with the classifier, who establishes and ranks taxa, the identifier's task is to associate the single specimen at hand with the correct established taxon.[3,4] The microanalytical entomologist is primarily an identifier who works with individual specimens to associate them with the correct taxon.

The goal of an identifier is to work a specimen down the classification hierarchy as far as possible, arriving at the least inclusive taxon to which the specimen can be reliably assigned. In practice, working a specimen through its kingdom, phylum and class is usually done by memory. Below that, the identifier increasingly depends on more reliable methods of identification. The next section describes the scientific methodologies used by identifiers, including microanalysts.

B. METHODS OF IDENTIFICATION

Before taking up the topic of how to identify filth in foods, it is necessary to discuss a common pitfall that may confuse or frustrate the identifier. One must always consider the possibility that a specimen may not be covered by the available identification keys, references or other resources. A specimen's true taxon may fall outside the scope of available reference materials. Alternately, there may be a natural variability among members of a taxon that is not accounted for in the reference materials. Besides variabilities in color and size, the identifier should be alert for variabilities between developmental stages of a taxon, sexual dimorphism and unusual variants such as dwarfs, hybrids and terata.[4] An ambiguous specimen may remain unidentified in the final report, but unidentified should not mean ignored. Unidentified specimens should be measured and characterized as accurately as possible. Descriptive terms such as "unidentified beetle adult," "early instar maggot" and "dermestid larva" are useful in this respect. Levels of taxonomic precision may vary also. One may be able to identify young moth larvae, for example, only to the family level, whereas more mature moth larva specimens develop morphological characteristics that support a genus- or species-level identification. It is incumbent on each identifier to recognize and observe the limitations imposed by their individual situations.[(2)]

1. Direct Comparison

Identification by direct comparison with already-identified specimens is the surest means of identifying filth from food products. This method of identification requires a reference collection of authentic specimens (see chapter 14). Direct comparison allows the identifier to compare the entire morphologies of the specimen to be identified with authentic specimens. The first step in this procedure is to match the morphological features described in the literature for a taxon with the features of both specimens. If all features match, the identifier examines all other morphological features of both specimens, using experience to decide whether any observed morphological differences are trivial or significant. If the two specimens are a match, the specimen from the food product is determined to belong in the same taxon as its matching authentic specimen.

2. Identification Keys

To avoid the tedium of trial-and-error sifting through reference collections for matching authentics, taxonomists have devised an efficient short-cut in the form of the identification key. An identification key consists of a sequential series of either–or choices that guides the identifier through feature-by-feature morphological comparisons to arrive at an identification. Each choice in a key is usually presented as a pair of mutually exclusive statements, called a couplet, about the morphology of the specimen to be identified. The identifier chooses which statement is correct for a specimen and follows that statement's direction to another couplet. The process is repeated until the identifier arrives at the correct taxon for the specimen. After the taxon is determined by process of elimination, the identification is easily confirmed by direct comparison because the identifier knows exactly which specimens to consult from the reference collection.

Identification keys are essentially methods for analyzing the morphology of a specimen. A key is a qualitative test in the sense that an unknown specimen is compared against a morphological standard to determine the identity of the unknown. As with any methodology, keys come in many variations, each suited to a particular set of conditions that are likely to confront the identifier. General keys are designed to differentiate major taxonomic groups and generally work for all adult forms, developmental stages and variant forms. A general key is a means of passing to more specialized keys that are more precise. General keys are available in most general textbooks dealing with insects.[(5,6)] Specialized keys for each major group of animals discussed in this book are included in that group's respective chapter.

There are two common types of specialized keys: taxonomic keys and field keys. Taxonomic keys attempt to reflect relationships between the taxa of a group as well as facilitating identification. They are characterized by their inclusivity of all taxa in a group and by complex statements that almost qualify as technical descriptions of the taxa. They are most useful to taxonomy experts working on classification research but are also usable to experienced identifiers. Field keys, also called analytical keys, are prepared by specialists of a group for use by identifiers. Field keys are often limited in scope to a specific geographic region or type of habitat. They use morphological characters that are easily observed and they do not attempt to fully reveal relationships between taxa. Most of the keys in this book are field or analytical keys for pests that infest stored food products.

3. Other Resources

In addition to reference specimens and identification keys, other resources are often available to assist the identifier. Pictures are often useful for verifying an identification, especially if an authentic specimen of a taxon is not available. Published technical descriptions

TABLE 1
Literature on Fragment Identification

Type of Fragment	Insect Group	Ref.
General	Insects	2,10,11,13,14,29–31
Antenna	Insects	15
	Roaches	16
	Moths	17
	Flies	32
Mouthparts	Insects	2,18
	Roaches	16
	Beetles	11,19–21,27,33,34
	Moths	11,16,35,36
Head capsule	Insects	2
	Roaches	16
	Moths	17,35,36
	Flies	32
Leg	Insects	37
	Roaches	16
Genitalia	Beetles	26–28
	Moths	2
Wing	Moths	38,39
Elytron	Beetles	22–25,40,41
Urogomphi	Beetles	2,11,42
Seta	Insects	2
	Flies	43

of a taxon can also be used to verify an identification. A specimen can be referred to a taxonomy specialist, but this option should be arranged well in advance because most specialists are in the business of classification, not identification, and requests to identify single specimens may interfere with business.

III. FRAGMENT IDENTIFICATION

The identification of insect fragments from processed food products is a unique aspect of microanalytical entomology. The premise of fragment identification is that certain integumental fragments of stored-product insects carry sufficient morphological information to infer the taxonomic identity of the insect, to the genus and species level in many cases. In addition to the myriad published technical descriptions of stored-product insects, which point out unique morphologies, there is a discrete body of research on insect fragment identification that is oriented toward the actual fragments found in processed foods. The literature reports of this latter research are summarized in Table 1. Not all fragments can be identified to precise taxonomic levels, but those that can should be identified to the most precise level possible.

A. WHOLE OR EQUIVALENT INSECTS

For purposes of quantifying the filth found in food products, the Association of Official Analytical Chemists (AOAC) has established a category of filth designated as "whole or equivalent" insects, often abbreviated in analytical reports as "w/e" insects.[9] The purpose of this filth category is to standardize the tabulation of insects that are damaged but not to the extent that the insect's body becomes totally disarticulated and randomly distributed throughout the food product. In practice, a w/e insect is an intact head with any other portion of the main body attached, up to and including intact specimens. There are a few ambiguities involving this categorization, however. Ecdysial fragments meet the criteria for w/e categorization but obviously do not represent a whole insect. These should be tabulated as "cast skins," "cast skin fragments" or, if the head portion only, "ecdysial head capsules." Another problematic question is whether fragments such as antennae, legs, elytra or abdomens that have apparently disarticulated from a particular w/e insect should be tabulated separately, or should an attempt be made to reassemble theoretical w/e insects from these scattered parts? In most cases, the analyst will have extra legs or too few antennae or other unaccounted parts for the number of heads involved so that the logical strategy is to tabulate separately but identify completely. Although it is theoretically possible to relate fragment evidence to original numbers of whole insects, it is seldom necessary.[10]

B. INSECT FRAGMENTS

1. General Recognition Criteria

The first step in identifying insect fragments is to distinguish them from similar-appearing components of the product where the fragments are found. The recognition characteristics[11,12] that distinguish minute to microscopic insect fragments from other material are

- Absence of cell structure
- Translucence
- Sutures and articulations
- Setal structures
- Surface sculpturing
- Recognizable, discrete features

The first and second characteristics, absence of cell structure and translucence, are found in all insect fragments. This is due to the nature of insect cuticle, which is a secreted substance. Insect cuticle lacks the rigid cellularity of plant tissue and possesses a characteristic translucence that is shared by only a few closely related terrestrial arthropods whose presence in food is also objectionable. The remaining characteristics are variable in occurrence, depending on the point of origin of a fragment on the insect body, and are confirmatory when they occur. Using these criteria, the analyst can recognize and quantify minute, nondescript insect fragments with confidence.

2. Diagnostic Insect Fragments

Generic recognition of insect fragments using the above criteria enables the analyst merely to differentiate minute bits of insects from other material. It is frequently necessary to perform more precise identifications of insect fragments to trace the etiology of adulteration from stored-product and commensal pests.[13] Although many insect fragments are nondescript, many bear morphological features that are uniquely diagnostic for a particular taxon.[2,14] It is these diagnostic fragments that the analyst must isolate and examine to more precisely identify and interpret the filth found in a product. Certain types of diagnostic fragments are found more frequently than others because their durability allows them to survive the food-processing technologies that break insects into fragments.[10,11] Among the most durable diagnostic fragments are the antenna, mouthparts, elytral fragments, caudal urogomphi and genitalia. Articulated antennae are usually diagnostic, but disarticulated individual segments are not.[15-17] Mandibles are excellent diagnostic fragments, bearing distinctive morphologies in the apical teeth, prosthecae, molars, articulation points, setation and color patterns.[17-21] Beetle elytra bear diagnostic sculpturing and setation that are identifiable even on minute elytral fragments.[22-25] Caudal urogomphi of beetle larvae and other insects are diagnostic if intact.[2] The genitalia of adult insects are widely recognized as diagnostic fragments, especially in beetles and moths.[2,26-28] These are the standard diagnostic fragments used by microanalytical entomologists to identify insect filth. Depending on the technology and insect involved, many other diagnostic fragments may occur in food products. In the following chapters these other diagnostics fragments are noted separately for each insect group.

Fragment identification requires a reference collection of authentic specimens, as required for the identification of whole insects. Fragment reference specimens are mounted on glass microscope slides in glycerin jelly for storage in the collection (see chapter 14). Direct comparison with authentic material from the insect is the surest means of identification. The references listed in Table 1 provide identification keys, pictures and technical descriptions for the more frequently encountered diagnostic fragments of stored-product insects. The procedures and guidance provided earlier in this chapter for the identification of whole insects also fully apply to the identification of insect fragments. Experience is the best teacher. Over time, an analyst will encounter certain types of diagnostic fragments so frequently that memorization is likely. In the interim, there are compendial references available as technical resources for fragment identification.(2)

REFERENCES

1. Gahan, A.B., The role of the taxonomist in present day entomology, *Proc. Entomol. Soc. Wash.*, 25, 68-78, 1923.
2. Kurtz, O.L. and Harris, K.L., *Microanalytical Entomology for Food Sanitation Control*, Assoc. Off. Agric. Chem., Washington, DC, 1-576, 1962.
3. Mayr, E., *Principles of Systematic Zoology*, McGraw-Hill, NY, 1-428, 1969.
4. Blackwelder, R.E., *Taxonomy: A text and Reference Book*, Wiley, NY, 1-698, 1967.
5. Borror, D.J., Triplehorn, C.A., and Johnson, N.F., *An Introduction to the Study of Insects*, 6th ed., Saunders Coll. Publ., Philadelphia, 1-875, 1989.
6. Gorham, J.R., Ed., *Insect and Mite Pests in Food*, Vols. 1 & 2, U.S. Dept. Agric., Agric. Handbook No. 655, 1-767, 1987/1991.
7. Schenk, E.T. and McMasters, J.H., *Procedure in Taxonomy*, 3rd ed., Stanford Univ. Press, Palo Alto, 1-149, 1956.
8. Gorham, J.R., Ed., *Principles of Food Analysis for Filth, Decomposition, and Foreign Matter*, 2nd ed., U.S. Food and Drug Admin. Tech. Bull. No. 1., Washington, DC, 1-286, 1981.
9. Boese, J.L. and Bandler, R., Extraneous Materials: Isolation, in *Official Methods of Analysis of the Association of Official Analytical Chemists*, 15th ed., Vol. 1, Helrich, K., Ed., Assoc. Off. Anal. Chem., Arlington, VA, 369-406, 1990.
10. Kurtz, O.L. and Harris, K.L., Identification of insect fragments: Relationships to the etiology of the contaminants, *J. Assoc. Off. Agric. Chem.*, 38, 1010-1019, 1955.
11. Kurtz, O.L., An approach to insect larval identification by structures isolated as fragments from cereal products, *J. Assoc. Off. Agric. Chem.*, 37, 167-169, 1954.
12. Gecan, J.S. and Brickey, P.M., Jr., *Cocoa Bean Histology and Comparative Micromorphology of Internal Bean Infesting Insects*, U.S. Food and Drug Admin., Washington, DC, 1-31, 1966.
13. Tilden, D.H., Measurement of filth in foods by microanalytical methods, *Adv. Chem. Ser.*, 3, 61-67, 1950.
14. Harris, K.L., Identification of stored product insects by the micromorphology of the exoskeleton, *J. Assoc. Off. Agric. Chem.*, 33, 878-933, 1950.
15. Jackson, M.M., Ratay, A.F., and Woznicki, E.J., Identification of stored product insects by the micromorphology of the exoskeleton. II. Adult antennal characters, *J. Assoc. Off. Agric. Chem.*, 39, 879-888, 1956.
16. Eiduson, H.P., Firestone, D., and Roe, J.E., Cockroaches and their fragments as food contaminants, *J. Assoc. Off. Agric. Chem.*, 41, 886-898, 1958.
17. Ensminger, L.G., Identification of stored product insects by the micromorphology of the exoskeleton. VIII. Moth larvae, *J. Assoc. Off. Agric. Chem.*, 41, 704-738, 1958.
18. Kurtz, O.L., Carson, N.A., and Van Dame, H.C., Identification of cereal insects in stored grain by their mandible characteristics, *J. Assoc. Off. Agric. Chem.*, 35, 817-826, 1952.
19. Jackson, M.M., Ratay, A.F., and Woznicki, E.J., Identification of stored product insects by the micromorphology of the exoskeleton. VI. Adult and larval beetle mandibles, *J. Assoc. Off. Agric. Chem.*, 41, 460-471, 1958.

20. Kvenberg, J.E., Scanning electron microscope study of adult stored product beetle mandibles, *J. Assoc. Off. Anal. Chem.*, 60, 1185-1209, 1977.
21. Kvenberg, J.E., Comparison of larval stored product beetle mandibles by light microscope and scanning electron microscope, *J. Assoc. Off. Anal. Chem.*, 64, 199-224, 1981.
22. Heuermann, R.F., Note on elytral patterns of storage insects, *J. Assoc. Off. Agric. Chem.*, 40, 662-664, 1957.
23. Heuermann, R.F. and Kurtz, O.L., Identification of stored product insects by the micromorphology of the exoskeleton. I. Elytral patterns, *J. Assoc. Off. Agric. Chem.*, 38, 766-781, 1955.
24. Khalaf, K.T., Micromorphology of beetle elytra, using simple replicas, *Fla. Entomol.*, 63, 307-340, 1980.
25. Kvenberg, J.E., Scanning electron microscope study of adult stored product beetle elytra, *J. Assoc. Off. Anal. Chem.*, 61, 76-87, 1978.
26. Olsen, A.R., Differentiating the saw-toothed grain beetle, *Oryzaephilus surinamensis* (L.), and the merchant grain beetle, *O. mercator* (Fauv.), by the morphology of the male genitalia, *J. Assoc. Off. Anal. Chem.*, 60, 1223-1226, 1977.
27. Zimmerman, M.L., Identification of *Ahasverus advena* (Waltl) (foreign grain beetle) and *Ahasverus rectus* (LeConte) (Coleoptera: Cucujidae) by the micromorphology of adult fragments including genitalia, *J. Assoc. Off. Anal. Chem.*, 70, 484-495, 1987.
28. Banks, H.J., Identification of stored product *Cryptolestes* spp. (Coleoptera: Cucujidae): A rapid technique for preparation of suitable mounts, *J. Austral. Entomol. Soc.*, 18, 217-222, 1979.
29. Clark, J.O., The detection and estimation of filth and decomposition in foods, *J. Assoc. Off. Agric. Chem.*, 31, 28-36, 1948.
30. Daly, A.W., Blomquist, V.H., and Hodges, F.A., Identification of stored product insects by the micromorphology of the exoskeleton. V. Identification of miscellaneous insects by the micromorphology of the exoskeleton, *J. Assoc. Off. Agric. Chem.*, 41, 206-220, 1958.
31. Van Dame, H.C., Reed, G.L., and Ikari, P.T., Identification of stored product insects by the micromorphology of the exoskeleton. IX. Head, thorax, and abdomen, *J. Assoc. Off. Agric. Chem.*, 41, 828-886, 1958.
32. Carson, N.A. and Martinez, E.F., Fly identification by the micromorphology of the head and head appendages, *J. Assoc. Off. Anal. Chem.*, 50, 1146-1193, 1967.
33. Spangler, P.J., Notes and pictorial key for separating khapra beetle (*Trogoderma granarium*) larvae from all other Nearctic species of the genus, *Coop. Econ. Insects Rep.*, 1161-1162, 1961.
34. Spilman, T.J., Identification of larvae and pupae of the larger grain borer, *Prostephanus truncatus* (Coleoptera, Bostrichidae) and the larger black flour beetle, *Cynaeus angustus* (Coleoptera, Tenebrionidae), *Proc. Int. Conf. Stored-Prod. Entomol.*, 3, 44-52, 1983.
35. Aitken, A.D., A key to the larvae of some species of Phycitinae (Lepidoptera, Pyralidae) associated with stored products, *Bull. Entomol. Res.*, 54, 175-188, 1963.
36. Capps, H.W., Keys for the identification of some lepidopterous larvae frequently intercepted at quarantine, *U.S. Dept. Agric. Agric. Res. Serv. Publ.*, 33-20-1, 1-34, 1963.
37. Thrasher, J.J. and Kurtz, O.L., Identification of stored product insects by the micromorphology of the exoskeleton. IV. Adult legs, *J. Assoc. Off. Agric. Chem.*, 40, 973-999, 1957.
38. Harris, M., Identification of stored product insects by the micromorphology of the exoskeleton. X. Common fragments of mill and grain moths, *J. Assoc. Off. Agric. Chem.*, 43, 444-458, 1960.
39. Khalaf, K.T., Differences between moths' and butterflies' scale base, *J. Econom. Sci.*, 21, 289-293, 1986.
40. Brickey, P.M. and Gecan, J.S., Introductory study of scanning electron microscopy of elytral fragments of stored product beetles, *J. Assoc. Off. Anal. Chem.*, 57, 1235-1247, 1974.
41. Rowley, J.Q., A simple method for the separation of *Gnatocerus* spp. and *Tribolium* spp. (Coleoptera: Tenebrionidae), *J. Stored Prod. Res.*, 19, 139-140, 1983.
42. Dobson, R.M., The species of *Carpophilus* associated with stored products, *Bull. Entomol. Res.*, 45, 398-402, 1954.
43. Sabatino,F.J., Identification of insect setae as an index of contamination in dairy products, *J. Assoc. Off. Agric. Chem.*, 37, 960-966, 1954.

Chapter 4

AMETABOLOUS AND HEMIMETABOLOUS INSECTS

Thomas H. Sidebottom

CONTENTS

I. Introduction 29

II. Thysanura 30
  A. Morphology 30
  B. Identification 30
  C. Discussion of Families 30
    1. Lepismatidae 30
    2. Nicoletiidae and Lepidotrichidae 31

III. Dermaptera 31
  A. Morphology 31
  B. Identification 31
  C. Discussion of Families 32

IV. Blattaria 33
  A. Morphology 33
  B. Identification 33
  C. Discussion of Families 33
  D. Diagnostic Fragments 34

V. Psocoptera 35
  A. Morphology 35
  B. Identification 35
  C. Discussion of Families 36
    1. Liposcelidae (Booklice) 36
    2. Lachellidae 37
    3. Trogiidae 37
    4. Psoquillidae 37
    5. Psyllipsocidae 37

References 37

I. INTRODUCTION

This chapter is primarily concerned with the common orders of ametabolous and hemimetabolous insects associated with food production and food storage, which include Thysanura (silverfish/firebrats), Blattaria (cockroaches), Dermaptera (earwigs) and Psocoptera (booklice). These groups contain insects known for their ability to transmit disease and to be prolific and gregarious invaders of buildings. They range in size from the minute (1 mm) to some that have been described as the size of small bats.[1-3] It is important for the microanalytical entomologist to be able to identify these insects as well as diagnostic fragments encountered in processed food.

II. THYSANURA

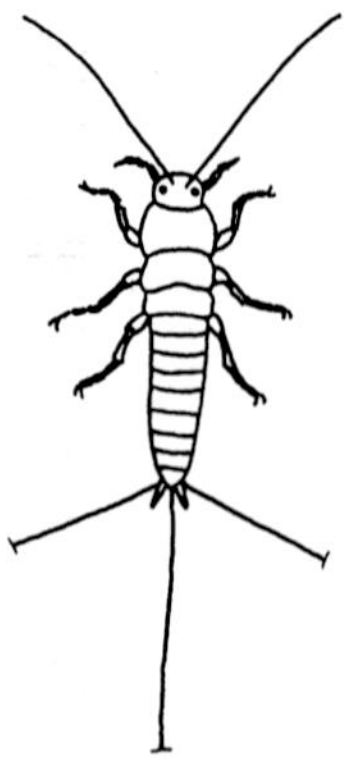

FIGURE 1. Silverfish (Thysanura: Lepismatidae), adult, dorsal view.

A. MORPHOLOGY

Silverfish and firebrats belong to the subclass Apterygota. Adult and immature Thysanura (Figure 1) are somewhat flattened, torpedo shaped, usually covered with silvery-gray to brown scales over the body. The scales are thin, flat and slightly round, similar to the scales of adult lepidopterans. Compound eyes are either reduced or absent. Chewing mouthparts are simple; mandibles are dicondylic, biting transversely as in ectognathous winged insects with the molar and incisor regions contiguous.[4] The maxillary palpi can be five or six segmented. Antennae are generally filiform and can extend the length of the body. In addition to the small scales covering the body, the thoracic nota and abdominal segments can have larger setae referred to as macrochaetae. The macrochaetae are often apically bifid and either smooth or spinose under higher magnification. The setae can be arranged in a single longitudinal series or in a grouped series. The grouped series form transverse bristlecombs on the body segments of the thoracic nota and abdominal tergites. Thysanura have three movable caudal appendages that can exceed the length of the body. The outer pair are referred to as cerci and the middle appendage is called the median caudal filament. The abdominal segments 2–9 can have smaller styles or stylets, which at a glance may appear like little legs. Dead, dehydrated specimens will easily lose a part of or all the caudal filaments. Silverfish molt many times, and some species have been recorded to go as high as 60 molts.[3]

B. IDENTIFICATION

Rapid Key to the Families of Thysanura

1. Compound eyes absent; (scales present or absent) .... Nicoletiidae
   Compound eyes present 2
2. Ocelli absent; scales present .... Lepismatidae
   Ocelli present; scales absent .... Lepidotrichidae

C. DISCUSSION OF FAMILIES

1. Lepismatidae

Lepismatidae include all the well-known species of synanthropic Thysanura encountered in filth work. The common silverfish, *Lepisma saccharina* L., and its kin are adapted to a wide range of microclimates including those found in buildings, warehouses, walls, bathrooms and kitchens. They are omnivorous nocturnal bandits that feed on a wide variety of foods, grains, dead insects, decomposing plant and animal matter, insect cast skins (particularly their own) and other materials or paper goods consisting of carbohydrates and proteins.[3-6] Tolerable microclimates can range from very humid to very dry with a moderate range in temperatures. Firebrats, *Thermobia domestica* (Packard), and gray silverfish, *Ctenolepisma longicaudata* Escherich, can absorb water from the atmosphere through their rectal wall. Silverfish can survive long periods of time without food or water. One lepismatid is known to have survived for more than 300 days in a laboratory jar without food or water.[3] It is not uncommon to find silverfish around the ovens of bakeries, where temperatures can be as high as 41°C. They have also been found living in the nests of birds and mammals.[3] On occasion, silverfish have even been found in bodybags by forensic entomologists.[5] Some members of this family are inquilines (cohabitants) in ant and termite nests and could be encountered during analyses of foods in such things as ant eggs for human consumption (see chapter 13).[4]

Lepisma saccharina **L.**—Common silverfish: Macrochaetae are smooth and usually bifid. The last abdominal tergite is elongate, with the posterior margin truncate.

Ctenolepisma lineata pilifera **(Lucas)[=** ***C. quadriseriata*** **(Packard)]**—Four-lined silverfish and *C. longicaudata* Escherich [= *Ctenolepisma urbana* Slabaugh]—Giant or gray silverfish: Macrochaetae are spinose and the thoracic nota has bristlecombs consisting of at least two macrochaetae. The maxillary palpi are five segmented. Abdominal tergites have a 3+3 bristlecomb arrangement. The last abdominal tergite has a posterior margin that is straight or can be slightly curved anteriorly. The median bristlecombs are absent from abdominal sternites.

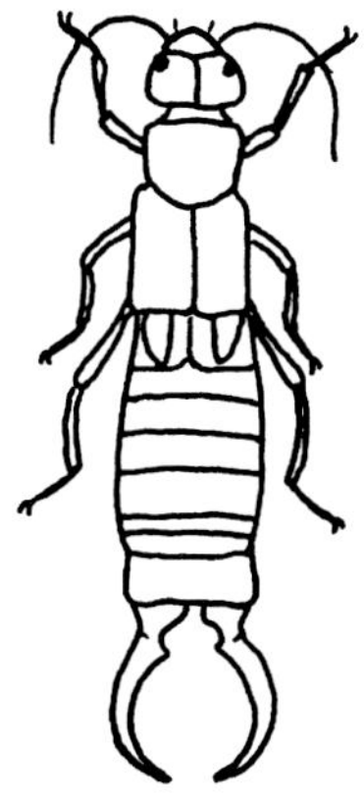

FIGURE 2. Earwig (Dermaptera), adult, dorsal view.

Thermobia domestica **(Packard)**—Firebrats: *T. domestica* is similar to the description above for *Ctenolepisma* spp., however, the abdominal tergites have a 2+2 bristlecomb arrangement and the maxillary palpi are six segmented. *Acrotelsia collaris* (Fabricius): The last abdominal tergite is distinctly pointed, with several bristlecombs on each side.

The fragments of silverfish that may be encountered include the mandibles, legs, antennae, cerci, scales and major body fragments. The identification of these fragments should be made in comparison with an authentic fragments from a reference collection of authentic specimens.

2. Nicoletiidae and Lepidotrichidae

Nicoletiidae and Lepidotrichidae prefer cooler, moist environments such as those found under rocks or in caves. Some are plant pests, others are inquilines. In Europe they are a common pest of greenhouses, and in the United States, Lepidotrichidae occur under the decaying bark of fallen trees.[7] Silverfish in this family are rarely encountered during analysis of foods.

III. DERMAPTERA

A. MORPHOLOGY

Earwigs (Figure 2) are rather elongated, somewhat flattened insects (5–20 mm). The fairly heavily sclerotized body is similar to that of the rove beetle (Staphylinidae); however, it is easily distinguished by a pair of forceps-type cerci projecting from the posterior end of the abdomen in both males and females.[1] Inner margins of the forceps of the male are usually distinctly curved, whereas those of females and immatures are usually straight. If winged, the forewing [tegmin (pl. tegmina)] is leathery and usually truncate, whereas the hindwing is membranous, fan-shaped, with radial venation. Mouthparts are chewing type; compound eyes are large, ocelli are indistinct or lacking; antennae are long and beadlike; tarsi are three segmented with the second tarsal segment subcylindrical, cylindrical or expanded.

B. IDENTIFICATION

Rapid Key to the Families of Dermaptera

1. Tarsus with second segment extending below the base of the third; antenna with 12–16 segments 2
   Tarsus with second segment not extending below the base of the third; antenna with 10–31 segments 3

2. Tarsus with second segment expanding laterally; color usually yellowish or brownish .... Forficulidae
   Tarsus with second segment not expanding, no wider than third; color dark brown to black .... Chelisochidae
3. Pronotum and tegmina light brown with two dark longitudinal stripes; length 20–30 mm .... Labiduridae
   Pronotum and tegmina uniformly colored; length <25 mm 4
4. Antenna with 10–16 segments, segments 4–6 together longer than the first segment; tegmina present or absent: if present they appear as rounded flaps not meeting along inner basal margins .... Labiidae
   Antenna with 14–24 segments, segments 4–6 together rarely longer than the first segment; tegmina present, meeting along entire midline .... Carcinophoridae

C. DISCUSSION OF FAMILIES

Dermaptera are cosmopolitan and feed on a variety of living and dead plant and animal matter, including cadavers.(5,8) The most well-known earwig is the european earwig, *Forficula auricularia* L. (Forficulidae). The other commonly encountered species (suborder Forficulina) are separated by a combination of characters that are described in the literature(3,4,6,9) and include:

Chelisochidae—Black earwigs:
  Chelisoches morio (Fabricius)
Labiduridae—Shore and striped earwigs:
  Labidura riparia (Pallas)—shore earwig
Labiidae—Little earwigs:
  Marava arachidis (Yersin)
Carcinophoridae—Seaside and ring-legged earwigs:
  Euborellia annulipes (Lucas)—Ring-legged earwig
  Anisolabis maritima (Bonelli)—Seaside earwig

E. annulipes (Lucas) and *M. arachidis* (Yersin) are tropical species that lack hindwings.(9) The ring-legged earwig, *E. annulipes* (Lucas), has yellowish legs that are banded with brown rings around the femur and tibia, and it is one of the most common pests in flour mills and meat-processing plants.(3) In Hawaii *E. annulipes* was recovered from decomposing human remains.(8) Birds, bats, shrew and other insectivores are common predators of earwigs, and some are even parasitized by Tachinidae flies (*Triarthria* spp.).(4)

There are two other families of earwigs: Arixeniidae and Hemimeridae. Earwigs in these families are found in association with rodents and therefore may be encountered in microanalytical analyses. Arixeniidae are robust, hairy, apterous, have rodlike forceps and lack eyes, making them easily distinguishable from the more common synanthropic families. In Malaysia, Indonesia, and the Philippines earwigs belonging to the family Arixeniidae have been recorded in association with velvet-furred free-tailed bats (Molossidae), which are known for their house-haunting habits.(4,10) Earwigs belonging to the family Hemimeridae in Africa are parasites under the fur of the giant rats (*Cricetomys* spp.) and the East African mouse (*Beamys* spp.).(4) The Hemimeridae, like the Arixeniidae, are eyeless, apterous and have rodlike forceps. They are distinguished by a depressed, smooth body and specialized large tarsal pads for gripping hairs on their hosts. The classification of these two families varies among authors.(4)

The most diagnostic earwig fragments in microanalytical analyses are the anal forceps (Figure 2). Because of their heavy sclerotization, the mandibles, tegmina, fan-shaped wings, legs and major body fragments are likely to be found in food. Earwig fragments should be compared with authentics for identification.

IV. BLATTARIA

A. MORPHOLOGY

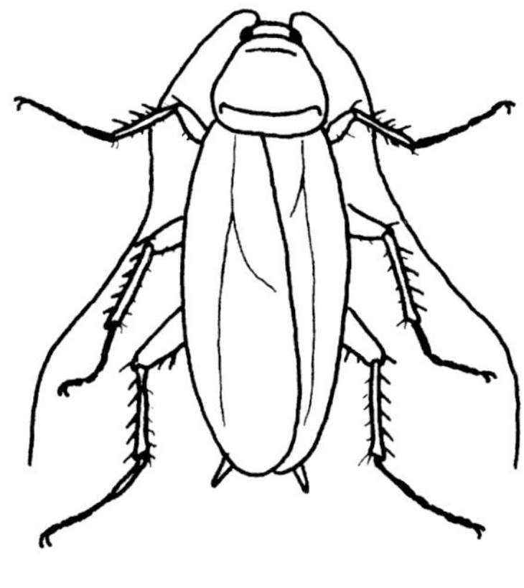

FIGURE 3. Cockroach (Blattaria), adult, dorsal view.

Cockroaches (Figure 3) are generally large (American and Australian cockroaches can be as big as a small bat), heavily sclerotized and dorsoventrally flattened.[1] The head is reposed beneath a large flattened pronotum with the chewing mouthparts directed downward. Antennae are elongated filaments. When present, forewings (tegmina) are leathery with a membranous hindwing underneath. The hindwings can be present, reduced or absent. Immatures show no signs of wings when they are younger; however, in the winged roaches, older immatures develop wing pads. The Blattidae and Blattellidae roaches have large, articulating, tactile leg setae. The second and third femora have these large setae along both ventral margins. The tibia have similar rows of articulating spines. Roaches have one pair of segmented anal cerci. Cockroach excrement is somewhat six sided, the shape apparently being formed in the rectum by the six internal podlike rectal glands.[11]

B. IDENTIFICATION

Rapid Key to the Families of Blattaria

1. Femora 2 and 3 lack strong spines on the ventral margin; length 3 mm or less, found in ant nests .... Polyphagidae
   Femora 2 and 3 with strong spines, similarly arranged along both ventral margins; length 3 mm 2
2. Length 17–50 mm; reddish to dark brown or nearly black; pronotum usually uniform in color, some with yellowish markings, but without dark longitudinal stripes ...... Blattidae
   Length 17 mm; light brown to straw-colored; pronotum usually with two dark longitudinal stripes (except *Supella* spp.) .... Blattellidae

C. DISCUSSION OF FAMILIES

The synanthropic species are nocturnal, opportunistic feeders that congregate in drains, sewers, small cracks and crevices in the floor and walls and even in electrical outlets and switches.[3] Roaches are placed in a high-risk category (see chapter 1) as vectors of food-borne disease and other microbial flora.[12-15] Bacteria have been isolated from their cuticle as well as within their digestive tract.[16] They destroy foodstuffs by eating them, and their excrement fouls what they walk on.[4] Allergens are present in cockroach feces, cast skins and vomit, which can be ingested in contaminated foods.[6] The family Blattellidae contains the well-known domiciliary pests: the German cockroach [*Blattella germanica* (L.)] and the brownbanded cockroach [*Supella longipalpa* (F.)]. *S. longipalpa* lacks the two longitudinal stripes on the pronotum that distinguish *B. germanica*, but it has two transverse brownish bands on the front wings. In the southeastern United States a newly introduced species, the Asian cockroach (*B. asahinai* Mizukubo), is morphologically very similar to the German cockroach. The family Blattidae includes the more well known species of *Periplaneta americana* (L.) (American

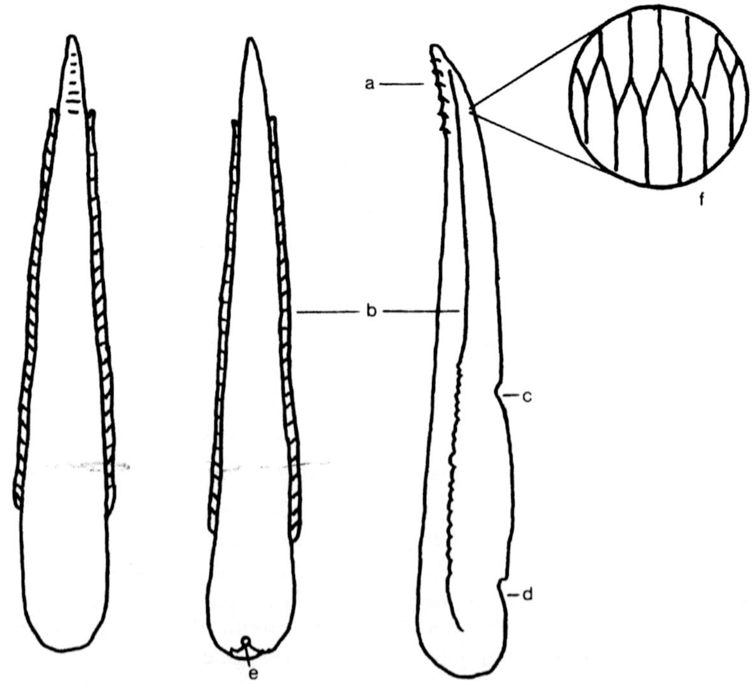

FIGURE 4. (Left) cockroach seta (ventral view).
FIGURE 5. (Middle) cockroach seta (dorsal view), showing (b) lateral ridge and (e) basal pit.
FIGURE 6. (Right) cockroach seta (lateral view), showing (a) distal serration, (b) lateral ridge, (c) dorsomedial depression, (d) dorsobasal depression and (f) detail of sculpturing. (Illustration by V.K. Avaritt.)

cockroach), *P. australasiae* (F.) (Australian cockroach), *P. fuliginosa* (Serville) (smokybrown cockroach) and *Blatta orientalis* L. (oriental cockroach).[2,7]

D. DIAGNOSTIC FRAGMENTS

Few roach fragments are as diagnostic or as easily recognized as the mandibles, antennae and legs. The mandible has a membranous tuft between the molar tooth and the base of the mandible that may be absent or shriveled when isolated from foods.[17] Incisor lobes of the right and left mandibles are not mirror images. The antennal segments are shorter and wider at the base and elongate as the segments taper toward the apex. The middle segments of the antennae are nearly square. The articulating leg setae, or strong spines, are not easily removed or shed from live or freshly killed roaches; however, desiccated specimens will easily lose their setae.[18]

Cockroach leg setae are distinctive from the setae of other closely related orthopterans. The most distinguishing feature of the seta under widefield magnification is the conical shape (Figure 4). The seta also has unique sculpturing and structures, which become more apparent after it is mounted on a microscope slide and viewed under higher magnification (Figures 4, 5 and 6). The distal end of the seta (lateral view) appears to have scalelike plates that overlap to form a short, serrated edge (Figure 6). The ventral view provides a slightly different appearance of short, vertically raised ridges in a series (Figure 4). The exterior edges from both the dorsal and the ventral views exhibit an overlapping pattern along the middle two-thirds of the seta (Figure 5). The lateral view of the same edge appears serrated and forms an anterior ridge. One or two slight depressions can be seen along the dorsal edge in the lateral view. The depression toward the basal end is more prominent (Figure 6). There is a small, single depression located at the basal end of the seta on the dorsal surface (Figure 5). This depression near the basal attachment to the leg is called the campaniform sensillum, which acts as a mechanoreceptor in controlling leg movements.(19) Sculpturing is a prominent feature on the surface of the articulating setae. The appearance of the sculpturing can vary from bricklike to angular depending on the orientation. In lateral view the sculpturing appears bricklike, but if the seta is rotated either dorsally or ventrally the patterns appear more angular (Figure 6).(18) A more technical description of roach fragments and their egg cases (ootheca) can be found in the literature.(17,20-22)

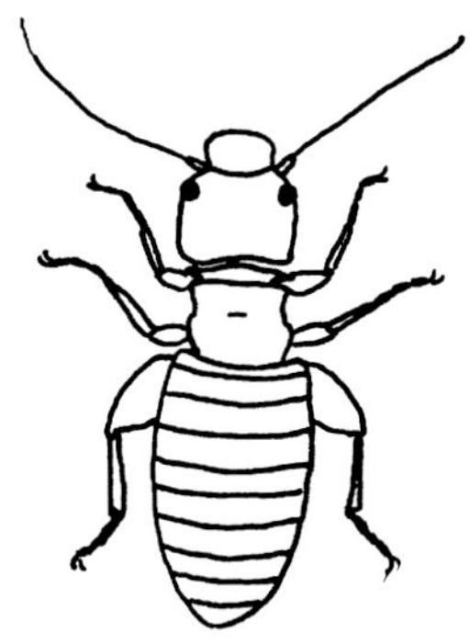

FIGURE 7. Booklouse (Psocoptera: Liposcelidae), *Liposcelis* sp., adult, dorsal view.

V. PSOCOPTERA

A. MORPHOLOGY

Booklice (Figure 7) and other psocids are small (<5 mm), soft-bodied insects, often white to light in color, with long filiform antennae of 12–50 segments. The antennae of booklice belonging to the suborder Troctomorpha have ringlike patterns or secondary annulations on the individual segments. The psocid head is relatively large, and the thoracic segments are somewhat compressed and narrower than the head and the wider abdomen. The mouthparts include asymmetrical mandibles, chitinous hypopharynx and maxillae with a rodlike lacinia, and the basal segment of the labrum (clypeus) is swollen or bulbous. Winged psocids usually have four wings held rooflike over the abdomen; however, the wings can be reduced with little or no venation or be completely absent. Psocids with reduced wings or wing pads are sometimes termed brachypterous. Adult tarsi can be two or three segmented; however, immatures always have two-segmented tarsi and a reduced number of antennal segments.(23) Psocids, unlike the preceding hemimetabolous insects covered in this chapter, do not have anal cerci. All three suborders of psocids contain families that are able to reproduce asexually (parthenogenesis).(23) When the virgin females reach sexual maturity (in some *Liposcelis* spp. maturity is reached as early as the fourth molt) they begin to lay fertilized eggs.(24) Parthenogenesis occurs in the families of stored-product psocids and, combined with optimum conditions, allows for the production of large colonies.

B. IDENTIFICATION

Rapid Key to Suborders and Families of Psocoptera

1. Antenna with 18–50 segments; labial palpus with two segments; hypopharynx with chitinous filaments separated throughout their length (Trogiomorpha) 4
   Antenna fewer than 18 segments; labial palpus with either one or two segments; hypopharynx not as above 2
2. Antenna with 11–17 segments; flagellum beyond segment 4 or 5 with secondary annulations; labial palpus with two segments; hypopharynx with chitinous filaments separated only in distal region (Troctomorpha) 6

3. Antenna with 13 or fewer segments, without secondary annulations; labial palpus with one segment; hypopharynx with chitinous filaments separated posteriorly (Psocomorpha) 6
4. Head long and anteriorly narrow; compound eyes subdorsal; maxillary palpus with sensillum on second segment .... Psyllipsocidae
   Head short and anteriorly broad; compound eyes contiguous with the dorsal surface; and maxillary palpus with sensillum on second segment 5
5. Forewings reduced to veinless pads or knobs (brachypterous) .... Trogiidae
   Forewings well developed or reduced but always with veins .... Psoquillidae
6. Hind femur enlarged or swollen, with a notch or lateral process, or without; wings usually absent .... Liposcelidae
   Hind femur not as above; wings present .... Lachesillidae

C. DISCUSSION OF FAMILIES

Booklice (*Liposcelis* spp.), are frequently encountered in buildings or domestic habitats. If you have ever opened an old, infested book (hence the name "booklice"), you are no doubt familiar with these minute intruders. Booklice, however, not to be confused with the chewing and sucking lice (Phthiraptera), are far from restricted to feeding on the glue and bindings of old books. At least 50 species are known to occur in foods and foodstuffs.(23,25) Booklice attack a wide variety of raw materials and foodstuffs, including grains, dried spices and dried seafood, and have also been intercepted in food processing facilities and warehouses.(26-29) They have even been found feeding on dead insects and cast skins.(23,24,29-31)

Stored-product psocids are usually less than 3 mm long and prefer dark, concealed areas where they often go unnoticed. Under favorable conditions stored-product psocids can be prolific breeders and form large gregarious populations. In large populations, psocids damage foods and adulterate products with their excreta, cast skins and dead bodies. Cheyletid mites are often predators of young immatures and eggs in large populations. Some psocids are able to tolerate up to two months without food.(32) It is a common misperception that booklice are primarily or exclusively attracted to moldy foods.(33) Although there are some mold-feeding species of Psocoptera, others species such as *Liposcelis*, have been observed feeding on dried capsicums. Under higher magnification the reddish-orange oils from the capsicums are quite conspicuous in the gut.

1. Liposcelidae (Booklice)

Of the five families of stored-product psocids covered in this section, the most frequently encountered are booklice, belonging to the family Liposcelidae. Three genera, *Belaphotroctes* spp., *Embidopsocus* spp. and *Liposcelis* spp., are encountered in stored-products. Of these, *Liposcelis* spp. are more prevalent as adulterants. Diagnostic fragments of *Liposcelis* spp. include the head, mandibles, hind leg and abdomen. The head has 2–8 ommatidia in each eye, and the dorsal cuticle is usually covered with arched ridges or minute tubercles. The mandibles are asymmetrical with no setae or pores (Figure 8).(34) The hind femur is enlarged or swollen, and, in *Liposcelis* spp., the femur has a notch or lateral process (Figure 9). The abdomen is also sculptured as the head. The following key separates the important genera.

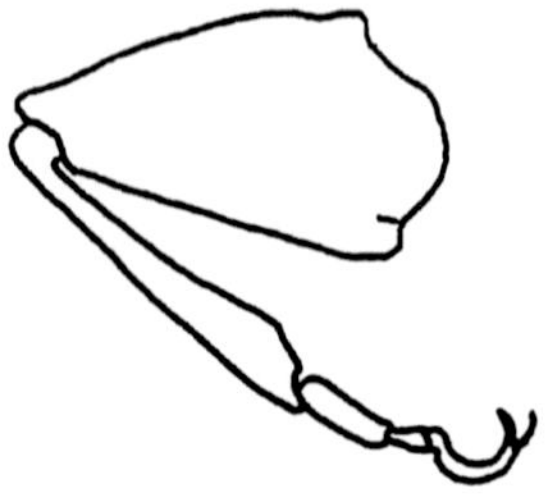

FIGURE 8. (Left) *Liposcelis* spp. mandible.
FIGURE 9. (Right) *Liposcelis* spp. leg III.

Rapid Key for the Identification of Common Genera of Liposcelidae

1. Hind femur with a notch or lateral process; dorsal cuticle with sculpturing of arched ridges or minute tubercles .... *Liposcelis* spp.
   Hind femur without notch; dorsal sculpturing faint or absent 2
2. Terminal segment of maxillary palpus swollen or bulbous; abdominal terga without a narrow, sclerotized transverse strip .... *Belaphotroctes* spp.
   Terminal segment of maxillary palpus not as above; abdominal terga with a narrow, sclerotized transverse strip .... *Embidopsocus* spp.

2. Lachesillidae

The remaining four families of stored-product psocids do not have a swollen hind femur. Lachesillidae have two pairs of wings and a two-segmented tarsus. The forewings are usually clear and do not have a prominent arrangement of setae on the wing margin. The cosmopolitan grain psocid, *Lachesilla pedicularia* (L.), as its name implies, is a common invader of grains. In *L. pedicularia* (L.) the forewing is clear, with brown borders on the descending portion of the distal veins.

3. Trogiidae

Trogiids have reduced, veinless forewings that resemble small pads or knobs. *Lepinotus* spp. and *Trogium* spp. may be found in grains as well as in dried plant and food materials.(35) *Trogium pulsatorium* (L.) is also known to inhabit the nests of birds and wasps in buildings.(23)

4. Psoquillidae

Psoquillidae have compound eyes that are even with the dorsal surface of the head and veined forewings that are well developed or reduced. The maxillary palpus has sensillum on the second segment. *Psoquilla* spp. and *Rhypsocus* spp. have been found in stored foods, infrequently. Most species of the psoquillids are associated with dead plant or packing materials. In the tropics they occur outside.(23,25)

5. Psyllipsocidae

Psyllipsocids have compound eyes that are subdorsal on the head and have a maxillary palpus with a sensillum on the second segment. *Dorypteryx* spp. and *Psyllipsocus* spp. are associated with caves, cellars and houses. *Psocatropus microps* (Enderlein) has been found in stored foods.(23,25)

REFERENCES

1. Ross, H.H., *A Textbook of Entomology*, 2nd ed., John Wiley & Sons Inc., NY, 1-519, 1956.
2. Burgess, N.R.H., *Public Health Pests: A Guide to Identification, Biology, and Control*, Chapman Hall, NY, 1-162, 1990.
3. Hickin, N.E., *Pest Animals in Buildings: A World Review*, Longman, NY, 1-385, 1985.
4. Naumann, I.D., Ed., *The Insects of Australia*, 2nd ed., CSIRO, 1-1075, 1991.
5. Catts, P.E. and Haskell, N.H., *Entomology and Death: A Procedural Guide*, Joyce's Print Shop, Clemson, SC, 29-48, 1990.
6. Goddard, J., *Physician's Guide to Arthropods of Medical Importance*, CRC Press, Boca Raton, FL, 1-332, 1993.
7. Borror, D.J., Triplehorn, C.A., and Johnson, N.F., *An Introduction to the Study of Insects*, 6th ed., Saunders Coll. Publ., Philadelphia, 1-875, 1989.
8. Goff, M.L., Comparison of insect species associated with decomposing remains recovered inside dwellings and outdoors on the island of Oahu, Hawaii, *J. Forensic Sci.*, 36, 748-753, 1991.

9. Freeman, P., Ed., *Common Insect Pests of Stored Food Products: A Guide to Their Identification*, 6th ed., Brit. Mus. (Natur. Hist.), London, 1-69, 1980.
10. Allen, G.M., *Where Bats Hide*, Dover Publ. Inc., NY, 1-368, 1939.
11. Harris, K.L. and Reynolds, H.L., Eds. *Microscopic-Analytical Methods in Food and Drug Control*, FDA Tech. Bull. No. 1, 1- 270, 1960.
12. Sramova, H., Daniel, M., Absolonova, V., Dedicova, D., Jedlickova, Z., Lhotova, H., Petras, P., and Subertova, V., Epidemiological role of arthropods detectable in health facilities, *J. Hosp. Infection*, 20, 281-292, 1992.
13. Domenichini, G., Fogliazza, D., and Pagani, M., Studies on the transmission of listeria by means of arthropods, *Ital. J. Food Sci.*, 4, 269-278, 1992.
14. Roth, M.L. and Willis, E.R., The medical and veterinary importance of cockroaches, *Smithson. Misc. Collect.*, 134, 1-147, 1957.
15. Singh, S.P., Sethi, M.S., Sharma V.D., The occurrence of salmonellae in rodent, and shrew, cockroach and ant, *Int. J. Zoonoses*, 7, 58-61, 1980.
16. Rivault, C., Le Guyader, A., and Cloarec, A., Transport de bacteries par les Blattes en milieu urbain, *Bull. Soc. Zool. Fr.*, 116, 235-241, 1991.
17. Kurtz, O.L. and Harris, K.L., *Microanalytical Entomology for Food Sanitation Control*, Assoc. Off. Agric. Chem., Washington, DC, 1-576, 1962.
18. Avaritt, V.K. and Zimmermann, M.L., Identification of the larger tactile setae on the legs of cockroaches, *Laboratory Information Bulletin* (FDA), 3281, 1-4, 1988.
19. Chapman, R.F., *The Insects: Structure and Function*, 3rd ed., Harvard Univ. Press, Cambridge, MA, 1-919, 1982.
20. Eiduson, H.P., Firestone, D., and Roe, J.E., Cockroaches and their fragments as food contaminants, *J. Assoc. Off. Agric. Chem.*, 41, 886-898, 1958.
21. Narasimham, A.U. and Sankaran, T., On some diagnostic characters of the oothecae of six common cockroaches (Blattidae), *Syst. Entomol.*, 5, 105-107, 1980.
22. Tanaka, A. and Kitamura, S., Spur patterns on the tibia and femur of the german cockroach (Blattodea: Blattellidae), *Ann. Entomol. Soc. Am.*, 85, 768-774, 1992.
23. Mockford, E.L., *North American Psocoptera (Insecta)*, Sandhill Crane Press, Gainesville, FL, 1-455, 1993.
24. New, T.R., Biology of the Psocoptera, *Oriental Insects*, 21, 1-109, 1987.
25. Gorham, J.R., Ed., *Insect and Mite Pests in Food: An Illustrated Key*, Vols. 1 & 2, U.S. Dept. Agric., Agric. Handbook No. 655, 1-767, 1987.
26. Olsen, A.R., List of stored-product insects found in imported foods entering the United States at southern California ports, *Bull. Entomol. Soc. Am.*, 27, 18-20, 1981.
27. Olsen, A.R., Mites and other filth in dried shrimp imported into United States from the orient, *J. Food Prot.*, 45, 1204-1207, 1982.
28. Olsen, A.R., Bryce, J.R., Lara, J.R., Madenjian, J.J., Potter, R.W., Reynolds, G.M., and Zimmerman, M.L., Survey of stored-product and other economic pests in import warehouses in Los Angeles, *J. Econ. Entomol.*, 80, 445-459, 1987.
29. Obr, S., Psocoptera of food-processing plants and storages, dwellings and collection of natural objects in Czechoslovakia, *Acta Entomol. Bohemoslov.*, 75, 226-242, 1978.
30. Ghani, M.A. and Sweetman, H.L., Ecological studies of the book louse, *Liposcelis divinatorius* (Mull.), *Ecology*, 32, 230-244, 1951.
31. Broadhead, E., A revision of the genus *Liposcelis* Motschulsky with notes on the position of this genus in the order corrodentia and on the variability of ten *Liposcelis* species, *Trans. Ry. Entomol. Soc. London*, 101, 335-393, 1950.
32. Turner, B.D. and Maude-Roxby, H., Starvation survival of the stored product pest *Liposcelis bostrychophilus* Badonnel (Psocoptera, Liposcelidae), *J. Stored Prod. Res.*, 24, 23-28, 1988.
33. Rees, D.P. and Walker, A.J., The effects of temperature and relative humidity on population growth of three *Liposcelis* species (Psocoptera: Liposcelidae) infesting stored products in tropical countries, *Bull. Entomol. Res.*, 80, 353-358, 1990.
34. Peace, D.M., *Key for Identification of Mandibles of Stored-Food Insects*, Assoc. Off. Anal. Chem., Arlington, VA, 1-169, 1985.
35. Fahy, E.D., Some factors in the ecology and dispersal of *Lepinotus patruelis* Pearman (Psocoptera), a pest of stored products, *J. Stored Prod. Res.*, 7, 107-123. 1971.
36. Sinha, R.N. and Watters, F.L., *Insect Pests of Flour Mills, Grain Elevators, and Feed Mills and Their Control*, Agric. Can. Publ. 1776, Ottawa, 1-290, 1985.

Chapter 5

BEETLES (INSECTA: COLEOPTERA)

James J. Madenjian

CONTENTS

I. Introduction 40
A. Scope 40
B. Categorization of Stored-Product Beetles 41

II. Significance of Food-Contaminating Beetles 41

III. Techniques and Equipment 41

IV. Morphology 42
A. Head and Head Appendages 42
1. Mandibles 42
2. Antennae 44
B. Thorax and Thoracic Structures 44
1. Legs 45
2. Elytra 45
C. Abdomen 45

V. Identification 45
A. Identification Key 46
B. List of Common Species 48

VI. Discussion of Families 48
A. Common Food-Contaminating Families 52
1. Anobiidae (Drugstore and Cigarette Beetles) 52
2. Anthicidae (Antlike Flower Beetles) 52
3. Anthribidae (Fungus Weevils) 53
4. Bostrichidae (False Powderpost Beetles) 53
5. Bruchidae (Seed Beetles) 53
6. Carabidae (Ground Beetles) 54
7. Cerylonidae (Minute Bark Beetles) 54
8. Ciidae (Minute Tree Fungus Beetles) 55
9. Cleridae (Checkered Beetles) 55
10. Corylophidae (Minute Fungus Beetles) 56
11. Cryptophagidae (Silken Fungus Beetles) 56
12. Cucujidae (Flat Bark Beetles) 56
13. Curculionidae (Weevils) 58
14. Dermestidae (Dermestid Beetles) 59
15. Erotylidae (Pleasing Fungus Beetles) 60
16. Histeridae (Hister Beetles) 60
17. Languriidae (Lizard Beetles) 60
18. Lathridiidae (Minute Brown Scavenger Beetles) 61
19. Lyctidae (Powderpost Beetles) 61

61
20. Mycetophagidae (Hairy Fungus Beetles) 62
21. Nitidulidae (Sap Beetles) 62
22. Ptiliidae (Featherwinged Beetles) 62
23. Ptinidae (Spider Beetles) 63
24. Scolytidae (Bark and Ambrosia Beetles) 63
25. Staphylinidae (Rove Beetles) 64
26. Tenebrionidae (Darkling Beetles) 64
27. Trogossitidae 65
B. Other Families 65

References 66

I. INTRODUCTION

This chapter addresses the beetles associated with stored food products. Species associated with stored food products are those with a history of being found in these products. Throughout this book, the term "stored-product" refers principally to stored food products. I specify "food" product only here and in the discussion of those families that also contain wood product pests to emphasize this usage. Beetles are the largest group of stored-product insects.[1] Estimates of the numbers of different kinds of beetles associated with all stored-products (including food, lumber, furniture, textiles and hides) range up to 500–600 species.[2-4] In this vein, Halstead included 40 families in his identification key to the beetles associated with stored-products.[3] More conservatively, only about 100 species of insects, beetles included, are thought to cause contamination or loss of stored food products.[4] On the basis of my experience analyzing food products, I have included 27 beetle families in this chapter as representative of what the microanalytical entomologist can actually and realistically expect to encounter in stored food products. Families that we in the FDA have not encountered in our samples of stored-products or have rarely encountered, or are represented only by field or incidental species, have not been treated. Likewise, within families that are included in this chapter, taxa such as genera and species that have not been encountered by us generally are not treated. In contrast, families that we have repeatedly recorded in our samples but that are not generally acknowledged as being associated with stored-products have been included.

A. SCOPE

This chapter is biased toward the experiences of the entomology laboratory of the U.S. FDA in Los Angeles during the last 10 years. During this period we have analyzed, for insects and other filth, roughly 1500–2500 food samples per year. The commonly sampled countries of origin include Thailand, the Peoples Republic of China, the Republic of China, Hong Kong, India, South Korea, the Philippines and Mexico. Other countries of origin include Indonesia, Malaysia, Sri Lanka, Singapore, Italy, Greece, Peru, Colombia, Paraguay and Guatemala. Southern and eastern Asia were well represented and Africa was poorly represented. Ports of entry of southern California and Arizona include airports, harbors and the California–Arizona and Mexico international border.

Products include dried nuts, dried vegetables and fruits, preserved vegetables and fruits in liquid, frozen vegetables and fruits, dried mushrooms, canned mushrooms, dried whole or ground spices, whole cereals and pulses, various flours, flour products such as pasta and bean thread, various vegetable sauces, dried fish, fresh and frozen fish, frozen shrimp, fish sauces, crab and shrimp pastes, whole and ground coffee and others. Included in this chapter as food are leafy wrappers such as corn husks, banana leaves and bamboo leaves. Although these wrappers may not be consumed, they contain material that is. Food also includes tea stock which is not directly consumed but is mixed in a liquid that is consumed. Examples include medicinal mushrooms, various dried twigs and leaves and dried lizards. Products excluded because they are regulated by agencies other than the FDA are beef, poultry, pork and their derivatives.

B. CATEGORIZATION OF STORED-PRODUCT BEETLES

The species associated with stored-products may not necessarily be able to develop in these products. For example, *Bruchus pisorum* (L.) (pea weevil) and *Curculio caryae* (Horn) (pecan weevil) are both field or preharvest pests. Neither can develop in stored-products. Pea weevil larvae develop in growing peas, but the adults may emerge in lots of stored peas. These adults do not normally breed in the stored peas. Pecan weevil larvae develop in growing pecans. The larvae remain in the dropped nuts and then burrow into the ground to pupate, although some larvae can be found in the harvested nuts. Pecan weevils cannot successfully breed in stored pecans. Other beetle species are pests only in wood structures of food warehouses and wood containers of stored food. These species only incidentally infest the food products.(5)

Some field pests [e.g., *Acanthoscelides obtectus* (Say) (bean weevil)] are also storage pests. They can survive harvest and produce many generations in storage. Extremely high densities of these pests can be reached after successive generations in storage.

Beetle species that can develop in stored-products may not feed on the actual product. The different categories are listed in Hinton.(6) Some species feed on mold on the product. Some species scavenge on the remains of dead insects, and others prey on live insects. There are probably many stored-product species that have more than one feeding (trophic) habit.

Some beetle species can develop in a huge variety of stored-products whereas other species have been recorded only in one species of stored-product. For a full discussion of the different perspectives on categorizing stored-product insects, including the beetles, see chapter 1.

II. SIGNIFICANCE OF FOOD-CONTAMINATING BEETLES

The aesthetic factor of beetles in food cannot be ignored. Their presence may be repugnant to some consumers. Live beetle infestations, in particular, carry a high consumer outrage factor. Beetles in food may be indicators of decomposition or mold.(6) Nitidulids, staphylinids and other families can occupy decaying organic material. Lathridiid adults and larvae feed exclusively on myxomycetes and fungi (particularly molds).

Phillips and Burkholder reviewed the potential hazards of beetle adults, larvae and their fragments in food to consumers and some employees of the stored-product industry.(7) Nine categories were discussed including mechanical irritation of skin and mucous membranes, different types of allergens, mechanical irritation from ingestion and invasion by live individuals from ingestion, chemical toxins, different types of vectors and mycotoxins.

The setae of dermestid larvae are hazardous in several ways. They can cause mechanical irritation to the skin and mucous membranes of dock workers who unload heavily infested product. They also have allergic properties and may cause mechanical injury if ingested. An infant suffered ulcerative colitis as the result of eating cereal infested with dermestid larvae.(8,9)

Tenebrionids, including *Tribolium castaneum* (Herbst), produce highly reactive alkylated benzoquinones as a defense against predation. Flour and grain in which these species are thriving are rendered unpalatable by the objectionable taste and odor of these chemicals. Quinones can cause dermatitis with skin discoloration, erythema (reddening) and the formation of papules and vesicles. Quinones are suspected of being carcinogens.(10)

III. TECHNIQUES AND EQUIPMENT

Most of the morphological characteristics mentioned below can be viewed under a dissecting microscope at 60× or less, but some must be viewed under a high-magnification microscope (up to 400×). In the latter case, the specimen should be mounted on a microscope slide.

Glycerin jelly is a very convenient mounting medium because it simply needs to be melted on a hot plate. In addition, a mount employing glycerin jelly can last a fairly long time (more than 10 years) if sealed with fingernail polish (see chapter 14). Examples of characteristics requiring high magnification are the whorls of hairs on the antennal segments of ptiliid adults; and the clavate, pectinate setae and dorsolateral gland openings of corylophid larvae. In both cases, I recommend mounting the whole individual.

Within some genera, the most important adult characteristics for species determinations are those of the genitalia. Two examples are *Cryptolestes* spp. (Cucujidae) and *Sitophilus* spp. (Curculionidae). For *Cryptolestes* spp. specimens, I recommend that the abdomen plus meso- and metathoraces (without elytra) be removed and cleared in 10% KOH. After rinsing, the fragment can be mounted on a slide and examined at 400×.[11] If the genitalia character(s) cannot be examined adequately, then further dissection is necessary. For *Sitophilus* spp. specimens, the genitalia can be isolated relatively easily through dissection under the dissecting microscope. The aedeagus of the male can be cut at about mid-length and the cross-section examined under the dissecting microscope in order to distinguish *S. zeamais* Motschulsky (maize weevil) from *S. oryzae* (L.) (rice weevil). The Y-shaped sclerite of the female genitalia should be mounted on a microscope slide and viewed under higher magnification. Within some genera, only the male genitalia are taxonomically useful [e.g., *Oryzaephilus* spp. (Cucujidae)].

IV. MORPHOLOGY

Beetles (Figures 1 and 2) are readily recognized by the specialized front wings, the elytra (sing. elytron), which form a hard or leathery dorsal shell over the mesothorax, metathorax and usually most or all of the abdomen. The beetle elytra meet in a straight line along the middle of the back without overlapping and they exhibit no venation. They sheathe the membranous hind wings that are intricately folded underneath the elytra when not in use for flying. The last abdominal tergite, the pygidium, is sometimes exposed beyond the tips of the elytra, and in some beetle groups the elytra are truncate, exposing up to seven abdominal tergites. The only other insect group with leathery elytra that meet in a straight line are the earwigs (Dermaptera), which are distinguished from the beetles by their unique pincerlike caudal forceps.

A. HEAD AND HEAD APPENDAGES

The adult beetle head (Figures 3 and 4) is typically ovoid with a large pair of lateral compound eyes and occasionally with a single, median ocellus. The margin of the compound eye may be entirely round or it may be notched slightly to deeply on the inner margin. In the weevils (Figure 5), the front of the head is elongated into a slender snout. Mouthparts (Figure 6) are the chewing type with the labrum, mandibles, maxillae and labium evenly developed and relatively unmodified. The mouthparts project forward, or downward if the head is deflexed. In the weevils the mouthparts (Figure 5A) occur at the tip of the snout. The head may be variously sculptured with punctations, rugosities, carinae (ridges) and smooth areas.

1. Mandibles

Mandibles are diagnostic for most genera of stored-product beetles, in both the adult and larval stages.[18-20] The beetle mandible has a typical pyramid shape with basal articulating condoyles and a broad molar area, in food-contaminating pests. The distal apex of the mandible usually bears two (bidentate), three (tridentate) or more apical teeth. The expanse between the molar area and the apical teeth, the mesal ridge, may be straight or concave. In beetle larvae especially, the mesal ridge or concavity is often armed with a fleshy or membranous projection, the prostheca, whose shape has taxonomic significance. Other mouthparts are also identifiable as filth fragments in food but do not survive food processing technology as

1

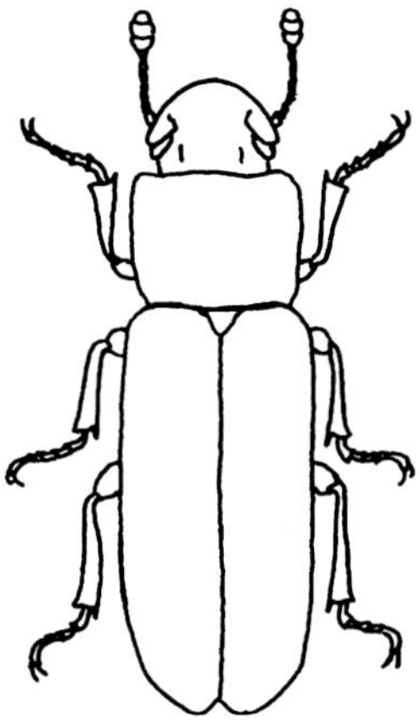

2

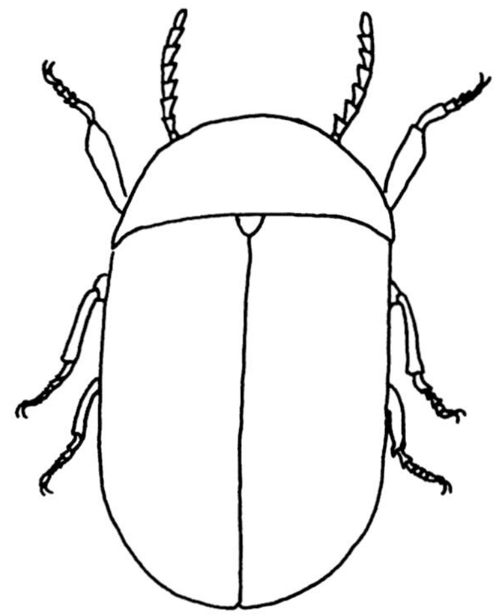

3

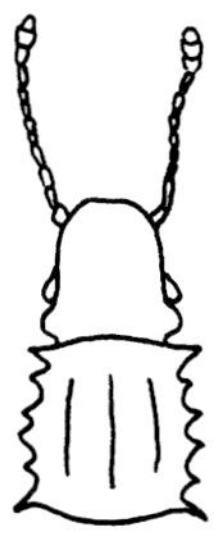

4

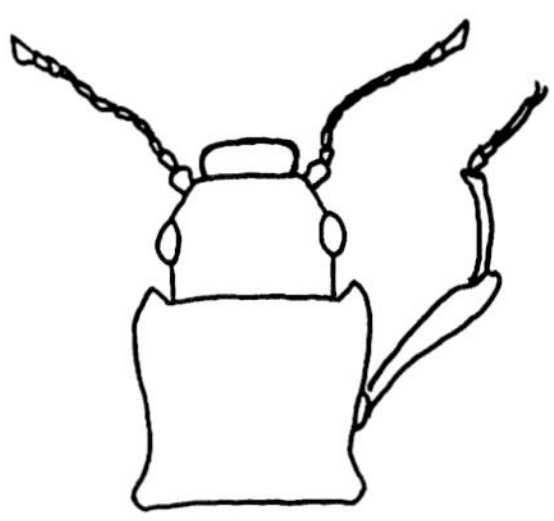

5

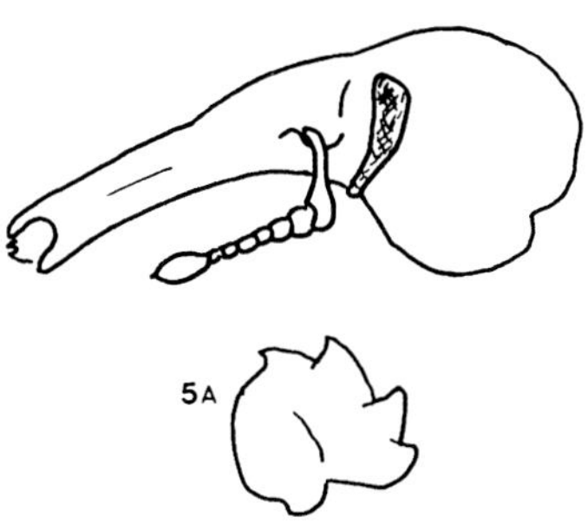

6

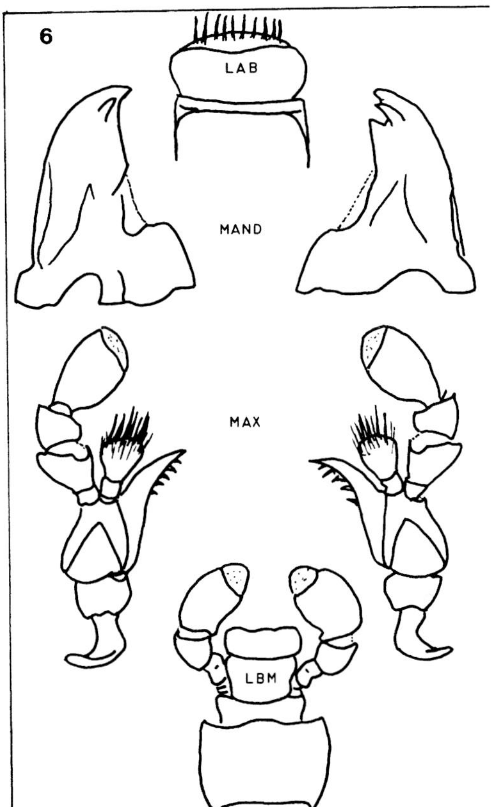

FIGURE 1. Adult red flour beetle, *Tribolium castaneum* (Tenebrionidae), showing general beetle morphology.

FIGURE 2. Adult cigarette beetle, *Lasioderma serricorne* (Anobiidae), showing morphology. Note deflexed head.

FIGURE 3. Sawtoothed grain beetle, *Oryzaephilus surinamensis* (Cucujidae), adult head and prothorax. Note lateral teeth on pronotum.

FIGURE 4. Squarenosed fungus beetle, *Lathridius minutus* (Lathridiidae), adult head and prothorax. Note three-segmented tarsus.

FIGURE 5. Weevil head, *Sitophilus* spp. (Curculionidae). (A) Detail of adult mandible.

FIGURE 6. Typical beetle mouthparts (*Tribolium sp.*). LAB, labrum; MAND, mandible; MAX, maxilla; LBM, labium.

7

8

9

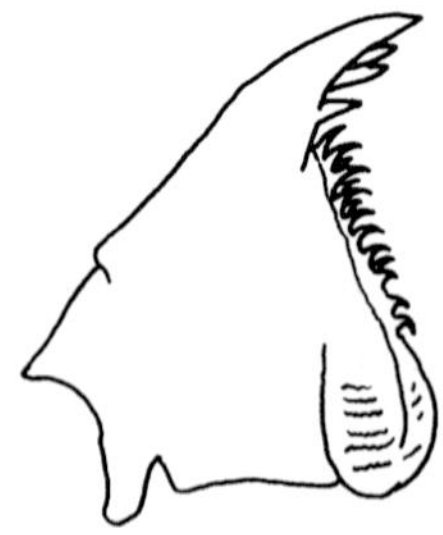

10

11

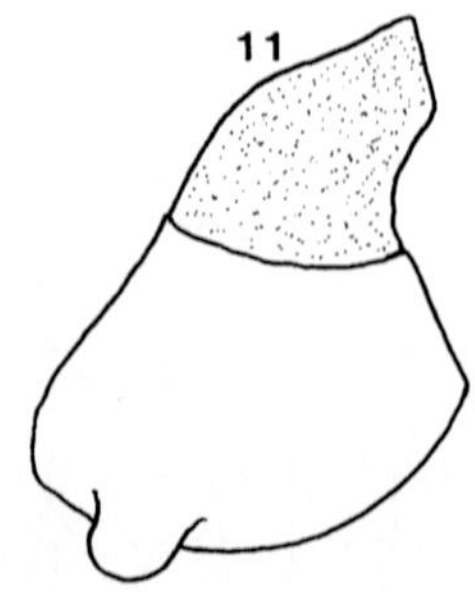

12

FIGURE 7. Larval mandible, lesser grain borer, *Rhyzopertha dominica* (Bostrichidae).
FIGURE 8. Larval mandible, redlegged ham beetle, *Necrobia rufipes* (Cleridae).
FIGURE 9. Larval mandible, driedfruit beetle, *Carpophilus* spp. (Nitidulidae).
FIGURE 10. Larval mandible, hide beetle, *Dermestes* spp. (Dermestidae).
FIGURE 11. Larval mandible, dermestid beetle, *Trogoderma* spp. (Dermestidae).
FIGURE 12. Larval mandible, flour beetle, *Tribolium* spp. (Tenebrionidae).

well as the rugged mandibles. Figures 7–12 illustrate some of the more commonly encountered larval mandibles.

2. Antennae

Beetle antennae vary greatly in form. Most stored-product pest species have adult antennae that are filiform, clavate (clubbed), serrate or geniculate. Clavate antennae may have a loose club, where the apical club segments are enlarged and loosely joined, or a compact club, where the enlarged club segments are tightly joined together and compacted. The number of segments in the antennae and the number of segments in the club, if present, are important taxonomic characters. Antennae, or substantial portions of antennae, are identifiable filth fragments when found in food.[21]

B. THORAX AND THORACIC STRUCTURES

The thorax is dominated dorsally by the shieldlike pronotum, which is usually flat and quadrate, sometimes rounded laterally. The corners, called angles, of the pronotum and its lateral margins may bear toothlike projections (Figure 3). The mesonotum and metanotum lie hidden beneath the elytra except for a small, triangular sclerite, the scutellum, which is located medially between the bases of the elytra. Ventrally, the thorax has three platelike sternites, each bearing the coxal attachment cavities for a pair of legs. The coxal cavity may be closed (entirely surrounded by the sternite) or open behind. The front and middle coxae articulate in these cavities but the hind coxae are flattened and immovably attached at the thorax, appearing more as part of the thorax than an appendage segment. The prosternum is sometimes flanked by a pair of lateral sclerites that abut the pronotum along the notopleural sutures. In most cases, however, the prosternum meets the rest of the prothorax directly and the notopleural suture is absent. The mesosternum and metasternum are each flanked by two lateral pairs of sclerites.

1. Legs

The legs are composed of five segments: coxa, trochanter, femur, tibia and tarsus. The shape and position of the coxae are sometimes taxonomically significant. Coxa shapes range from flat and elongate to round or globular. The two coxae of a thoracic segment can be separated or contiguous. Beetle trochanters are not directly aligned between coxa and femur, being offset to a greater or lesser degree. In one group of beetles (suborder Adephaga), the trochanter on the hind leg is large and so strongly offset toward the midline that the coxa and femur are nearly touching. In the extreme case, the ground beetles (family Carabidae), the trochanters appear to hang suspended from the hind coxae, like a pair of beans, completely independent of the rest of the legs. The femur and tibia may exhibit diagnostic shapes or structures.

Beetle tarsi can vary in number of secondary segments, or tarsomeres, to the point that various combinations of numbers of segments are reliable taxonomic guides to various groups. The tarsal formula of a group is a triad of numbers that indicate the number of tarsal segments on each leg, in order from prothorax to metathorax. For example, a tarsal formula of 5-5-4 (family Tenebrionidae) (Figure 1) indicates five segments on each front tarsus, five on each middle tarsus and four on each hind tarsus. Care must be taken in evaluating the tarsal formula of beetle specimens. Some groups of beetles have pseudotetramerous tarsi where the third segment is bilobed, concealing a minute fourth segment at the crotch. This condition is referred to as "apparently four segmented" by some taxonomists.

2. Elytra

The elytra (sing. elytron) are diagnostic fragments, even if broken into many tiny pieces. The surface sculpturing of beetle elytra is composed of numerous microscopic repetitive units or "cells" that are recognizable on an individual basis even if separated from the original elytron by mechanical damage from food processing technology.[22] The recognition of these patterns on elytral fragments is well-established in the literature for most of the common stored-product species of beetles.[23-26] Elytral fragments are such a distinctive food contaminant that they are used as a standard spike material for comparison and validation of analytical methods for the recovery of insect fragments from foods.[27]

C. ABDOMEN

The beetle abdomen consists of three to eight visible segments (usually five or six), each supporting a tergite and sternite with membranous pleural areas bearing a pair of spiracles. The genitalia are retracted inside the abdomen. The tergites are usually covered by the elytra. The taxonomically important structures from the standpoint of fragment identification are the elytra (actually attached to the mesothorax)(see above) and the genitalia. Genitalia, and even fragments of some genitalia, can be identified to the species level for many stored-product insects. Dissected genitalia are used extensively to identify whole specimens of beetles to the species level.[11,28,29]

V. IDENTIFICATION

I have strictly followed the family classification of Arnett.[12] Halstead's scheme differs in a number of ways.[3] For example, Arnett retained the Laemophloeinae, Passandrinae and the Silvaninae as subfamilies of the Cucujidae whereas Halstead elevated these three taxa to family level. I have used the taxonomic characteristics and morphological terminology from several sources. Many adult characteristics were taken from Halstead[3] and Freeman.[2] Others were found in Arnett[13] and Arnett et al.[14] Larval terminology can be found in Lawrence.[15] I have not mentioned the overall form of the larva unless it is grublike or C-shaped (scarabaeiform).

Common names of beetle families and species have been abundantly supplied. The most important sources are the various contributors of Gorham[16] and of Stehr.[17] Taxonomic characteristics in the following identification key and discussion of families may be limited in scope. Family characters may not represent all members of the family, but rather only those members associated with stored food products.

A. IDENTIFICATION KEY

Rapid Key to the Families of Common Food-Contaminating Beetle Adults[2]

1. Head elongated in front of eyes to form a narrow snout or beak .... Curculionidae
   Head without a distinct snout or beak 2
2. Hindwing fringed with hairs that often project beyond the elytra; lenght 0.5–5.0 mm 3
   Hindwing without fringe of hairs 4
3. Fringe long; antenna 11 segmented with whorls of long hair on each segment; length 0.5–1.0 mm .... Ptiliidae
   Fringe short; antenna clubbed, without hairs and with basal two segments enlarged .... Corylophidae
4. Prothorax with a short, narrow apical neck so that head is completely exposed and never partly received into the anterior part of the thorax .... Anthicidae
   Prothorax without an apical neck and head always with at least base fitting into thorax 5
5. Body oval, convex; antenna ten segmented, with one-segmented club and received into a cavity of the prothorax .... (*Murmidius* spp.) Cerylonidae
   Body not both oval and convex 6
6. Elytra leaving at least one abdominal segment exposed dorsally 7
   Elytra completely covering all abdominal segments dorsally 12
7. Elytra very short; abdomen flexible with at least six segments exposed ventrally and with six or more segments exposed dorsally .... Staphylinidae
   Elytra moderately long; abdomen hard and not flexible, with not more than five segments exposed ventrally, and never with more than three segments exposed dorsally 8
8. Antenna elbowed; head below and in front of eye with a groove that receives the first antennal segment when the latter is retracted .... Histeridae
   Antenna not elbowed and head without grooves for reception of the first antennal segment 9
9. Antenna ten segmented; antennal club large and one segmented; form slender; length 1.8–2.5 mm .... (*Monotoma* spp.) Rhizophagidae
   Antenna 11 segmented; antennal club with more than one segment, or without a distinct club; form not slender; length usually 2–5 mm 10
10. Antenna with a large, distinct, compact three-segmented club; abdomen with two or three segments exposed dorsally; elytra not striate .... Nitidulidae
    Antenna not distinctly clubbed, although sometimes with three apical segments enlarged; abdomen with only last segment vertically inclined; elytra deeply striate 11
11. Antenna with three apical segments distinctly larger than other segments; eyes entire, nowhere emarginate .... (coffee bean weevil) Anthribidae

Antenna with three apical segments not distinctly larger than other segments; eye usually with a broad and deep margination that extends backward from the base of the antenna .... Bruchidae

12. First segment of abdomen completely divided ventrally by hind coxae .... Carabidae
First segment of abdomen not divided ventrally 13

13. Elytra with a blue or blue-green metallic luster on at least apical half .... Cleridae
Elytra never blue or blue-green 14

14. Species 14–24 mm long (tarsal formula 5-5-4) .... (*Blaps* spp., *Tenebrio* spp.) Tenebrionidae
Species 11 mm long or less 15

15. Prothorax with six or more robust teeth on each side; disk sometimes with three longitudinal ridges 16
Prothorax with not more than two large teeth on each side, or without teeth (denticles may be present); disk of prothorax never with three longitudinal ridges ... 17

16. Prothorax with six large teeth on each side; disk of prothorax with three longitudinal ridges .... (*Oryzaephilus* spp.) Cucujidae
Prothorax with more than six teeth, usually nine or ten; small and more or less equal teeth on each side; disk of prothorax without longitudinal ridges .... (*Henoticus* spp.) Cryptophagidae

17. Prothorax with an entire sublateral ridge or carina on each side, more or less parallel with or near lateral margin; head not deflexed 18
Prothorax without sublateral ridges, or, if ridges present, head deflexed 19

18. Antenna usually more than half as long as body and without an apical club; body very flat; length 1.3–4.0 mm .... (*Cryptolestes* and related spp.) Cucujidae
Antenna much less than half as long as body and with loose three-segmented club .... (*Mycetaea* spp.) Endomychidae

19. Head deflexed; not visible from above; antenna clubbed or serrate; small brown or black beetles 2–5 mm long 20
Head not deflexed; visible, at least partly, from above 22

20. Antenna elbowed (geniculate) and short; with small, compact cylindrical club .... Scolytidae
Antenna not elbowed, club loose or absent 21

21. Prothorax very densely and, particularly anteriorly, coarsely tuberculate; antenna with a loose, lopsided club .... Bostrichidae
Prothorax punctate; antennal club loose or absent .... Anobiidae

22. Head partly concealed by pronotum; antenna with loose, symmetrical club; eyes entire and oval; tarsal formula 4-4-4 .... Ciidae
Not as above 23

23. Prothorax with apical angles distinctly toothed or with a very much thickened, nearly smooth, oval, disklike area at apical angle; antenna with a two- or three-segmented club 24
Prothorax with apical angles not toothed or thickened 25

24. Prothorax with a distinct tooth on about middle of lateral margin .... (*Cryptophagus* spp.) Cryptophagidae
Prothorax without a tooth on middle of lateral margin; lateral denticles may be present .... (*Ahasverus* and similar spp.) Cucujidae

25. Antenna not thickened apically, without a club and usually more than half as long as body; 11 segmented; antennae inserted very close together in front of head; distance between basal sockets less than length of first antennal segment; all tarsi five segmented .... Ptinidae

Antenna with apical segments always thicker than second segment and often forming a distinct club; less than half as long as body; antennae inserted laterally in front of eyes; distance between basal sockets greater than length of first antennal segment 26

26. Prothorax with a short and narrow but very distinct basal neck; antennal club three segmented and all tarsi five segmented ...... (*Thaneroclerus buqueti*) Cleridae

Prothorax without a distinct neck 27

27. Antenna with a distinct two segmented club; brown beetles with all tarsi five segmented although the basal segment is so minute that they appear four segmented; body narrow, with parallel sides .... Lyctidae

Antenna with club usually three or more segmented, or without distinct club; if club two segmented, then of different appearance or all tarsi three segmented 28

28. Head with median ocellus; body convex or oval, with either hair or scales .... Dermestidae (except *Dermestes* spp.)

Head without median ocellus 29

29. Tarsi all three segmented .... Lathridiidae

Tarsi other than three segmented 30

30. Dorsum distinctly hairy (at 15×); body convex, elongate-oval 31

Dorsum without fine hairs, or, if with a few hairs (scarcely visible at 15×) then body distinctly flattened 32

31. All tarsi five segmented; length 5.5–10.0 mm .... Dermestidae (*Dermestes* spp.)

Tarsal formula 4-4-4 in females, 3-4-4 in males; length 2.5–4.0 mm ...... Mycetophagidae

32. Tarsal formula 5-5-4 .... Tenebrionidae

Tarsal formula not 5-5-4 33

33. Pronotum produced forward at sides; tarsal formula 5-5-5, but first tarsal segment so small that all tarsi appear four segmented .... Trogossitidae

Pronotum not produced forward at sides; first tarsal segment not smaller than second 34

34. Tarsi all distinctly five segmented; fourth segment not hidden; antennal club compact .... (*Dacne* spp.) Erotylidae

Tarsi with fourth segment small, hidden; antennal club loose .... Languriidae

B. LIST OF COMMON SPECIES

Table 1 provides a list of the most commonly encountered stored-product beetles. The species marked by an asterisk (*) are included in the table based on the experience of FDA laboratories, especially the Los Angeles laboratory.[30,31] They are the species most likely to be found as contaminants in food products. The remaining species were included based on references in the scientific literature to their status as pests of stored food products.

VI. DISCUSSION OF FAMILIES

In describing geographic distributions, I have frequently used two terms. *Cosmopolitan* species are those found in most parts of the world and under varied ecological conditions. *Tropicopolitan* species occur throughout the tropics.

In providing product records, I have usually employed the common name only, such as "chilies" for the fruit of *Capsicum* or "tamarind" for the fruity pulp (usually with seeds) of *Tamarindus indica* L. We have analyzed many samples of "red beans" and "black beans" from Asia, but the species identities are not known. An effort has been made to identify some mushrooms to genus or species, and this information may be specified in cases of beetle species associated with a particular type of mushroom.

TABLE 1
List of Common Food-Contaminating Beetles

Family/Scientific Name	Common Name
Anobiidae	Drugstore and cigarette beetles
Lasioderma serricorne (F.)	Cigarette beetle
Stegobium paniceum (L.)	Drugstore beetle
Tricorynus herbarius (Gorham)	Mexican book beetle
Anthicidae	Antlike flower beetles
Anthicus floralis (L.)	Narrownecked grain beetle
*Undetermined spp.	
Anthribidae	Fungus weevils
Araecerus fasciculatus (De Geer)	Coffee bean weevil
Bostrichidae	False powderpost beetles
Prostephanus truncatus (Horn)	Larger grain borer
Rhyzopertha dominica (F.)	Lesser grain borer
Dinoderus minutus (F.)	Bamboo powderpost beetle
Bruchidae	Seed beetles
Acanthoscelides obtectus (Say)	Bean weevil
Bruchus ervi Froelich	
Bruchus pisorum (L.)	Pea weevil
Bruchus rufimanus Boheman	Broadbean weevil
Callosobruchus analis (F.)	
Callosobruchus chinensis (L.)	Southern cowpea weevil
Callosobruchus maculatus (F.)	Cowpea weevil
Caryedon serratus (Olivier)	Groundnut bruchid
Zabrotes subfasciatus (Boheman)	Mexican bean weevil
Carabidae	Ground beetles
*Miscellaneous opportunistic predators	
Cerylonidae	Minute bark beetles
Murmidius ovalis (Beck)	Oval grain beetle
Murmidius segregatus Waterhouse	
Ciidae	Minute tree fungus beetles
Cis asiaticus Lawrence	
Cis chinensis Lawrence	
Orthocis auriculariae Lawrence	
Cleridae	Checkered beetles
Necrobia ruficollis (F.)	Redshouldered ham beetle
Necrobia rufipes (De Geer)	Redlegged ham beetle
Necrobia violacea (L.)	Blacklegged ham beetle
Thaneroclerus buqueti (Lefevre)	
Corylophidae	Minute fungus beetles
*Undetermined spp.	
Cryptophagidae	Silken fungus beetles
Atomaria linearis Stephens	
Cryptophagus acutangulus Gyllenhal	Acute-angled fungus beetle
Cryptophagus pilosus Gyllenhal	
Cryptophagus cellaris (Scopoli)	Cellar beetle
Henoticus californicus (Mannerheim)	
Cucujidae	Flat bark beetles
Ahasverus advena (Waltl)	Foreign grain beetle
Cathartus quadricollis (Guérin-Méneville)	Squarenecked grain beetle
Cryptolestes capensis (Waltl)	Cape grain beetle
Cryptolestes cornutus Thomas & Zimmerman	Horned grain beetle
Cryptolestes ferrugineus (Stephens)	Rusty grain beetle
Cryptolestes klapperichi Lefkovitch	
Cryptolestes pusilloides (Steel & Howe)	
Cryptolestes pusillus (Schönherr)	Flat grain beetle

TABLE 1 *(Continued)*
List of Common Food-Contaminating Beetles

Family/Scientific Name	Common Name
**Cryptolestes turcicus* (Grouvelle)	Turkish grain beetle
Cryptolestes ugandae (Steel & Howe)	
**Monanus concinnulus* Walker	
**Oryzaephilus mercator* (Fauvel)	Merchant grain beetle
**Oryzaephilus surinamensis* (L.)	Sawtoothed grain beetle
**Silvanus* spp.	
Curculionidae	Weevils
Caulophilus oryzae (Gyllenhal)	Broadnosed grain weevil
**Sitophilus granarius* (L.)	Granary weevil
**Sitophilus linearis* (Herbst)	Tamarind weevil
**Sitophilus oryzae* (L.)	Rice weevil
**Sitophilus zeamais* Motschulsky	Maize weevil
Dermestidae	Dermestid beetles
Anthrenocerus australis (Hope)	Australian carpet beetle
**Anthrenus flavipes* LeConte	Furniture carpet beetle
Anthrenus museorum (L.)	Museum beetle
Anthrenus oceanicus Fauvel	
**Anthrenus scrophulariae* (L.)	Carpet beetle
**Anthrenus verbasci* (L.)	Varied carpet beetle
**Attagenus fasciatus* (Thunberg)	Wardrobe beetle
Attagenus pellio (L.)	Fur beetle
**Attagenus unicolor* (Brahm)	Black carpet beetle
**Dermestes ater* De Geer	Black larder beetle
**Dermestes carnivorus* F.	
**Dermestes frischii* Kugelan	
Dermestes haemorrhoidalis Kuster	Sheepskin dermestid
**Dermestes lardarius* L.	Larder beetle
**Dermestes maculatus* De Geer	Hide beetle
**Dermestes marmoratus* Say	
Dermestes peruvianus Laporte de Castlenau	Peruvian larder beetle
**Dermestes undulatus* Brahm	
**Thorictodes heydeni* Reitter	
**Thylodrias contractus* Motschulsky	Odd beetle
**Trogoderma glabrum* (Herbst)	Glabrous cabinet beetle
**Trogoderma granarium* Everts	Khapra beetle
Trogoderma inclusum LeConte	Larger cabinet beetle
**Trogoderma sternale* Jayne	
**Trogoderma variabile* Ballion	Warehouse beetle
Endomychidae	Handsome fungus beetles
Mycetaea subterranea (F.)	Hairy cellar beetle
Erotylidae	Pleasing fungus beetles
**Dacne japonica* Crotch	
**Dacne picta* Crotch	
Histeridae	Hister beetles
**Carcinops pumilio* (Erichson)	
Carcinops troglodytes (Paykull)	
Languriidae	Lizard beetles
**Cryptophilus* spp.	
Leucohimatium arundinaceum (Forskål)	
**Pharaxonotha kirschi* Reitter	Mexican grain beetle
Lathridiidae	Minute brown scavenger beetles
Adistemia watsoni (Wollaston)	
Aridius nodifer (Westwood)	
**Cartodere constricta* (Gyllenhal)	Plaster beetle
Corticaria punctulata Marsham	

TABLE 1 *(Continued)*
List of Common Food-Contaminating Beetles

Family/Scientific Name	Common Name
**Dienerella* spp.	
Enicmus consimilis Mannerheim	
**Eufallia seminvenus* (Motschulsky)	
**Holoparamecus depressus* Curtis	
Lathridius anthracinus Mannerheim	
**Lathridius minutus* (L.)	Squarenosed fungus beetle
Lathridius pseudominutus (Strand)	
**Lathridius* spp.	
Thes bergrothi (Reitter)	Ridgewinged fungus beetle
Lyctidae	Powderpost beetles
**Lyctus brunneus* (Stephens)	
Mycetophagidae	Hairy fungus beetles
**Berginus pumilus* LeConte	
Mycetophagus quadriguttatus Muller	Spotted hairy fungus beetle
**Typhaea stercorea* (L.)	Hairy fungus beetle
Nitidulidae	Sap beetles
**Carpophilus dimidiatus* (F.)	Corn sap beetle
**Carpophilus hemipterus* (L.)	Driedfruit beetle
Carpophilus lugubris Murray	Dusky sap beetle
Carpophilus marginellus Motschulsky	Shiny sap beetle
**Carpophilus mutilatus* Erichson	Confused sap beetle
**Carpophilus obsoletus* Erichson	
Carpophilus spp.	
Haptoncus luteolus (Erichson)	Yellowbrown sap beetle
**Nitidula bipunctata* (L.)	Bone beetle
Urophorus humeralis (F.)	Pineapple beetle
Ptiliidae	Featherwinged beetles
*Undetermined spp.	
Ptinidae	Spider beetles
**Gibbium psylloides* (de Czenpinski)	Hump beetle
Mezium affine Boieldieu	Shiny spider beetle
**Mezium americanum* (Laporte)	American spider beetle
Niptus holoeucus (Faldermann)	Golden spider beetle
Ptinus clavipes Panzer	Brown spider beetle
**Ptinus fur* (L.)	Whitemarked spider beetle
Ptinus ocellus Brown	Australian spider beetle
Ptinus villiger (Reitter)	Hairy spider beetle
**Trigonogenius globulum* (Solier)	Globular spider beetle
Scolytidae	Bark and ambrosia beetles
**Hypothenemus areccae* (Hornung)	
**Hypothenemus eruditus* Westwood	
**Hypothenemus hampei* (Ferrari)	Coffee berry borer
Staphylinidae	Rove beetles
Oligota granaria Erichson	
*Undetermined spp.	
Tenebrionidae	Darkling beetles
**Alphitobius diaperinus* (Panzer)	Lesser mealworm
**Alphitobius laevigatus* (F.)	Black fungus beetle
Alphitophagus bifasciatus (Say)	Twobanded fungus beetle
Blaps mucronata Latreille	Churchyard beetle
Cynaeus angustus (LeConte)	Larger black flour beetle
**Gnatocerus cornutus* (F.)	Broadhorned flour beetle
**Gnatocerus maxillosus* (F.)	Slenderhorned flour beetle
**Latheticus oryzae* Waterhouse	Longheaded flour beetle
**Palorus ratzeburgi* (Wissmann)	Smalleyed flour beetle

TABLE 1 ***(Continued)***
List of Common Food-Contaminating Beetles

Family/Scientific Name	Common Name
**Palorus subdepressus* (Wollaston)	Depressed flour beetle
Platydema americanum (Laporte)	
Platydema ruficorne (Sturm)	Redhorned grain beetle
Tenebrio molitor L.	Yellow mealworm
Tenebrio obscurus F.	Dark mealworm
Tribolium audax Halstead	American black flour beetle
**Tribolium castaneum* (Herbst)	Red flour beetle
**Tribolium confusum* Jacquelin du Val	Confused flour beetle
Tribolium destructor Uyttenboogaart	Dark flour beetle
Tribolium madens (Charpentier)	Black flour beetle
Trogossitidae	
**Lophocateres pusillus* (Klug)	Siamese grain beetle
**Tenebroides mauritanicus* (L.)	Cadelle beetle

* Species most likely to be found as contaminants in food products.

A. COMMON FOOD-CONTAMINATING FAMILIES

1. Anobiidae (Drugstore and Cigarette Beetles)

Freeman reported that a worldwide total of about 15 species of anobiids have been recorded in stored food products, warehouses, etc. Some of these [e.g., *Anobium punctatum* (De Geer) (furniture beetle)] bore into the woodwork and are therefore only incidentally associated with food products.[(2)] At least four species actually attack stored food products. Most anobiid larvae feed on dead or dying wood, but some species feed on conifer cones, seeds, mushrooms or stored-products of both animal and plant origin.[(32)] Adults of at least the two most common stored-product pests do not feed.[(2)]

The two most common pests are *Lasioderma serricorne* (F.) (cigarette beetle) (Figure 2) and *Stegobium paniceum* (L.) (drugstore beetle). Both are cosmopolitan and have been recorded from a great variety of plant products and some animal products.[(31,33,34)] Our laboratory has recorded only these two species, but both have been encountered frequently.[(31)]

Anobiid adults are identified by the prothorax strongly hooded over the head; the serrate antenna or the antenna with a loose, lopsided, three-segmented club; and the pronotum not rugose (Figure 2). *L. serricorne* has a serrate antenna, whereas *S. paniceum* has an antenna with a loose, three-segmented club. Anobiid larvae are scarabaeiform with developed legs and they may have a mandible with two apical teeth. The *L. serricorne* larva has the posterior margin of the frons distinctly marked, and the *S. paniceum* larva does not.[(5)]

2. Anthicidae (Antlike Flower Beetles)

I suspect that the knowledge of stored-product anthicids is alarmingly incomplete. Hinton[(6)] discussed only three species, but our laboratory has recorded at least five (unpublished data). Habitats are quite varied. Adults can be found on branches and foliage of trees and shrubs, under decaying vegetation, beneath cover in sandy soil, or on flowers. Larvae have been found in heaps of decaying vegetation.[(6)] Trophic habits are also varied, but Chandler (personal communication) believes that the predominant role of stored-product species is one of microscavenger. Werner observed that adults of North American *Anthicus* spp. feed readily on dead insects and believed that larvae of most species probably subsist on dead insects and soil fungi.[(35)] Adults of *Notoxus monodon* (F.) have been found feeding on moth pupae, and larvae of *Anthicus heroicus* Casey apparently prey on dobsonfly eggs.[(6,35)]

Our laboratory only occasionally finds adult anthicids. Specimens have been recorded in preserved cabbage, brown sugar, dried chilies and dried apricots.

Adult anthicids can be identified by the prothorax with a short, narrow, apical neck; the head completely exposed; and the elytra entire.

3. Anthribidae (Fungus Weevils)

Araecerus fasciculatus (De Geer) (coffee bean weevil) is the only stored-product species in the Anthribidae (fungus weevils). The larvae feed on a variety of plant products including coffee.[2] This species is cosmopolitan.[34]

Our laboratory only occasionally encounters coffee bean weevil.[31] Products of record include coffee beans, dried chilies, dried sweet potato slices and tamarind. We find adults more frequently than larvae.

Coffee bean weevil adults can be identified by incomplete elytra exposing the pygidium; the three broad, apical antennal segments preceded by slender segments 3–8; and the pseudotetramerous tarsi with the third segment bilobed or apparently so. The larvae are scarabaeiform with undeveloped legs and mandible with three incisors and a mola.

4. Bostrichidae (False Powderpost Beetles)

A number of bostrichids attack commercial wood products.[36] Only four species are definitely known to feed on stored food products: *Rhyzopertha dominica* (F.) (lesser grain borer), *Prostephanus truncatus* (Horn) (larger grain borer), *Dinoderus minutus* (F.) (bamboo powderpost beetle) and *D. bifoveolatus* (Wollaston).[34,37] Bostrichids are mostly wood borers, but a few feed in the fruiting bodies of bracket fungi.[15] Many Dinoderinae (including *Dinoderus* spp.) prefer to attack monocotyledons such as bamboos or canes.

Adults and larvae of the cosmopolitan *R. dominica* and the adults of the cosmopolitan *D. minutus* have been recorded in a great variety of food products of plant origin, but the larvae of *D. minutus* thrive only in the stems of bamboos and canes.[33,34] However, according to Spilman, *D. minutus* breeds in dried, stored food in North America.[37] *P. truncatus* may have more limited geographic and host plant ranges, although some authorities feel that it is cosmopolitan.[33,38]

Our laboratory commonly encounters only *R. dominica*, on a great variety of plant products. We have also recorded *P. truncatus* once and have rarely encountered *Dinoderus* spp. adults and adult fragments.[31]

Bostrichid adults can be identified by the prothorax strongly hooded over the head; the rugose pronotum; and the loose, lopsided, three-segmented antennal club. *R. dominica* larvae can be identified by their C-shape, developed legs, and the distinctive mandible with three apical teeth and a protruding mola (Figure 7).

5. Bruchidae (Seed Beetles)

More than 20 species of bruchids in six genera have been reported in stored-products.[39-42] However, a few of these beetle species may be limited to the seeds of cover and forage crop species or ornamental species as opposed to species used for human consumption.

Bruchid larvae develop inside seeds, principally leguminous seeds but seeds of 33 other plant families can function as suitable hosts.[43] Adults do not feed at all or are restricted primarily to feeding on pollen and nectar from diverse families of plants.

Kurtz and Harris defined two categories (Groups I and II) of bruchid species associated with stored-products. The larvae of Group I start developing on immature seeds in the field, although adults may not emerge until subsequent storage of the seeds. Possibly because of the hardness of the seed, these larvae can not attack mature, stored seeds. The larvae of Group II can attack stored, mature seeds.[5] *Bruchus pisorum* (L.) (pea weevil) is a Group I species because it normally cannot attack mature, stored seeds.[44] Group II species include *Callosobruchus maculatus* (F.) (cowpea weevil), *Callosobruchus chinensis* (L.) (southern cowpea weevil), *Acanthoscelides obtectus* (Say) (bean weevil), *Zabrotes subfasciatus* (Boheman) (Mexican bean weevil) and *Caryedon serratus* (Olivier) (groundnut bruchid).[44,45]

Kingsolver and Pfaffenberger discussed the common species.[39,44] The cosmopolitan *B. pisorum* is encountered in and limited to peas, whereas the cosmopolitan *C. maculatus* and *C. chinensis* are found in a variety of beans and other pulses. The cosmopolitan *A. obtectus* can attack a variety of beans and other pulses, and the widespread *Z. subfasciatus* can attack pinto, haricot and kidney beans. *C. serratus* is widespread, although its range does not include the continental United States. This species can be found in tamarind and African peanuts.

Our laboratory regularly finds *C. maculatus* and *C. chinensis* in certain types of "beans," but *Callosobruchus analis* (F.) is rarely encountered (unpublished data). We have never recorded *Z. subfasciatus* and rarely, if ever, find *A. obtectus*. Adult fragments and whole first-instar larvae of *C.serratus* are commonly encountered in light filth extractions of processed tamarind products.

Adult bruchids can be identified by the incomplete elytra leaving the pygidium exposed, antenna not clubbed (serrate or pectinate), and a pseudotetramerous tarsi with the third segment bilobed or appearing so. I recommend Kingsolver's keys for differentiating the genera and for differentiating the *Bruchus* spp.[39,42] Haines's and Kingsolver's keys for differentiating the *Callosobruchus* species are also recommended.[41,46]

Mature larvae are scarabaeiform with undeveloped legs. The mandible has a characteristic shape including a gouge-shaped apex. The first-instar larva has a sclerotized X- or H-shaped prothoracic shield.

6. Carabidae (Ground Beetles)

Twenty species are mentioned in Hinton's treatment of the Carabidae (ground beetles) associated with stored-products.[6] He summarized the trophic habits as follows. Carabids are predominantly predaceous as larvae and adults, but probably most species supplement their diet with plant material. Species of a few groups can be largely phytophagous, but will readily feed on other invertebrates.

Bousquet's summary is slightly different. Adults of most species are polyphagous, feeding on both animal and plant matter. The adults of some species are exclusively carnivorous and those of other species are phytophagous. The larvae of most species are probably carnivorous.[47]

We have rarely encountered ground beetles in our imported food samples. Adults are recognized by the first visible abdominal sternite being completely divided by the hind coxae and the absence of fringe on the hind legs.

7. Cerylonidae (Minute Bark Beetles)

Only two genera (*Murmidius* and *Euxestoxenus*) of cerylonids (minute bark beetles) are associated with stored-products.[3] Two species of *Murmidius* are reported by Hinton. Cerylonids typically occur in leaf litter, in rotten wood or under bark and probably feed on fungal hyphae and spores.[15] Cultured adults of *Murmidius ovalis* (Beck) (oval grain beetle) have been observed feeding on mold, and mold is believed to be essential for larval development.[48] *M. ovalis* is a cosmopolitan species that has been found in rice, seeds, dried fruit, shelled corn and stored wheat.[6]

Our laboratory has only rarely encountered *Murmidius segregatus* Waterhouse. We have recorded it in dried chilies and jishengcha (tea stock) from China. China may be a new geographic record (Slipinski, personal communication). We have no records of any other species in this family.

The adults of *Murmidius* spp. can be identified by the antennal cavities on the pronotum near the anterior angles; the ten-segmented antennae with a one-segmented club; tarsi 4-4-4; the oval, shining, strongly convex body; and the small size (<1.5 mm). The larvae can be identified by the broadly oval body, which completely conceals the head dorsally, and by the flat projections bearing spinulose setae on all the tergites.

8. Ciidae (Minute Tree Fungus Beetles)

This family is represented by 550 described species and at least 400 undescribed species.[49] Although it is mentioned by both Hinton and Halstead, it has not been treated in detail, with respect to stored-products, until recently.[3,6,50,51] Three species in two genera have been reported in commercial, stored-products.[50]

Ciids spend most of their lives within the sporophores (mushrooms) of Polyporaceae and other Basidiomycetes and are entirely dependent on them for food and shelter. Under natural conditions, ciids usually concentrate on mushrooms that have shed their spores and are starting to decompose.[51a] They are one of the most important agents of decomposition of mushrooms.[52] *Hadraule blaisdelli* (Casey) has been recorded as a pest of dried mushrooms in herbarium collections.[49] *Cis chinensis* Lawrence has been found developing, in high density, in dried, commercial mushrooms.[50] The origin of ciid infestations could be pre- or postharvest, but I believe that the attainment of very high densities occurs after harvest.

Some degree of specialization may exist in the utilization of mushroom taxa by the three species found in commercial products.[50] *Orthocis auriculariae* Lawrence has been found only in the edible ear fungus *Auricularia* spp, whereas *C. chinensis* has been found only in a medicinal mushroom known as "Ling Chih" [*Ganoderma lucidum* (Leysser ex Fries) Karsten] from the People's Republic of China. The only commercial species of mushroom in which *Cis asiaticus* has been found is the edible shiitake mushroom *Lentinus polychrous* Léveillé from Thailand, although it has also been recorded in Thailand from *Trametes corrugata* (Persoon) Cunningham.[51] *C. asiaticus* has not been found in the more common, commercial shiitake mushroom *Lentinus edodes* (Berkeley) Singer. All of our laboratory's records represent product from Asia.

Adult ciids can be identified by the pronotum partially concealing the head, but not strongly hooded over it; the oval and entire eyes; the symmetrical, loose, two- or three-segmented antennal club; and tarsi 4-4-4 in both sexes.

Ciid larvae have a pair of hooklike urogomphi and a very characteristic two-segmented antenna, which includes a long, sensory appendage near the base of the second segment and a long apical seta. The antenna is best seen by mounting the larva on a microscope slide and viewing under high magnification (400×).

9. Cleridae (Checkered Beetles)

Freeman stated that about ten species of clerids are associated with stored food products, and he treated seven species in five genera. Nearly all of the approximately ten species are thought to be predaceous.[2] *Thaneroclerus buqueti* (Lefevre) and *Paratillus carus* (Newman) specialize on anobiid beetles. In addition to being predaceous, *Necrobia rufipes* (De Geer) (redlegged ham beetle), *Necrobia ruficollis* (F.) (redshouldered ham beetle) and *Necrobia violacea* (L.) (blacklegged ham beetle) feed on oilseeds, old cheeses and cured meats such as ham, bacon and smoked pork.[2] Haines and Rees reported that the adults and larvae of *N. rufipes* commonly feed on dried fish and that *N. ruficollis* and *N. violacea* are rarely found on dried fish.[53] Kingsolver reported that adults and larvae of the three species feed on larvae of other insects and on molds.[54]

Our laboratory commonly finds *Necrobia* spp. larvae and larval fragments in fish products, but we also have recorded the genus in rice sticks, shrimp paste, crab paste and dried snakes.[31] Adults and adult fragments are recorded less frequently. We very occasionally find adults of *T. buqueti* in a variety of plant products, including dried aniseed, sweet basil leaves, watermelon seeds, dried chilies and jishengcha (tea stock). The three *Necrobia* spp. are cosmopolitan,[54] and we have recorded this genus in Asia, the Philippines and Central America.[31] In addition, we have recorded *T. buqueti* from several Asian countries.

Checkered beetle adults have bristly setae on the margins of the prothorax and elytra. The elytra of *Necrobia* spp. have a blue or blue-green metallic hue on at least the apical half. *T.*

buqueti has the base of its prothorax considerably constricted. A key for identifying species can be found in Freeman.[2]

Necrobia spp. larvae have biforous spiracles, widely separated urogomphi and characteristically shaped mandibles (Figure 8).

10. Corylophidae (Minute Fungus Beetles)

I know of no treatment of corylophids associated with stored-products, although Hinton included this family in his identification key.[6] Our laboratory only occasionally encounters undetermined adults or larvae in both plant and animal products.

The adults and larvae are primarily spore feeders and usually can be found in leaf litter and refuse heaps, or on surfaces covered with sporulating fungi of various kinds.[15] Species of *Sericoderus* and *Orthoperus* are often associated with molds. Gross (personal communication) has found corylophids in horse dung in San Mateo County, California.

The adults we encounter are small (ca. 1 mm) and have a hood-like pronotum that completely covers a proportionately small head. Other characteristics include six visible abdominal sternites; a large, loosely segmented antennal club; 4-4-4 tarsi; and the hindwings fringed with hairs. The larvae we encounter have clavate, pectinate setae on the thorax and abdomen; and paired dorsolateral gland openings toward the anterior and posterior ends of the abdomen.

11. Cryptophagidae (Silken Fungus Beetles)

Hinton stated that 25 species (in three genera) of Cryptophagidae have been found indoors and he included 19 in his key. Of the three genera, *Cryptophagus* is best represented with 19 species. Adults and larvae are found in moldy plant and animal substances; in flowers; in fungi; under bark; in caves; and in the nests of ants, social wasps and bees, birds and small mammals. Species in stored food products almost certainly live principally on the spores and hyphae of molds. However, *Atomaria linearis* Stephens can feed on the leaves, roots and germinating seeds of sugarbeet and mangold. Cryptophagids have been reported in a variety of plant products, including dried fruits, tobacco, peanuts and flour.[6] Several *Cryptophagus* spp. are geographically widespread.[6,55]

Our laboratory has rarely encountered cryptophagids. *Henoticus* sp., probably *californicus* (Mannerheim) was found in three different samples of dried chilies from Mexico.[31]

The identification of cryptophagid adults is complicated by their diversity. *Henoticus* spp. have the lateral margin of the pronotum serrate (nine to ten small teeth). The pronotum of *Cryptophagus* spp. has thickened and obliquely truncate anterior angles and a lateral tooth on both sides. *Atomaria* spp. have a simple pronotum, but the antennae are inserted close together on the frons. All three genera are pubescent and have a loose, three-segmented antennal club, procoxal cavities open behind and tarsi 5-5-5, except for some males with tarsi 5-5-4.

The mandible of the larva has two apical teeth, a well-developed mola and an elongate retinaculum.[6]

12. Cucujidae (Flat Bark Beetles)

Very recently the cucujoid group associated with stored-products has been thoroughly revised and discussed by Halstead, who splits the group into three families: Laemophloeidae, Passandridae and Silvanidae.[56] I have strictly followed Arnett's classification, which treats these three groups as subfamilies of Cucujidae.[12] This family is one of the richest, with 47 species in 18 genera associated with stored-products. Most cucujid species can be found under bark or in leaf litter and humus; others may be on flowers or leaves or in the galleries of wood-boring beetles.[15,56] They are primarily mycophagous.

Stored-product species can feed on a huge variety of plant products and can function as primary pests of whole grains. One common species, *Ahasverus advena* (Waltl) (foreign grain beetle), is generally considered to feed on mold on the product.[56] Other stored-product species are also believed to feed on mold. Pal (personal communication) stated that adults and larvae of *Monanus concinnulus* (Walker) feed on mold spores in haystacks in India. Dyte suspected that mold growth might provide a nutritional supplement for *Cryptolestes turcicus* (Grouvelle) (Turkish grain beetle) in static flour residue.[57] *Cryptolestes ferrugineus* (Stephens) (rusty grain beetle) can develop in various storage fungi.[58]

Cucujids can also function as predators or parasitoids. The adults and larvae of *C. ferrugineus* have been observed feeding on cultured eggs and larvae of *Sitotroga cerealella* (Olivier) (Angoumois grain moth).[59] Larvae of passandrines are ectoparasitoids of cerambycids and bostrichids, and adults are probably mainly predaceous on these wood borers.[56]

The only common stored-product genus of the Laemophloeinae is *Cryptolestes*. Halstead reported nine species, seven of which have been encountered in our laboratory.[31,56] *Cryptolestes pusillus* (Schönherr) (flat grain beetle) and *C. ferrugineus* are the most frequently encountered and most important pests. Both species are geographically widespread (although the former cannot survive outdoors in cooler temperate regions) and found in a great variety of plant products. Four other *Cryptolestes* spp.—*C. turcicus*, *C. pusilloides* (Steel and Howe), *C. klapperichi* Lefkovitch and *C. capensis* (Waltl) (cape grain beetle)—are less common, but have a fairly widespread to widespread geographic distribution and have been found on a variety of plant products. The other three *Cryptolestes* spp. are markedly less common and apparently have more limited known distributions. *C. cornutus* Thomas and Zimmerman (horned grain beetle) has only been found in dried chilies from Thailand, and *C. ugandae* (Steel and Howe) is limited to Africa. *C. divaricatus* has been found in illipe nuts from Malaysia and dessicated coconut from Sri Lanka. Mixed populations of *Cryptolestes* spp. in stored-products can occur, and up to four different species have been found in the same sampled lot (unpublished data).

Halstead noticed the following correlation in the group consisting of *C. pusillus*, *C. ferrugineus*, *C. turcicus*, *C. pusilloides*, *C. capensis* and *C. ugandae*. The species (*C. ferrugineus* and *C. capensis*) with the male antenna only slightly longer than the female's and with an additional lateral tooth on the male mandible can develop at humidities well below 50% r.h., whereas the species (*C. pusillus*, *C. turcicus*, *C. pusilloides* and *C. ugandae*) with the male antenna much longer than the female's and with the mandibles undifferentiated are unable to develop at humidities much below 50% r.h.[56]

Three species of silvanines are very commonly encountered in stored-products: *Ahasverus advena*, *Oryzaephilus surinamensis* (L.) (sawtoothed grain beetle) and *O. mercator* (Fauvel) (merchant grain beetle). All three are cosmopolitan or nearly so and each has been found in a huge variety of plant products, although *O. surinamensis* is principally a pest of cereals and cereal products, whereas *O. mercator* is most frequently found on oilseeds and their derivatives.[56] Two other *Oryzaephilus* spp. and two other *Ahasverus* spp. have been found in stored-products.[56,60]

Monanus concinnulus Walker is a silvanine species with a widespread geographic distribution, but it is less common than the three previously discussed species. It has been recorded from a variety of plant products.[31,56]

Our laboratory occasionally encounters *Cathartus quadricollis* (Guérin-Méneville) (squarenecked grain beetle). Halstead considered its geographic distribution virtually tropicopolitan and listed a variety of plant products from which it had been recorded.[56] Our laboratory has repeatedly found it in corn husks, dried chilies and tamarind products from Mexico.[31]

Eight other silvanine genera associated with stored-products are discussed in Halstead, but some of the species covered are not known to develop in stored-products.[56] Within these genera, our laboratory has found, but only rarely, adults of *Silvanoprus cephalotes* (Reitter), *Silvanoprus longicollis* (Reitter), *Silvanus lewisi* Reitter and *Silvanus proximus* Grouvelle.[31]

13

14

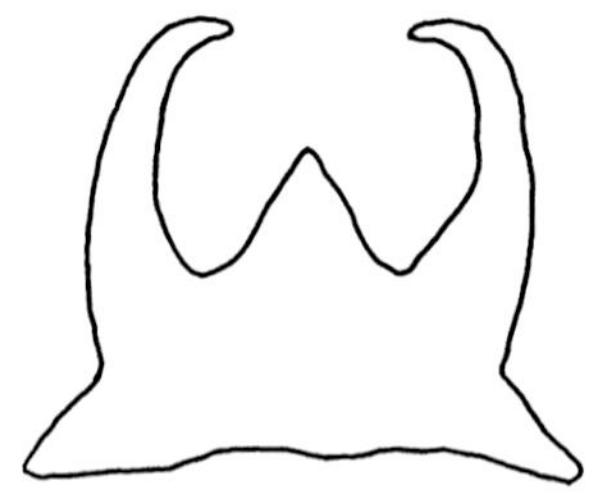

FIGURE 13. Larval anal urogomphi, flat grain beetle, *Cryptolestes* spp. (Cucujidae).

FIGURE 14. Larval anal urogomphi, Siamese grain beetle, *Lophocateres pusillus* (Trogossitidae).

Cryptolestes spp. adults, larvae and their fragments are extremely frequently recorded in stored-products. The sexually dimorphic adults are recognized by their dorsoventrally flattened appearance, the elytron with four longitudinal ridges, the pair of sublateral carinae on the pronotum and the lack of a distinct antennal club (front cover). Adult species taxonomy is thoroughly discussed by Halstead, who recommends examination of the genitalia to those who only occasionally need to make identifications.(56) Larvae have distinctly shaped anal urogomphi (Figure 13) and distinctive, subparallel hypostomal rods on the head venter.

Silvanine adults have a great diversity in form, which can make the family diagnosis very difficult. The only comprehensive taxonomic treatment is Halstead's.(56) *A. advena* is recognized by the developed, rounded lobes at the anterior angles of the pronotum. *Oryzaephilus* spp. have six large, acutely pointed teeth on both margins of the pronotum (Figure 3). The ratio of eye length to temple length is considerably greater for *O. mercator* than for *O. surinamensis*. *Monanus concinnulus* is identified by the approximately eight denticles (each with an anteriorly directed apical seta) in addition to the acute anterior angle on both sides of the pronotum, and by the dark transverse elytral band (not present on some individuals). The sexually dimorphic *C. quadricollis* is identified by the absence of pronotal lateral teeth or denticles; the closely joined apical and subapical antennal segments; and the pronotal length equal to (females) or considerably greater than (males) the pronotal width.

The larvae of at least the more common silvanine genera have a characteristically pointed prostheca on the mandible. Cutler's key differentiates *A. advena*, *O. surinamensis*, *O. mercator* and *C. quadricollis*.(61)

13. Curculionidae (Weevils)

Curculionidae is the largest beetle family in the world. About 30 species have been recorded in stored food products.(2) Whitehead's summary included more than 50 species in 29 genera, but he specified that most species cause only preharvest injury and cannot develop in stored-products.(62)

Weevils are phytophagous. The larvae of most species live inside the tissues of their plant hosts, but a few feed externally on leaves. Host tissues include those of the roots, stems, wood, leaves, fruits and nuts. Adults typically feed on leaves, pollen, flowers, developing fruit and fungi. A few burrow into wood.(63)

Whitehead selected the following five species which he felt were the most likely to cause losses in storage: *Sitophilus zeamais* Motschulsky (maize weevil), *S. oryzae* (L.) (rice weevil), *S. granarius* (L.) (granary weevil), *S. linearis* (Herbst) (tamarind weevil) and *Caulophilus oryzae* (Gyllenhal) (broadnosed grain weevil). The first three are cosmopolitan, whereas *S. linearis* is tropicopolitan. *C. oryzae* is found in the Caribbean region, in the southeastern

United States and perhaps elsewhere. Whitehead also summarized host data for these five species: *S. zeamais*, *S. oryzae* and *S. granarius* develop in whole grains; *S. linearis* develops in the fruits of tamarind; *C. oryzae* develops in corn, but also can develop in certain fruits and in root products.(62)

Our laboratory commonly encounters *S. zeamais* and *S. oryzae* in a variety of plant products.(31) *S. linearis* is commonly encountered in tamarind products. In cooler wheat-growing areas, *S. granarius* is found by FDA labs in samples from grain elevators and its fragments in processed farinaceous products (A. Olsen, personal communication). We rarely encounter *C. oryzae*.

Curculionid adults are identified by their distinct beaks (Figure 5) and their geniculate antennae. The species morphology and taxonomy are discussed by Whitehead.(62) I recommend that the examination of genitalia be employed to differentiate *S. zeamais* from *S. oryzae*.

Weevil larvae are scarabaeiform with undeveloped legs and a pair of labral rods. The typical *Sitophilus* mandible has a single apical tooth. *S. zeamais*, *S. oryzae* and *S. granarius* can be differentiated by head characteristics.(64)

14. Dermestidae (Dermestid Beetles)

Freeman stated that about 55 species of dermestids are injurious to stored-products.(2) Hinton reported 52 species found in stored-products and discussed 48 of these. His thorough treatment of the family provides a useful summary of the ecology.(6) Dermestid habitats include burrows of various wood-boring beetles, in or near carrion, and the nests of bees, wasps, gregarious caterpillars, spiders, birds and mammals.

There are three established larval feeding categories: species that can maintain themselves on animal matter; species that normally live on animal matter but can also develop on plant material exclusively; and one species, *Trogoderma granarium* Everts (khapra beetle), that appears normally to be restricted to plant matter.(6) The majority of the stored-product species are in Group 1, including all species of *Thylodrias* and *Dermestes*, as well as nearly all species of *Attagenus*, *Anthrenus* and *Trogoderma*. Group 2 includes *Anthrenus verbasci* (L.) (varied carpet beetle), *Trogoderma sternale* Jayne and *Trogoderma versicolor* (Creutzer).

Group 1 can be divided into two overlapping subgroups: species that normally confine their feeding to more or less raw animal products such as dried fish and meats (all species of *Dermestes* and some species of *Attagenus* and *Anthrenus*), and species that may occasionally feed on raw products but are serious pests of highly processed products such as carpets, woolen clothes and silk [*Attagenus unicolor* (Brahm) (black carpet beetle), *Anthrenus vorax* Waterhouse and *Anthrenus scrophulariae* (L.) (carpet beetle)]. A difference in the tolerance to low moisture content may be the primary factor for this division.

Dermestid larvae may feed on other types of animal matter. They can feed on whole, dead insects, spiders, birds and mammals. Their presence in plant products may indicate feeding not on the product but on dead insects in the product. In addition to scavenging for animal matter, dermestid larvae may also function as predators on insects, insect eggs, young spiders, spider eggs and even nestling birds.(6)

Dermestes spp. can be found in a great variety of animal products such as furs, feathers, leather, skins, dried meats and fish, cheese and glue.(2) Haines and Rees reported seven species of *Dermestes* in cured fish.(53) *Dermestes* spp. may also be found in products of vegetable origin, but frequently not feeding on the actual product.(6,65) Full-grown larvae leave their food and reach pupation chambers by tunneling through compact material such as wood and metal.(6) Our laboratory commonly encounters whole larvae and larval fragments in dried fish and fish products.

Most species of *Attagenus*, *Anthrenus* and *Trogoderma* can be found in both plant and animal products, although *T. granarium* is limited to plant products.(6) Our laboratory has found individuals of these genera in various plant products such as fennel seeds, coffee beans, rice and mung beans. We have also very occasionally found adults and adult fragments of *Thorictodes heydeni* Reitter in plant products such as rice sticks, patongo mix and flour mix.[31]

The identification of dermestid adults is complicated by their diversity.[3] *Dermestes* spp. can be identified by their relatively large size (5.5–12.5 mm); the absence of a median ocellus; the densely pubescent cuticle; and the short antennae with a large, compact, three-segmented club. *T. heydeni* can be identified by the concealed head, the inconspicuous eye encircled by the dorsal cuticle of the head, the 5-5-5 tarsi, and the antennae with a compact club. The other genera (i.e., *Attagenus*, *Anthrenus*, *Trogoderma*, etc.) can be identified by the presence of the median ocellus, the body usually covered with scales or setae, the cavity usually present for the reception of the antenna, and the hind coxa usually with a cavity for the reception of the femur. Hinton's key is recommended for species identification because it is the most comprehensive, although *T. heydeni* is not included.[6]

Dermestid larval genera are characterized by their numerous, large setae. The mandibles of *Dermestes* have three apical teeth (Figure 10), whereas the mandibles of *Attagenus*, *Anthrenus* and *Trogoderma* are bicolored and have a single, blunt apex (Figure 11). The larvae of *Dermestes* spp. have a pair of urogomphi that are undeveloped in early instars. Hinton's key is again recommended for species identification.[6]

15. Erotylidae (Pleasing Fungus Beetles)

The Erotylidae are not widely recognized as a family containing stored-product species. Lawrence mentioned that *Dacne* spp. are pests in commercial, dried mushrooms. Erotylids apparently feed exclusively on mushrooms.[15] Our laboratory has found *D. picta* Crotch and *D. japonica* Crotch in stored-products. Savary found high densities of *D. picta* developing in dried mushrooms in a San Francisco warehouse.[66]

Our laboratory has occasionally found larvae and adults of *D. picta* in dried shiitake mushrooms [*Lentinus edodes* (Berkeley) Singer] from Korea, and we have found larvae and adults of *D. japonica* in one sample of dried shiitake mushrooms (*L. edodes*) from China. We have also found *D. picta* in dried straw mushrooms [*Volvariella volvacea* (Bulliard ex Fries) Singer].

Adults are identified by their red markings dorsally; the absence of teeth, lobes or callosities on the pronotum; the procoxae closed behind; the three-segmented antennal club; and the 5-5-5 tarsi with the first three segments broad and pubescent beneath.

The larvae have a pair of hook-like urogomphi and numerous asperities on the dorsum of the thorax and abdomen. The mandible is tridentate. The distal edge of the prostheca lies parallel to the base of the mandible and forms a right angle with its mesal edge, which extends to the well-developed mola.

16. Histeridae (Hister Beetles)

Fourteen species are covered in Hinton's treatment of the Histeridae (hister beetles) associated with stored-products.[67] Habitats include carrion, dung, decaying vegetable matter and bird and mammal nests. All species are probably predaceous, as larvae and adults, on mites, insect larvae and primitive insects. *Carcinops pumilio* (Erichson) is the commonest hister beetle found in stored food products, where both adults and larvae prey on stored-product mites and insects.[2]

We have encountered hister beetles rarely in food samples, but we have found them in both plant and animal products. Adults are identified by the distinctly elbowed antennae and the one or two abdominal tergites exposed by incomplete, truncate elytra.

17. Languriidae (Lizard Beetles)

The species of languriids associated with stored-products include the following: *Cryptophilus integer* (Heer), other *Cryptophilus* spp., *Pharaxonotha kirschi* Reitter (Mexican grain beetle) and *Leucohimatium arundinaceum* (Forskål).[3,6]

The biology and ecology of the family were reviewed by Hinton and Lawrence. The majority of species are usually associated with decaying plant tissue. Hinton concluded that both adults and larvae of these species feed on fungi, but Lawrence suggested only an association with molds for many of these species.[6,15] Mold and/or high moisture content may be required for *P. kirschi,* as it apparently cannot develop in healthy, dry whole grain.[68] Larvae in the atypical subfamily Languriinae develop in the living stems of various herbaceous plants, and the adults can be found on the host plants.

C. integer has been found in dried figs, and *P. kirschi* has been found in several types of plant products.[6] The distribution of the latter is North America, Central America, South America and Germany.[6] The widespread *L. arundinaceum* has been recorded in oats and barley.[15]

Our laboratory occasionally encounters adults of *Cryptophilus* spp. in a variety of Asian products such as hot bean sauce, dried chilies, herbal tea, dehydrated cole and dried squid. We find *P. kirschi* sporadically in dried chilies from Mexico and have never recorded *L. arundinaceum.*[31]

The identification of lizard beetle adults is complicated by the family's diversity.[3] Both *Cryptophilus* spp. and *P. kirschi* lack lateral callosities, laterally produced anterior angles and serrate lateral margins on the pronotum, and have pronounced pubescence on the elytra. In addition, they share 5-5-5 tarsi and a three-segmented antennal club with all three segments loosely attached. *P. kirschi* has a pair of short basal carinae on the pronotum that *Cryptophilus* spp. lack. *L. arundinaceum* is identified by the lateral callosities on the pronotum and the elongate body clothed with grayish, lanceolate setae.

The larva of the Mexican grain beetle has pairs of pregomphi and urogomphi, and numerous asperities on the dorsum of the thorax and abdomen. The mandibles of the Mexican grain beetle are similar to those of *Dacne* spp., but the former is bidentate and the latter is tridentate.

18. Lathridiidae (Minute Brown Scavenger Beetles)

I have retained Arnett's classification, which includes *Holoparamecus* spp. in this family, although Halstead placed that genus in the Merophysiidae and Lawrence placed it in the Endomychidae.[3,12,15] The Lathridiidae are rich with species associated with stored-products. Thirty-five species have been recorded in warehouses, granaries, and so forth, and in stored-products, 11 other species that had been found only in houses or cellars.[6] Our laboratory has found several additional species in stored-products (see Table 1).[31]

Hinton provided data on the numerous habitats of various stored-product species and defined their trophic niches. Lathridiids are found in moldy plant and animal substances; in myxomycetes and fungi; under bark and stones; and in ant, wasp, bee and bird nests. Both adults and larvae feed only on myxomycetes and fungi (particularly molds). They are able to transmit molds. Many species of stored-product lathridiids are widespread. Hinton found nine species with cosmopolitan or nearly cosmopolitan distributions.[6]

Lathridiid adults (Figure 4) are identified by tarsi 3-3-3, 2-3-3, or 2-2-3; antennae 8–11 segmented with a two- or three-segmented, elongate club (Figure 4); and by the coxae not largely hidden. Hinton's key is recommended for species identification because of its comprehensive nature, although the nomenclature is largely outdated. Larvae are identified by the characteristic mandible, which consists of a sclerotized basal portion and an entirely or mostly membranous apical portion.[6]

19. Lyctidae (Powderpost Beetles)

The larvae of Lyctidae bore into wood. I am aware of only three species of lyctids found in stored food products.[31,36] *Lyctus brunneus* (Stephens) has been found in a variety of food products of plant origin, but Freeman believed that its occurrence in warehouses is probably due to infected wood.[2,34]

Our laboratory has recorded the following three species: *Lyctoxylon* sp., probably *japonum* Reitter; *Lyctus* sp., probably *brunneus*; and *Minthea rugicollis* (Walker).[31] *L. japonum* apparently has an oriental distribution although it may have become established in the New World. *L. brunneus* is cosmopolitan, and *M.rugicollis* is tropicopolitan.[69]

Adult lyctids are identified by the exposed head; the elongate shape; and the symmetrical, two-segmented antennal club.

20. Mycetophagidae (Hairy Fungus Beetles)

Hinton discussed four species of stored-product mycetophagids, summarizing their habitats and trophic habits.[6] Mycetophagids can be found inside fungi, under bark, in vegetable refuse and in plant seed heads. The larvae and adults feed on fungi, except for *Berginus tamarisci* Wollaston, which feeds on pollen and stamens.

Our laboratory commonly encounters mycetophagid larvae and larval fragments in a variety of plant products, including species not mentioned in the literature. Most, if not all, of these specimens represent *Typhaea stercorea* (L.) (hairy fungus beetle), and the adults and adult fragments of this species are commonly encountered.[31] We have also recorded *Berginus pumilus* LeConte once from dried tomatoes from Mexico.

Mycetophagid adults are identified by the pubescent dorsal surface; the pronotum at base as wide as the elytra; and the three- or four-segmented antennal club. Other characteristics include tarsi 4-4-4 or 3-4-4 and front coxal cavities open behind. The larvae of *T. stercorea* have a pair of pointed urogomphi with basal transverse ridges and asymmetrical mandibles.

21. Nitidulidae (Sap Beetles)

This family is rich with species associated with stored-products. It contains approximately 30 stored-product species in seven genera.[6,70-73] Nitidulids can occupy a variety of habitats.[6] They have been found feeding on the sap of trees and the juice of fruits, particularly when partly fermented. They can live on flowers, fungi and carrion. The trophic habits of both adults and larvae are also varied.[6] Several species of *Carpophilus* plus *Haptoncus luteolus* (Erichson) (yellowbrown sap beetle) feed on decaying or fermenting plant material and on ripe or nearly ripe fruit. Apparently *Carpophilus dimidiatus* (F.) (corn sap beetle) is facultatively predaceous, as adults have been found feeding on aphids on maize. *Nitidula* spp. feed on carrion. Stored-product species can feed on a great variety of products, but those products that are contaminated with fungi are most attractive.[2] Adults can function as mechanical vectors of yeasts and bacteria.[6]

Carpophilus [including *C. hemipterus* (L.) (driedfruit beetle), *C. dimidiatus* (F.), *C. obsoletus* Erichson and *C. mutilatus* Erichson (confused sap beetle] is the only genus our laboratory commonly encounters.[31] We have recorded individuals on a great variety of plant products from many different countries. Larvae (particularly early instar) are found more frequently than adults. A total of 19 *Carpophilus* spp. have been reported associated with stored-products.[70-73] Many of these species are cosmopolitan or tropicopolitan.[6,73]

Carpophilus spp. can be recognized fairly easily, but differentiating the species is very difficult. Adults of *Carpophilus* spp. can be identified by the abrupt, compact antennal club consisting of three flattened, strongly tranverse segments; the incomplete, truncate elytra; and the transverse, exposed pygidium. Larvae can be identified by the unbranched apices of the urogomphi and by the multi-serrated prostheca of the mandible (Figure 9). To identify adults, I recommend Hinton or Connell and for larvae, Connell.[6,73] An earlier work by Connell provides illustrations of the genitalia of the commonest *Carpophilus* spp.[74]

22. Ptiliidae (Featherwinged Beetles)

The Ptiliidae occupy a variety of habitats, including different kinds of decaying organic matter [e.g., forest litter, dung, carrion and rotting fungi]; along bodies of water in wet moss, under seaweed, etc.; and in the thin spore tubes of some mushrooms (Sörensson, personal communication). Dybas believed that adults and larvae feed mainly on fungal spores and hyphae and on soft organic materials containing microorganisms.[75]

We only occasionally find featherwinged beetles, and most records represent an undescribed genus (near *Ptiliola*) in dried fish and shrimp fry from Asia and the Philippines. We also have a record of *Nanosella* sp. in dried forest mushrooms and a record of *Actidium* sp. in frozen fish.

Adults are identified by their minute size (<1 mm) and their fringed hindwings, as well as by the whorls of hairs at the apices of the antennal segments.

23. Ptinidae (Spider Beetles)

Ptinidae is a family rich with stored-product species. Howe discussed 28 species in ten genera that may be found in stored food and storage premises. Habitats include decaying wood, but more commonly, bee, ant, spider, bird and mammal nests, and the excrement of other animal species. The nest environment can provide for the typical ptinid, which is primarily a scavenger, feeding on plant and animal material. Ptinids can feed on leaves, seeds, pollen, excrement or carrion. They depend on water to lay eggs and multiply. Water is available in urine, moist excrement, honey and other fluids. Many spider beetle species are widespread and feed on a variety of products (both animal and plant). Some are sensitive to the higher temperatures of the tropics and cannot colonize there.[76]

Our laboratory rarely encounters ptinids in samples of imported food products. We have a record of *Mezium americanum* (Laporte) (American Spider Beetle) in shark fin and a record of *Ptinus fur* (L.) (whitemarked spider beetle) in dried, edible green. It has also been found infesting chocolate slabs in the Pacific northwest. Other common spider beetle pests of stored-products include *Gibbium psylloides* (de Czenpinski) (hump beetle) and *Trigonogenius globulum* (Solier) (globular spider beetle).[16,77]

Adults are identified by their filiform antennae, which are inserted close together on the front of the head, and by their spiderlike appearance. *Ptinus* contains the most stored-product species, some of which are sexually dimorphic. Species identification keys for adults can be found in Papp and Okumura, Hinton, and Spilman.[77-79] Larvae are scarabaeiform with developed legs and a characteristic mandible that has a single apical tooth and a small, mesal tooth. A species key can be found in Manton.[80]

24. Scolytidae (Bark and Ambrosia Beetles)

To the best of my knowledge, there is no comprehensive treatment of scolytid species associated with stored food products. Wood (personal communication) mentioned about 20 species (in four genera) that are expected to be associated with stored food products. Of these, the majority are in the genus *Hypothenemus*. *H. hampei* (Ferrari) (coffee berry borer) is well known to most stored-product entomologists, but other potential food contaminants include *Hypothenemus eruditus* Westwood, *H. areccae* (Hornung), *H. columbi* Hopkins, *H. crudiae* (Panzer), *H. obscurus* (F.) (apple twig beetle) and *H. seriatus* (Eichoff). Some scolytid species are limited to commercial wood products and have not been included in the approximately 20 species mentioned above. Ebeling discussed some of these wood product species and noted that they are incapable of developing in aged, dry bark or wood.[36]

The adults and larvae of the typical North American scolytid tunnel under the bark (bark beetles) or deep in the wood (ambrosia beetles) of a variety of conifers and deciduous trees.[81] Bark beetles feed on phloem of the inner bark, whereas ambrosia beetles feed on fungi growing on the gallery walls. Other feeding habits include herbiphagy (feeding on nonwoody plants), spermophagy (feeding on seeds and the protective fruit) and myelophagy (feeding on the pith of small stems). Some species, such as *H. eruditus*, are capable of living in almost any available source of plant material.[82]

Hypothenemus spp. exhibit consanguineous polygyny.[82] The males are dwarfed, deformed and flightless, and females are partly parthenogenetic. If mating does occur, it is usually between mother and son or sister and brother. Apparently some outbreeding occurs.

A number of *Hypothenemus* spp. are widespread and attack a variety of plant parts of various hosts. *H. eruditus*, *H. areccae*, *H. crudiae*, *H. obscurus* and *H. seriatus* develop in a variety of fruits, twigs and weeds, and each is tropicopolitan or nearly so. However, the tropicopolitan *H. hampei* is limited to the developing fruits of coffee (*Coffea* spp.). This species does not normally develop in mature, stored coffee beans.[82]

Our laboratory has very limited data on species records because of the taxonomic difficulties of the group. We have recorded *H. eruditus* in a processed tamarind product and in preserved cabbage from Thailand; in dried mushrooms from Korea; and in whole tamarind pods from Mexico. We have recorded *H. areccae* in orange peel plum from Hong Kong and dried chilies from Thailand. We have recorded *H. hampei* in coffee beans from the Philippines.

Hypothenemus spp. adults are identified by their small size (1/2–2 mm); the compact, symmetrical antennal club; the slight notch in the eye at the antennal base; the rows of scalelike setae on the elytron; and the group of teeth along the anterior rim of the pronotum. The species taxonomy is extremely difficult, but a key is provided by Wood.[82]

Hypothenemus spp. larvae are scarabaeiform, lack developed legs, have a pair of labral rods, and have mandibles with three apical teeth.

25. Staphylinidae (Rove Beetles)

Thirty-four species are covered in Hinton's treatment of the Staphylinidae associated with stored-products. Habitats include a variety of decaying plant material situations, carrion, dung and mushrooms. Most species are predaceous, as larvae and adults, chiefly on other small arthropods. Some species may feed on dung or decomposing plant and animal materials. Adults of 2 of the 34 species have been found to feed on seedlings of cucumber, melon, squash, pepper, beet and chrysanthemum.[6] Two other species in the group are possible mold feeders. Some staphylinid species feed on developing mushrooms.[83] Their presence in mushroom products may indicate preharvest contamination.

Our laboratory has found a variety of rove beetles from a variety of plant and animal products (unpublished data). The adults are encountered fairly commonly in food products and can be identified by the short, incomplete elytra exposing at least three tergites and by the six or seven visible abdominal sternites.

26. Tenebrionidae (Darkling Beetles)

The Tenebrionidae are richly represented among stored-product beetles. Freeman stated that worldwide there are over 100 species associated with stored food products, but he provided no details or references.[2] Spilman's discussion of 33 species in 16 genera is the most current comprehensive treatment. Of the 33 species discussed by Spilman, 12 are cosmopolitan.[84]

Lawrence and Spilman reviewed the larval habitats of the entire family. They split the family into two basic groups: xylophilous larvae, which occur in rotten wood and associated cambium and subcortical spaces, and geophilous larvae, which occur in the soil and in leaf litter. Subgroups of xylophilous larvae include those that occur in the galleries of wood-boring beetles; in fermenting tree wounds; in bamboo and palm fronds; in mushrooms; on algae, lichens and mosses growing on wood and bark; and in the nests of birds, ants or termites. The last subgroup contains many taxa that have become stored-product species. All of the important stored-product species have been derived from xylophilous members.[85]

Trophic habits are quite variable, but tenebrionids are primarily saprophagous, feeding on a variety of dead plant and animal matter.[85] *Alphitobius diaperinus* (Panzer) (lesser mealworm) will feed on dead chickens and breed in poultry manure.[2,85] Some species are primarily mycophagous, including those that inhabit and feed on mushrooms [e.g., *Platydema americanum* (Laporte)].[85] Freeman stated that adults and larvae of *Alphitophagus bifasciatus* (Say) (twobanded fungus beetle) feed on molds.[2] Some species of *Tribolium* and *Palorus* have been classified as semipredators.[85] Spilman believed that the exact food preferences of most of the stored-product species have not been determined and speculated that both the adults and larvae might subsist mainly on mold.[84]

Our laboratory has found tenebrionids in a great variety of plant products, although some species have also been recorded in animal products.[31,84] Darkling beetles are frequently found in grains. Because most of them cannot gnaw through the hard outer covering, they usually attack broken or damaged grain. In wheat, we have seen *Tribolium* spp. attacking the germ ends of otherwise sound kernels, causing considerable damage. Our laboratory has encountered tenebrionids from a number of countries.[31] The tenebrionid that we most commonly encounter is *Tribolium castaneum* (Herbst) (red flour beetle). We have occasionally found *Tenebrio* spp., *Alphitobius* spp., *Gnatocerus* spp., *Palorus* spp., *Latheticus oryzae* Waterhouse (longheaded flour beetle) and *Tribolium confusum* Jacquelin du Val (confused flour beetle).[31]

Darkling beetle adults can be identified by the nonlobed 5-5-4 tarsi and the genae more or less concealing the antennal insertions (Figure 1). *T. castaneum* adults can be identified by the 1:1 ratio [distance between eyes (ventral aspect):width of one eye (ventral aspect)] and by the longitudinal, short, elytral ridges. Most tenebrionid larvae have a characteristically shaped mandible with two apical teeth and a distinct mola (Figure 12). Some genera have a pair of urogomphi and others a single, median process. *Tribolium* spp. larvae have a pair of large, unicolored urogomphi. I recommend Spilman's keys for the species identifications of both adults and larvae.[84] Sokoloff provides detailed identification keys for *Tribolium* spp. and a complete morphological treatment with descriptions of the terrata caused by quinones, which the beetles emit, or by other processes.[86]

27. Trogossitidae

The following two species are fairly commonly encountered in stored-products: *Lophocateres pusillus* (Klug) (Siamese grain beetle) and *Tenebroides mauritanicus* (L.) (cadelle beetle).[31] Barron listed 17 other *Tenebroides* spp. in America north of Mexico, but Halstead indicated that other species are rarely found in stored-products.[3,87]

Trogossitids can be found beneath bark, in rotten wood or in the galleries of wood-boring insects.[88] Neither *L. pusillus* nor *T. mauritanicus* are limited to feeding on plant material.[2] The adults and larvae of the latter feed on plant material, but the larvae are also predaceous on other insect pests. In addition, mature larvae may damage woodwork by boring through it to reach pupation sites. The adults and larvae of *L. pusillus* feed on plant material and on dried insect remains.

Both *L. pusillus* and *T. mauritanicus* can be found in a variety of plant products and both are widespread geographically. However, the former is limited to warmer areas (tropicopolitan), whereas the latter is cosmopolitan.[2,34]

The larvae and adults of both species can be readily identified. Adults of *L. pusillus* have a depressed elytron with an explanate margin and longitudinal ridges, a large and globose antennal scape and a three-segmented antennal club. The larvae of *L. pusillus* have a pair of apically converging urogomphi and a single, median projection between the urogomphal bases. The adults of the *T. mauritanicus* can be distinguished by their relatively large size (5–11 mm) and the distinct separation of the prothorax from the elytra; the larvae can be identified by the large, characteristically shaped mandible.

B. OTHER FAMILIES

Other families of beetles are reported as occasional pests of stored-products. The hairy cellar beetle, *Mycetaea subterranea* (F.) (= *M. hirta* [Marsham]) belongs to the family Endomychidae (handsome fungus beetles). It is a convex, hairy beetle approximately 1.5 mm long that feeds on mold.[2] It has relatively short antennae with a loose three-segmented club. The slightly larger *Monotoma* spp. (family Rhizophagidae) are also mold feeders that are occasionally found on moldy products. The elytra of these small dark beetles are slightly truncate, exposing the tip of the abdomen. In *Monotoma* and related species, the head is constricted into a narrow neck just posterior to the eyes. The antennal club consists of an enlarged apical segment. In California, *Aglenus brunneus* (Gyllenhal)(Colydiidae) is reported from grain and animal feed.[89,90] Colydiids are hard, shiny beetles with a two- or three-segmented antennal club and tarsal formula of 4-4-4.

REFERENCES

1. Hill, D.S., *Pests of Stored Products and Their Control*, CRC Press, Boca Raton, FL, 1-274, 1990.
2. Freeman, P., Ed., *Common Insect Pests of Stored Food Products: A Guide to Their Identification*, 6th ed., Brit. Mus. (Natur. Hist.), London, 1-69, 1980.
3. Halstead, D.G.H., Keys for the identification of beetles associated with stored products. I. Introduction and key to families, *J. Stored Prod. Res.*, 22, 163-203, 1986.
4. Sinha, R.N. and Watters, F.L., *Insect Pests of Flour Mills, Grain Elevators, and Feed Mills and Their Control*, Agric. Can. Publ. 1776, Ottawa, 1-290, 1985.
5. Kurtz, O.L. and Harris, K.L., *Microanalytical Entomology for Food Sanitation Control*, Assoc. Off. Agric. Chem., Washington, DC, 1-576, 1962.
6. Hinton, H.E., *A Monograph of the Beetles Associated with Stored Products*, Brit. Mus. (Natur. Hist.), London, 1-443, 1945 (reprinted 1963, Johnson Reprint Co., London).
7. Phillips, J.K. and Burkholder, W.E., Health Hazards of Insects and Mites in Food, in *Insect Management for Food Storage and Processing*, Bauer, F.J., Ed., Am. Assoc. Cereal Chem., St. Paul, MN, 1-384, 1984.
8. Okumura, G.T., A report of canthariasis and allergy caused by *Trogoderma* (Coleoptera: Dermestidae), *Vector Views*, 14, 19-24, 1967.
9. Scott, H.G., Human myiasis in North America (1952-1962 inclusive), *Fl. Entomol.*, 47, 503-505, 1964.
10. Wirtz, R.A. and Fruin, J.T., Mutagenicity of tenebrionid flour beetle secretions using *Drosophila melanogaster* sex-linked recessive lethal test, *J. Food Prot.*, 45, 1084-1086, 1982.
11. Banks, H.J., Identification of stored product *Cryptolestes* spp. (Coleoptera: Cucujidae): A rapid technique for preparation of suitable mounts, *J. Austral. Entomol. Soc.*, 18, 217-222, 1979.
12. Arnett, R.H., Jr., *American Insects: A Handbook of the Insects of America North of Mexico*, Van Nostrand Reinhold, NY, 1-850, 1985.
13. Arnett, R.H., Jr., *The Beetles of the United States (A Manual for Identification)*, Am. Entomol. Inst., Ann Arbor, MI, 1-1112, 1973.
14. Arnett, R.H., Jr., Downie, N.M., and Jaques, H.E., *How to Know the Beetles*, 2nd ed., Wm. C. Brown, Dubuque, IA, 1-424, 1980.
15. Lawrence, J.F., Order Coleoptera, in *Immature Insects*, Vol. 2, Stehr, F.W., Ed., Kendall/Hunt, Dubuque, IA, 144-658, 1991.
16. Gorham, J.R., Ed., *Insect and Mite Pests in Food: An Illustrated Key*, U.S. Dept. Agric., Agric. Handbook No. 655, 1-767, 1987.
17. Stehr, F.W., Ed., *Immature Insects*, Vol. 2, Kendall/Hunt, Dubuque, IA, 1-975, 1991.
18. Kurtz, O.L., Carson, N.A., and Van Dame, H.C., Identificationof cereal insects in stored grain by their mandible characteristics, *J. Assoc. Off. Agric. Chem.*, 35, 817-826, 1952.
19. Jackson, M.M., Ratay, A.F., and Woznicki, E.J., Identification of stored product insects by the micromorphology of the exoskeleton. VI. Adult and larval beetle mandibles, *J. Assoc. Off. Agric. Chem.*, 41, 460-471, 1958.
20. Kvenberg, J.E., Scanning electron microscopic study of adult stored product beetle mandibles, *J. Assoc. Off. Anal. Chem.*, 60, 1185-1209, 1977.
21. Jackson, M.M., Ratay, A.F., and Woznicki, E.J.,Identification of stored products insects by the micromorphology of the exoskeleton. II. Adult antennal characters, *J. Assoc. Off. Agric. Chem.*, 39, 879-888, 1956.
22. Heuermann, R.F., Note on elytral patterns of storage insects, *J. Assoc. Off. Agric. Chem.*, 40, 661-664, 1957.
23. Heuermann, R.F. and Kurtz, O.L., Identification of stored products insects by the micromorphology of the exoskeleton. I. Elytral patterns, *J. Assoc. Off. Agric. Chem.*, 38, 766-781, 1955.
24. Brickey, P.M., Jr. and Gecan, J.S., Introductory study of scanning electron microscopy of elytral fragments of stored product beetles, *J. Assoc. Off. Anal. Chem.*, 57, 1235-1247, 1974.
25. Kvenberg, J.E., Scanning electron microscopic study of adult stored product beetle elytra, *J. Assoc. Off. Anal. Chem.*, 61, 76-87, 1978.
26. Khalaf, K.T., Micromorphology of beetle elytra, using simple replicas, *Fl. Entomol.*, 63, 307-340, 1980.

27. Brickey, P.M., Jr., Gecan, J.S., Thrasher, J.J., and Vazquez,A.W., Notes on microanalytical techniques in the analysis of foods for extraneous materials. Part II, *J. Assoc. Off. Anal. Chem.*, 55, 51-56, 1972.
28. Olsen, A.R., Differentiating the saw-toothed grain beetle, *Oryzaephilus surinamensis* (L.), and the merchant grain beetle, *O. mercator* (Fauv.), by morphology of the male genitalia, *J. Assoc. Off. Anal. Chem.*, 60, 1223-1226, 1977.
29. Zimmerman, M.L., Identification of *Ahasverus advena* (Waltl) (foreign grain beetle) and *Ahasverus rectus* (LeConte) (Coleoptera: Cucujidae) by the micromorphology of adult fragments including genitalia, *J. Assoc. Off. Anal. Chem.*, 70, 484-495, 1987.
30. Olsen, A.R., List of stored-product insects found in imported foods entering the United States at southern California ports, *Bull. Entomol. Soc. Am.*, 27, 18-20, 1981.
31. Zimmerman, M.L., Coleoptera found in imported stored-food products entering southern California and Arizona between December 1984 through December 1987, *Coleopt. Bull.*, 44, 235-240, 1990.
32. Lawrence, J.F. and de Viedma, M.G., Anobiidae (Bostrichoidea), in *Immature Insects.*, Vol. 2, Stehr, F.W., Ed., Kendall/Hunt, Dubuque, IA, 441-444, 1991.
33. Anderson, D.M., Larval beetles (Coleoptera), in *Insect and Mite Pests in Food: An Illustrated Key*, Gorham, J.R., Ed., U.S. Dept. Agric., Agric. Handbook No. 655, 95-114, 1987.
34. Kingsolver, J.M., Adult beetles (Coleoptera), in *Insect and Mite Pests in Food: An Illustrated Key*, Gorham, J.R., Ed., U.S. Dept. Agric., Agric. Handbook No. 655, 75-94, 1987.
35. Werner, F.G., A revision of the North American species of *Anthicus*, s. str. (Coleoptera: Anthicidae), *Misc. Publ. Entomol. Soc. Am.*, 14, 195-242, 1964.
36. Ebeling, W., *Urban Entomology*, Univ. Calif. Press, Berkeley,1-695, 1975.
37. Spilman, T.J., False powderpost beetles of the genus *Dinoderus* in North America, *Coleopt. Bull.*, 36, 193-196, 1982.
38. Hodges, R.J., The biology and control of *Prostephanus truncatus* (Horn) (Coleoptera: Bostrichidae) — a destructive storage pest with an increasing range, *J. Stored Prod. Res.*, 22, 1-14, 1986.
39. Kingsolver, J.M., Seed beetles (Bruchidae, Coleoptera), in *Insect and Mite Pests in Food: An Illustrated Key*, Gorham, J.R., Ed., U.S. Dept. Agric., Agric. Handbook No. 655, 215-222, 1987.
40. Southgate, B.J., Biology of the Bruchidae, *Annu. Rev. Entomol.*, 24, 449-473, 1979.
41. Haines, C.P., Observations on *Callosobruchus analis* (F.) in Indonesia, including a key to storage *Callosobruchus* spp. (Col., Bruchidae), *J. Stored Prod. Res.*, 25, 9-16, 1989.
42. Kingsolver, J.M., *A Preliminary Key to Species of the Genus* Bruchus *(Bruchidae) Commonly Intercepted in U.S.D.A. Quarantine Inspections*, U.S. Natl. Mus., Washington, DC, 1-6, 1965.
43. Johnson, C.D., Biosystematics of the Arizona, California, and Oregon species of the seed beetle genus *Acanthoscelides* Schilsky (Coleoptera: Bruchidae), *Univ. Calif. Berkeley Publ. Entomol.*, 59, 1-116, 1970.
44. Pfaffenberger, G.S., Bruchidae (Chrysomeloidea), in *Immature Insects*, Vol. 2, Stehr, F.W., Ed., Kendall/Hunt, Dubuque, IA, 561-568, 1991.
45. Davey, P.M., The groundnut bruchid, *Caryedon gonagra* (F.), *Bull. Entomol. Res.*, 49, 385-404, 1958.
46. Kingsolver, J.M., A key to the species of *Callosobruchus* (Bruchidae) intercepted in USDA plant quarantine inspections, *USDA Plant Quarantine Memorandum No. 690*, 1-14, 1969.
47. Bousquet, Y., Carabidae (Adephaga), in *Immature Insects*, Vol.2, Stehr, F.W., Ed., Kendall/Hunt, Dubuque, IA, 306-310, 1991.
48. Halstead, D.G.H., Observations on the biology of *Murmidius ovalis* (Beck) (Coleoptera: Cerylonidae), *J. Stored Prod. Res.*, 4, 13-21, 1968.
49. Lawrence, J.F., Revision of North American Ciidae (Coleoptera), *Bull. Mus. Comp. Zool. Harv.*, 142, 419-522, 1971.
50. Madenjian, J.J., Eifert, J.D., and Lawrence, J.F., Ciidae: Newly recognized beetle pests of commercial dried mushrooms, *J. Stored Prod. Res.*, 29, 45-48, 1993.
51. Lawrence, J.F., Three new Asiatic Ciidae (Coleoptera:Tenebrionoidea) associated with commercial, dried fungi, *Coleopt. Bull.*, 45, 286-292, 1991.

51a. Lawrence, J.F., Host preference in ciid beetles (Coleoptera: Ciidae) inhabiting the fruiting bodies of Basidiomycetes in North America, *Bull. Mus. Comp. Zool. Harv.*, 145, 163-212, 1973.

52. Graves, R.C., Ecological observations on the insects and other inhabitants of woody shelf fungi (Basidiomycetes: Polyporaceae) in the Chicago area, *Ann. Entomol. Soc. Am.*, 53, 61-78, 1960.
53. Haines, C.P. and Rees, D.P., *A Field Guide to the Types of Insects and Mites Infesting Cured Fish*, FAO Fisheries Tech. Paper No. 303, 1-33, 1989.
54. Kingsolver, J.M., Checkered beetles (Coleoptera, Cleridae), in *Insect and Mite Pests in Food: An Illustrated Key*, Gorham, J.R., Ed., U.S. Dept. Agric., Agric. Handbook No. 655, 149-150, 1987.
55. Kingsolver, J.M., Cryptophagid beetles (Cryptophagidae, Coleoptera), in *Insect and Mite Pests in Food: An Illustrated Key*, Gorham, J.R., Ed., U.S. Dept. Agric., Agric. Handbook No. 655, 175-178, 1987.
56. Halstead, D.G.H., Keys for the identification of beetles associated with stored products. II. Laemophloeidae, Passandridae and Silvanidae, *J. Stored Prod. Res.*, 29, 99-197, 1993.

57. Dyte, C.E., Studies on the abundance of *Cryptolestes turcicus* (Grouv.) (Coleoptera, Cucujidae) in different machines of an English flour mill, *J. Stored Prod. Res.*, 1, 341-352, 1966.
58. Loschiavo, S.R. and Sinha, R.N., Feeding, oviposition and aggregation by the Rusty Grain Beetle, *Cryptolestes ferrugineus* (Coleoptera: Cucujidae) on seed-borne fungi, *Ann. Entomol. Soc. Am.*, 59, 578-585, 1966.
59. Sheppard, E.H., Notes on *Cryptolestes ferrugineus* Steph., a cucujid occurring in the *Trichogramma minutum* parasite laboratory of Colorado State College, *Colo. Agric. Exp. St. Tech. Bull.*, 17, 1-20, 1936.
60. Halstead, D.G.H., A revision of the genus *Oryzaephilus* Ganglbauer, including descriptions of related genera, *Zool. J. Linn. Soc.*, 69, 271-374, 1980.
61. Cutler, J.R., A key for distinguishing the larvae of *Ahasverus advena* (Waltl), *Cathartus quadricollis* (Guér.), *Oryzaephilus surinamensis* (L.) and *Oryzaephilus mercator* (Fauv.) (Coleoptera: Silvanidae), *J. Stored Prod. Res.*, 7, 125-127, 1971.
62. Whitehead, D.R., Weevils (Curculionidae, Coleoptera), in *Insect and Mite Pests in Food: An Illustrated Key*, Gorham, J.R., Ed., U.S. Dept. Agric., Agric. Handbook No. 655, 223-230, 1987.
63. White, R.E., *A Field Guide to the Beetles of North America*, Houghton Mifflin, Boston, 1-368, 1983.
64. Hunkapiller, P.D. and O'Donnell, A.E., A taxonomic separation of larvae of the genus *Sitophilus* by head capsule morphology (Coleoptera: Curculionidae), *J. Kansas Entomol. Soc.*, 40, 435-440, 1967.
65. Kingsolver, J.M., Dermestid beetles (Dermestidae,Coleoptera), in *Insect and Mite Pests in Food: An Illustrated Key*, Gorham, J.R., Ed., U.S. Dept. Agric., Agric. Handbook No. 655, 115-136, 1987.
66. Savary, W.E., *Dacne picta*: A recently introduced pest of stored, dried shiitake mushrooms (Coleoptera: Erotylidae), *Pan-Pac. Entomol.* (in press).
67. Hinton, H.E., The Histeridae associated with stored products, *Bull. Entomol. Res.*, 35, 309-340, 1945.
68. Chittenden, F.H., The Mexican grain beetle (*Pharaxonotha kirschi* Reitt.), *U.S. Dep. Agric. Bur. Entomol. Bull.*, 96, 8-13, 1911.
69. Gerberg, E.J., A revision of the New World species of powderpost beetles belonging to the family Lyctidae, *U.S. Dep. Agric. Tech. Bull.*, 1157, 1-55, 1957.
70. Dobson, P.M., The species of *Carpophilus* Stephens (Col., Nitidulidae) associated with stored products, *Bull. Entomol. Res.*, 45, 389-402, 1954.
71. Williams, R.N., Fickle, D.S., Kehat, M., Blumberg, D., and Klein, M.G., Bibliography of the genus *Carpophilus* Stephens (Coleoptera: Nitidulidae), Ohio Agric. Res. Develop. Center, Res. Circular 278, 1-95, 1983.
72. Archibald, R.D. and Maddison, P.A., A summary of stored product Coleoptera in New Zealand and neighboring Pacific countries, *N.Z. Entomol.*, 11, 1-6, 1988.
73. Connell, W.A., Sap beetles (Nitidulidae, Coleoptera), in *Insect and Mite Pests in Food: An Illustrated Key*, Gorham, J.R., Ed., U.S. Dept. Agric., Agric. Handbook No. 655, 151-174, 1987.
74. Connell, W.A., Nitidulidae of Delaware, *Univ. Del. Tech. Bull., 318*, 1-67, 1936.
75. Dybas, H.S., Ptiliidae (Staphylinoidea), in *Immature Insects*, Vol. 2, Stehr, F.W., Ed., Kendall/Hunt, Dubuque, IA, 322-324, 1991.
76. Howe, R.W., Studies on beetles of the family Ptinidae. XVII. Conclusions and additional remarks, *Bull. Entomol. Res.*, 50, 287-326, 1959.
77. Papp, C.S. and Okumura, G.T., A preliminary study of the Ptinidae of California, *Calif. Dep. Agric. Bull.*, 48, 228-248, 1959.
78. Hinton, H.E., The Ptinidae of economic importance, *Bull. Entomol. Res.*, 31, 331-381, 1941.
79. Spilman, T.J., Spider beetles (Ptinidae, Coleoptera), in *Insect and Mite Pests in Food: An Illustrated Key*, Gorham, J.R., Ed., U.S. Dept. Agric., Agric. Handbook No. 655, 137-148, 1987.
80. Manton, S.M., The larvae of the Ptinidae associated with stored products, *Bull. Entomol. Res.*, 35, 341-365, 1945.
81. Bright, D.E., Scolytidae (Curculionoidea), in *Immature Insects*, Vol. 2, Stehr, F.W., Ed., Kendall/Hunt, Dubuque, IA, 613-116, 1991.
82. Wood, S.L., *The Bark and the Ambrosia Beetles of North and Central America (Coleoptera: Scolytidae): A Taxonomic Monograph*, Great Basin Naturalist Memoirs No. 6, Brigham Young Univ., Provo, UT, 1-1359, 1982.
83. Klimaszewski, J. and Peck, S.B., Succession and phenology of beetle faunas (Coleoptera) in the fungus *Polyporellus squamosus* (Huds.: Fr.) Karst (Polyporaceae) in Silesia, Poland, *Can. J. Zool.*, 65, 542-550, 1987.
84. Spilman, T.J., Darkling beetles (Tenebrionidae, Coleoptera), in *Insect and Mite Pests in Food: An Illustrated Key*, Gorham, J.R., Ed., U.S. Dept. Agric., Agric. Handbook No. 655, 185-214, 1987.
85. Lawrence, J.F. and Spilman, T.J., Tenebrionidae (Tenebrionoidea), in *Immature Insects*, Vol. 2, Stehr, F.W., Ed., Kendall/Hunt, Dubuque, IA, 520-528, 1991.
86. Sokoloff, A., *The Biology of Tribolium*, Vol. 1, Clarendon Press, Oxford, 1-300, 1972.
87. Barron, J.R., A revision of the Trogositidae of America north of Mexico (Coleoptera: Cleroidea), *Mem. Entomol. Soc. Can.*, 75, 1-143, 1971.

88. Foster, D.E. and Lawrence, J.F., Trogossitidae (Cleroidea) in *Immature Insects*, Vol. 2, Stehr, F.W., Ed., Kendall/Hunt, Dubuque, IA, 448-450, 1991.
89. Strong, R.G. and Okumura, G.T., Insects and mites associated with stored foods and seeds in California, *Bull. Calif. Dep. Agric.*, 47, 233-249, 1958.
90. Okumura, G.T. and Strong, R.G., Insects and mites associated with stored foods and seeds in California. Part II, *Bull. Calif. Dep. Agric.*, 54, 13-23, 1965.

Chapter 6

FLIES (INSECTA: DIPTERA)

Alan R. Olsen

CONTENTS

I. Introduction ... 72

II. Morphology ... 73
- A. General Morphology ... 73
- B. Wing Venation ... 75

III. Identification ... 77
- A. Identification Key ... 77
- B. Significance of Food-Contaminating Families ... 80

IV. Discussion of Families ... 80
- A. Common Food-Contaminating Families ... 80
  1. Anthomyiidae ... 80
  2. Calliphoridae (Blow Flies) ... 80
  3. Carnidae ... 81
  4. Ceratopogonidae (Biting Midges) ... 81
  5. Chironomidae (Midges) ... 81
  6. Chloropidae (Eye Gnats) ... 82
  7. Drosophilidae (Pomace Flies) ... 82
  8. Ephydridae (Shore Flies) ... 82
  9. Milichiidae ... 82
  10. Muscidae (House Flies) ... 83
  11. Phoridae (Humpbacked Flies) ... 84
  12. Piophilidae (Skipper Flies) ... 84
  13. Psychodidae (Sewer Flies) ... 84
  14. Platystomatidae (Picture-Winged Flies) ... 85
  15. Sarcophagidae (Flesh Flies) ... 85
  16. Scathophagidae (Dung Flies) ... 85
  17. Scatopsidae (Minute Black Scavenger Flies) ... 85
  18. Scenopenidae (Window Flies) ... 86
  19. Sciaridae (Dark-Winged Fungus Gnats) ... 86
  20. Sepsidae (Black Scavenger Flies) ... 86
  21. Sphaeroceridae (Small Dung Flies) ... 87
  22. Stratiomyidae (Soldier Flies) ... 87
  23. Syrphidae (Rat-Tailed Maggots) ... 87
- B. Other Families ... 87

V. Remarks ... 88

References ... 88

I. INTRODUCTION

The flies constitute one of the largest orders (Diptera) of insects in terms of numbers of species and families.[1,2] Among this diversity, there is a discrete group of families, some of whose members have adapted to flourish in association with humans and their civilizations. These flies, termed synanthropic flies, are attracted to the food and shelter that we unintentionally provide for them in our farms, villages, towns and cities around the world.[3] Synanthropic flies differentially survive and increase in population in the altered environments of human settlements better than they do in their natural habitats.[4,5] Some synanthropic flies, such as the cheese skippers and windowpane flies, are true stored-product pests in that they are chiefly found infesting food products and rarely seen outside of food processing and storage facilities. Most synanthropic flies, however, breed in the filth and offal that are byproducts of civilization.[3,6-8] Both types of synanthropy concern the microanalytical entomologist. Stored-product flies are a concern because they infest stored food products and adulterate them with their infestation. Filth flies are an added concern because they account for direct adulteration when they visit food products and are potential vectors for various food-borne pathogens.

The association of synanthropic flies with the transmission of disease is well established.[9-12] In relation to food-borne pathogens, their chief role is that of a passive vector, physically transporting pathogens from filthy reservoirs to foods where the pathogens' potential for causing illness is greatly amplified.[13-16] Table 1 summarizes the major food-borne pathogens that have been isolated from natural populations of flies. Although flies may not be efficient enough vectors to always deliver the quantities of pathogens that will immediately provoke illness in humans, they are generally capable of transporting pathogens and depositing them on food in a viable state. If further mishandled in ways that promote the growth of a pathogen, the food that was inoculated by flies may become hazardous to consume.[10]

Other hazards associated with synanthropic flies are the introduction of agents of decomposition and the causing of myiasis, or the invasion of the body by fly larvae. Agents of food decomposition (molds, yeasts and spoilage bacteria) are transported by the same mechanical means as pathogens.[10,11] In myiasis, the fly itself becomes an agent of disease. There are three categories of myiasis.[10,17,18] Obligatory myiasis is caused by flies that are obligatory endoparasites and cannot survive without a living host. Examples are the primary screw-worms (Calliphoridae), bot flies and warble flies (Oestridae).[17,19] Facultative myiasis is caused by flies that usually feed on carrion or other dead, decayed material. These flies may opportunistically oviposit on wounds or orifices of living hosts, including humans, and the hatchling larvae invade the host. Examples include the secondary screw-worms (Calliphoridae) and many other carrion-feeding flies.[18] Accidental myiasis occurs when fly larvae are accidently eaten while alive.[10,17] About 50 species of flies have been reported as causative of accidental myiasis. Cases where the larvae merely survive the rigors of the host's gastrointestinal system, as opposed to feeding and growing there, are termed pseudomyiasis.[17,18] In either case, clinical symptoms are sometimes severe, including pain, vomiting, diarrhea, vertigo, nausea, depression and secondary infections. About 25% of human myiasis cases involve accidental ingestion.[20] Table 2 lists the fly species that have been documented as causing enteric myiasis in humans.

TABLE 1
Selected Examples of Microbial Pathogens Isolated from Natural Populations of Flies

Pathogen	Ref.
Campylobacter jejuni	103
Entoamoeba histolytica	104
Escherichia coli	105
Salmonella	106–108
Shigella	107, 109, 110
Staphylococcus aureus	111
Vibrio cholera	16

Modified from Gorham.[102]

TABLE 2
Flies That Cause Human Food-Borne Enteric Myiasis

Fly	Ref.
Calliphora vicina (Calliphoridae)	18, 20
Eristalis tenax (Syrphidae)	10, 17, 18
Fannia canicularis (Muscidae)	10, 17, 18, 20
F. scalaris (Muscidae)	10, 17, 18, 20
Hermetia illucens (Stratiomyidae)	10, 17, 18
Leptocera spp. (Sphaeroceridae)	20
Megaselia spp. (Phoridae)	17, 18
Musca domestica (Muscidae)	10, 17, 18
Muscina stabulans (Muscidae)	10, 17, 18
Phaenicia spp. (Calliphoridae)	20
Piophila casei (Piophilidae)	10, 17, 18, 20
Sarcophaga haemorrhoidalis (Sarcophagidae)	17, 18, 20
Stomoxys calcitrans (Muscidae)	17, 18
Sylvicola fenestralis (Anisopodidae)	17, 18

The fly taxa encountered by the FDA in food products include more than 70 species from 42 families. About half of these (21 families) are filth flies. The remainder are stored-product pests and incidental flies, mostly the latter. Table 3 summarizes the significance of the families of flies that are most frequently encountered in food products. This chapter deals with the common food-contaminating filth flies and stored-product flies with emphasis on the adult flies that are potential vectors of food-borne disease.

II. MORPHOLOGY[1,21-23]

Flies (order Diptera) are easily recognized by the possession of only one pair of wings connected to a thorax which is highly modified to support the musculature and articulation of the single pair of wings.

A. GENERAL MORPHOLOGY

The head (Figure 1) is dominated by a pair of compound eyes that take various forms. On the vertex of the head between the compound eyes there are up to three ocelli, occurring as either three, one or none. The ocelli are usually situated on a distinct, often raised, triangular area. This ocellar triangle is usually bare. Posterior to the ocellar triangle are the taxonomically important postvertical bristles. A full complement of three ocelli is arranged in a triangle on the vertex with two posterolateral eyes and one anteromedial eye. The front of the head bears the antennae, which may be filiform, plumose or aristate, occasionally subclavate (e.g., Scatopsidae). The aristate antenna is unique to flies, consisting of only three segments with the apical segment bearing a setiform process called a style or arista. A style is located apically and is relatively rigid. An arista is located dorsobasally on the third segment and is somewhat flexible. Mouthparts are located ventrally and may be of the piercing type, sponging type or a combination of both. In some groups the mouthparts are nonfunctional or vestigial. To either side of the oral cavity are the bristlelike oral vibrissae. The presence, absence and sizes of the oral vibrissae are taxonomically significant.

The thorax consists mostly of the well-developed mesothorax, which bears the wings. The mesonotum is divided into the anterior notum and the smaller, usually triangular, posterior scutellum. In the calyptrate flies (and a few other groups), the notum is further divided by a transverse suture midway from the anterior margin. Laterally (Figure 2), the mesothorax

TABLE 3
Significance of Families of Food-Contaminating Flies

Family	Common name	Significance[a]
Anisopodidae	Wood gnats	f,m
Anthomyiidae	Anthomyiid flies	i
Calliphoridae	Blow flies	f,m,v
Carnidae	Carnid flies	f
Ceratopogonidae	Biting midges	f,i
Chironomidae	Midges	f,i
Chloropidae	Eye gnats	f
Drosophilidae	Pomace flies	f
Ephydridae	Shore flies	f,i
Heleomyzidae	Heleomyzid flies	i
Milichiidae	Milichiid flies	f,v
Muscidae	Muscid flies	f,m,v
Phoridae	Humpbacked flies	f,m,s
Piophilidae	Skipper flies	f,m,s
Psychodidae	Sewer flies	f
Platystomatidae	Picture-winged flies	f
Sarcophagidae	Flesh flies	f,m,v
Scathophagidae	Dung flies	f,v
Scatopsidae	Minute black scavenger flies	f
Scenopenidae	Window flies	s
Sciaridae	Dark-winged fungus gnats	s
Sepsidae	Black scavenger flies	f
Sphaeroceridae	Small dung flies	f,m,v
Stratiomyidae	Soldier flies	f,s
Syrphidae	Rat-tailed maggots	f,m

[a] f, filth fly; i, incidental fly; m, enteric myiasis-producing fly; s, stored-product pest; v, vector of food-borne pathogens.

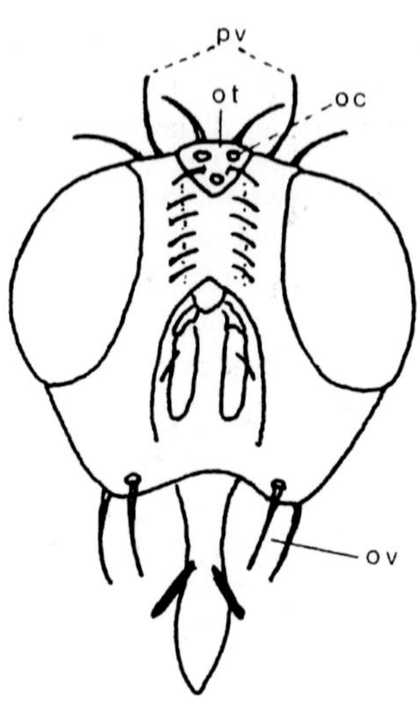

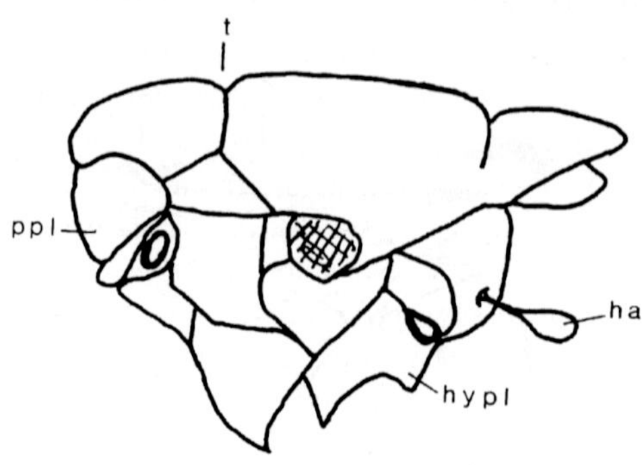

FIGURE 1. (Left) diagram of fly head (Muscomorpha) showing ocellus (oc); ocular triangle (ot); oral vibrissa (ov); and postvertical bristles (pv).

FIGURE 2. (Right) diagram of fly thorax, lateral view, showing halter (hal); hypopleuron (hypl); propleuron (ppl); and transverse suture (t).

resembles a fan or wedge, and one can discern the narrow prothoracic and metathoracic areas to which the prothoracic and metathoracic legs are attached before and behind the mesothoracic legs. The pleural areas are entirely sclerotized with a number of distinct sclerites. The setation of the sclerite (the propleuron) anterior to the prothoracic spiracle is often significant for identification. Of particular interest is the hypopleural sclerite (hypopleuron) located between the coxa and spiracle on the metathorax. A vertical row of bristles on the hypopleuron distinguishes two groups of calyptrate flies. The metathorax also bears the knobbed halteres, vestigial wings.

The abdomen consists of five to eleven external segments that are visible. In most cases where few segments are visible, the remaining segments are retracted into the abdomen and modified for reproductive purposes. The retracted segments can be exerted for copulation or oviposition. Each external segment bears a pair of lateral spiracles on the membranous pleural areas, a dorsal tergite and ventral sternite.

The six legs are usually uniform in size and structure. The tarsus bears a pair of apical claws, a pair of fleshy pulvilli (pads) beneath the claws and a central structure called an empodium. The empodium may be padlike, resembling the pulvilli, or it may be bristlelike or absent. The legs are rarely modified for any purpose other than walking. The body and legs are armed with numerous setae. The chaetotaxies of the head, thorax and legs are important taxonomic features as well as the isolated setae found in foods.[1,3,17,24-26]

B. WING VENATION

The wings of flies are membranous and either clear, tinted or patterned with transparent colors. One form of coloration is a stigma (pl. stigmata), or dark spot, on an otherwise clear wing. The wing shape is paddlelike with a narrow base. The posterior wing margin near the base often bears a deep incision, creating a basal flap, the alula. In some flies there are two additional flaps, the calypters, that are partially attached to the thorax, and one usually overlaps the other. The calypters lie below the plane of the wing when at rest, whereas the slightly distal alula is in the plane.

The most important taxonomic feature of flies is the intricate pattern of veins and cells on the membranous wing. There are numerous systems of nomenclature for the wing venation of flies. This chapter uses McAlpine's system (Figure 3).[22] In most cases, wing venation is diagnostic to the family level or lower, and the identification of flies is greatly simplified by learning to recognize characteristic patterns, using a standard nomenclature system to do so.

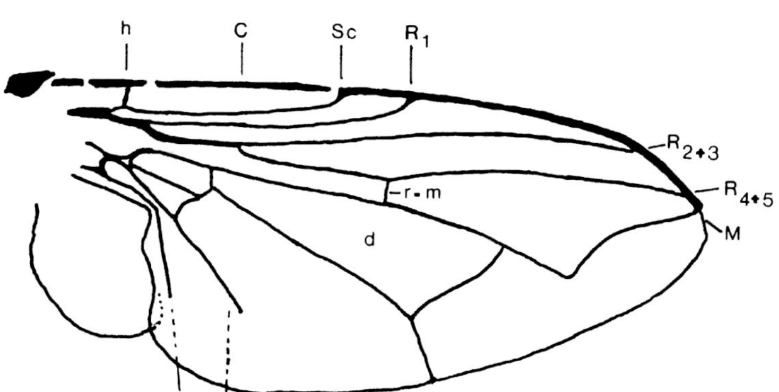

FIGURE 3. Diagram of a housefly wing showing principal veins. Branches of anal vein (A); costa (C); anterior branches of cubitus (CuA); media (M); branches of radius (R); humeral cross vein (h); radial-medial cross vein (r-m); discal cell (d); subcosta (Sc).

The principal veins are longitudinal, originating at the base of the wing and extending toward its tip (apex). These longitudinal veins are named, from anterior to posterior:

Costa (C)
Subcosta (Sc)
Radius (R)
Median (M)
Cubitus (Cu)
Anal (A)

The costa proceeds along the anterior margin of the wing and defines this margin for the length of the vein. It is not branched. The other veins traverse the membrane of the wing and may or may not fork into numerous branches from a single basal vein. Branches may proceed to the wing margin, end abruptly before the margin, intersect and coalesce in various patterns. In addition to the longitudinal veins, cross veins that connect two longitudinals may occur. The principal cross veins and the longitudinal veins (or branches) that they connect are

(h) (humeral vein) connects C and Sc or R basally
(r-m) connects R and M in middle of wing
(dm-cu) connects M and Cu in middle of wing
(bm-cu) connects M and Cu basally
(CuA) connects Cu and A basally

The system can daunt one with its apparent complexity. The key to deciphering the morphology of complex wing venation patterns is to locate the r-m cross vein near the exact center of the wing. The anterior end of this small cross vein terminates at the last branch of the radius (R) and thus orients the eye to the front half of the wing containing the branches of the radius (R), the subcosta (Sc), when present, and the costa (C), which form the anterior margin of the wing. The posterior end of r-m terminates at the first branch of the median (M) and orients the eye to the back half of the wing containing, among other things, the taxonomically important discal cell (dm), upon whose anterior margin the r-m vein terminates as well as the basal medial cell (bm) and posterior cubital cell (cu*p*) just basad of the discal cell. The r-m cross vein serves as a convenient orientation landmark that allows the observer to determine where a particular vein or cell is without tracing veins and branches from the wing base where vein origins can be crowded and obscure. When a vein or cell is located, its orientation relative to r-m is important for comparing positions, shapes and relationships of venation patterns among different groups of flies. Diagrams of typical wing patterns from literature sources are also more easily interpreted by first locating the r-m cross vein.

Cells, the open areas between veins, are often difficult to decipher because the shape and position can vary widely for the same cell between groups of flies. Cells can be open or closed. An open cell is one where the apical margin of the cell is defined by an area of wing margin past the end of the costa (i.e., no vein closes the cell's apical margin). A closed cell is one bounded on all sides by veins. The most useful cells for identificaton purposes are the discal cell (dm) just behind the posterior end of r-m and the posterior cubital cell (cu*p*) just posterior and basad of the discal cell. The presence, absence, size, shape and position of these cells are taxonomically significant.

Figures 4 and 5 illustrate the venation patterns of the wings of the most frequently encountered food-contaminating filth flies. Other families discussed in this chapter have patterns similar to the ones depicted. The analytical key to fly families and the discussions of each family provide the diagnostic venation characteristics necessary to differentiate each family.

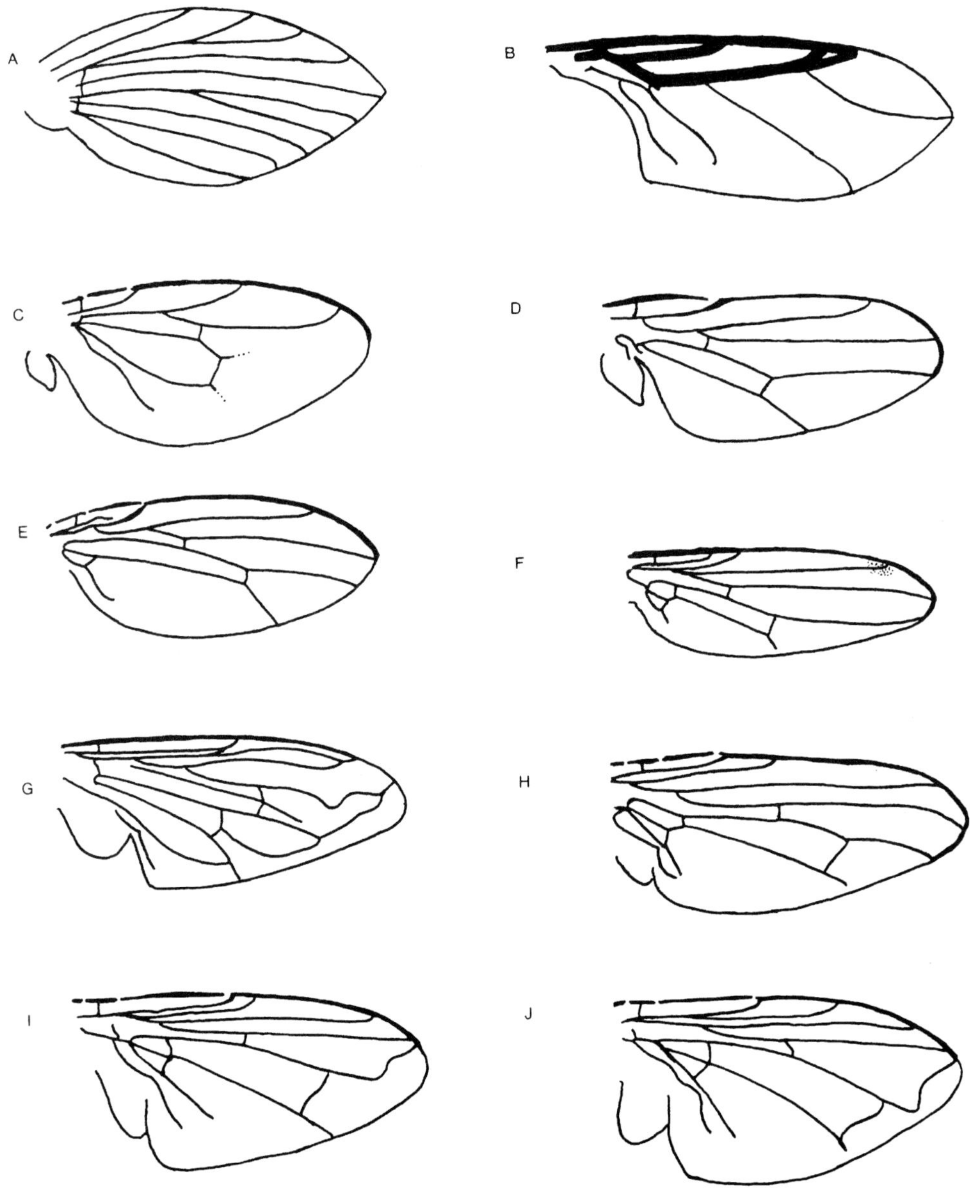

FIGURE 4. Diagrams of wings of common food-contaminating flies. (A) Psychodidae, *Psychoda* sp.; (B) Phoridae, *Megaselia* sp.; (C) Sphaeroceridae, *Leptocera* sp.; (D) Chloropidae, *Cadrema* sp.; (E) Drosophilidae, *Drosophila* sp.; (F) Sepsidae, *Sepsis* sp.; (G) Syrphidae, *Eristalis* sp.; (H) Muscidae, *Fannia* sp.; (I) Muscidae, *Musca* sp.; (J) Calliphoridae, *Chrysomya* sp.

III. IDENTIFICATION

A. IDENTIFICATION KEY

Rapid Key for Major Families of Food-Contaminating Flies (Adults)

1. Antenna with six or more segments; discal cell, when present, always open 2
   Antenna with five or fewer segments; discal cell, when present, usually 6
2. Wing pointed apically; C continuing around apex; body and wings densely clothed with fine hairs or scales .... Psychodidae

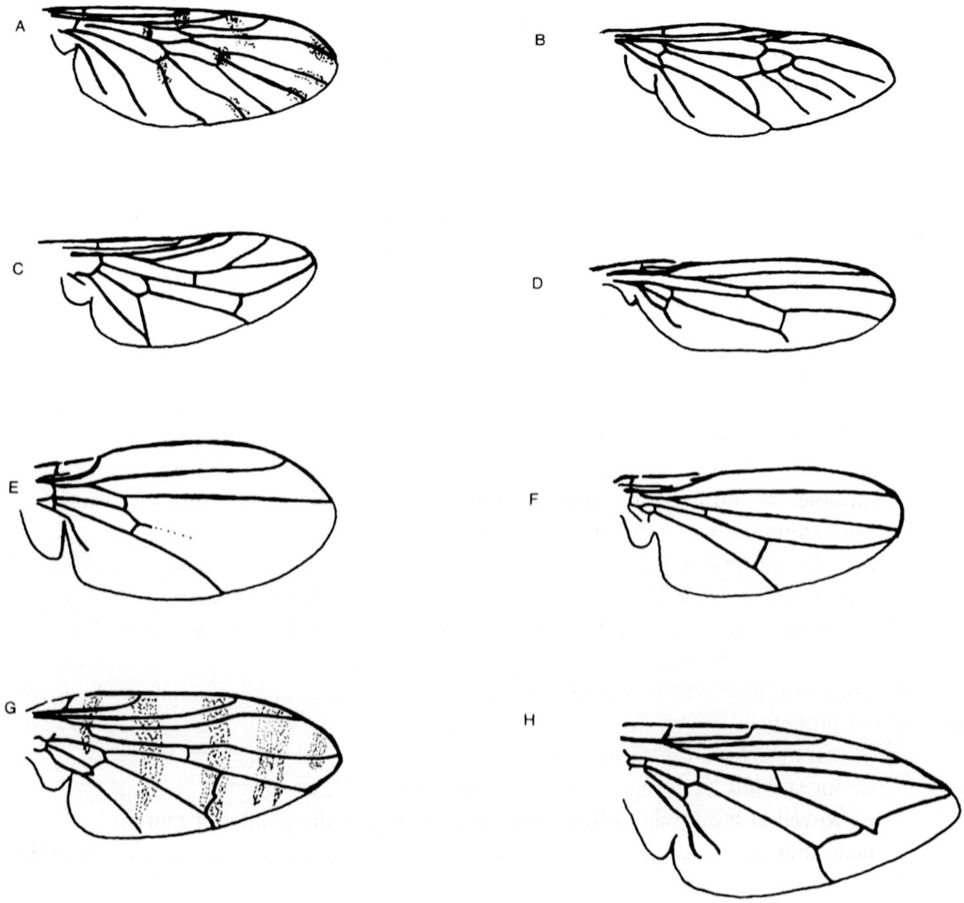

FIGURE 5. Diagrams of wings of common food-contaminating flies. (A) Anisopodidae, *Sylvicola* sp.; (B) Stratiomyidae, *Hermetia* sp.; (C) Scenopenidae, *Scenopinus* sp.; (D) Scathophagidae, *Scathophaga* sp.; (E) Carnidae, *Meoneura* sp.; (F) Milichiidae, *Milichiella* sp.; (G) Platystomatidae, *Scholastes* sp.; (H) Sarcophagidae, *Sarcophaga* sp.

Wing rounded apically; body not densely clothed with hairs or scales 3

3. Ocelli present; compound eyes meet above antennae; M always present and forked at or beyond the middle of the wing; posterior half of wing lacking closed cells 4

   Ocelli absent; compound eyes variable; M present or absent, but when present usually forked basad of wing middle 5

4. C ending near middle of anterior wing margin; posterior venation weak; tibia without apical spur; antennae and palpi short .... Scatopsidae

   C ending near apex of wing; r-m cross vein parallel to R_1; tibia with apical spur; antennae, palpi long .... Sciaridae

5. All branches of R reaching anterior wing margin at or before the middle of the margin; posterior veins indistinct; compound eyes nearly meeting above antennae; mouthparts short, piercing type .... Ceratopogonidae

   Some or all branches of R terminating beyond middle of wing; compound eyes separated above; mouthparts not piercing type .... Chironomidae

6. Anterior wing veins close together near front margin; discal cell present, compact, not elongate; empodium padlike .... Stratiomyidae

   Venation variable; discal cell elongate or absent; empodium bristlelike 7

7. M_{1+2} ending at or before apex of wing; CuA_2 arced broadly, closing cu*p* past the midpoint of A_1; antenna three segmented, peglike and lacking a style or arista .... Scenopenidae
   Antenna three-segmented with third segment bearing an arista; venation variable but usually without CuA_2 forming a broad arc 8
8. C ending near middle of wing; strong anterior veins concentrated near basal front margin; other veins indistinct; R forked just before terminating at the tip of C .... Phoridae
   C extending nearly to wing apex or beyond, some or all posterior veins distinct 9
9. Wing with spurious vein crossing r-m but not reaching wing margin; R_{4+5} cell petiolate with veins R_{4+5} and M_{1+2} coalescing before reaching wing margin; R_{4+5} undulate in middle after r-m cross vein .... Syrphidae
   Wing lacking spurious vein 10
10. C with three breaks; Sc incomplete; M and CuA_1 ending at or just beyond cross vein; first segment of hind tarsus enlarged .... Sphaeroceridae (part)
   C variable; M and CuA_1 reaching wing margin or nearly so, or bending forward but extending well beyond the cross vein in any case; cu*p* rounded or truncate apically, or absent 11
11. Sc complete, reaching C independent of R_1; calypters present or absent 12
   Sc incomplete, not reaching C, sometimes terminating at R_1; calypters absent 21
12. C unbroken or with one break 13
   C broken in three places (basally, at h and Sc) 16
13. C not broken 14
   C broken near Sc; oral vibrissae present 15
14. Posterior spiracle (near halter) with one or more bristles; abdomen elongate, narrowed at base; wing often with a single stigma along anterior margin near apex .... Sepsidae
   Posterior spiracle without bristles; wing patterned; R_1 setulose above along entire length; M_3+CuA much shorter than dm-cu cross vein .... Platystomatidae
15. C spinose; postocellar bristles convergent; antennae small; one or more dorsal preapical bristles on tibia .... Heleomyzidae
   C bare; stigma present between Sc and R_1; A_1 not reaching wing margin; postocellar bristles divergent; legs yellow, lacking dorsal bristles on tibia .... Piophilidae
16. Calypters present, rounded apically; M straight or nearly so 17
   Calypters present, truncate apically; M curved sharply forward after dm-cu cross vein 19
17. A_1 reaching wing margin, at least as a fold 18
   A_1 not reaching wing margin; A_1 shorter than A_2; A_2 curving slightly forward but not enough to cross A_1; Sc parallel to R_1, curving forward only near termination at C .... Muscidae (Fanniinae)
18. Scutellum with fine, erect hairs underneath .... Anthomyiidae
   Scutellum bare beneath; frontal bristles smooth, not cruciate; C not spinose .... Scathophagidae
19. Curve of M is even or angular, but M never terminates in R_{4+5}; hypopleuron without vertical series of bristles .... Muscidae (Muscinae)
   Curve of M angular, sometimes terminating in R_{4+5}; hypopleuron with vertical series of bristles 20
20. Bend of M closer to wing margin than to dm-cu cross vein; dm-cu slightly sigmoid; body usually metallic .... Calliphoridae
   Bend of M closer to dm-cu than to wing margin; dm-cu straight; body gray to black, rarely metallic .... Sarcophagidae

21. C with one break; CuA_1 with a slight kink or offset thickening midway; ocellar triangle large, shiny; oral vibrissae hairlike .... Chloropidae
C with more than one break 22
22. cu*p* absent; face convex; oral vibrissae absent .... Ephydridae
cu*p* usually present; if absent, arista slightly pubescent, not bare or plumose; face flat or concave 23
23. cu*p* small, compact; genal bristles, if present, in lower half of gena only; proboscis geniculate, usually slender .... Milichiidae
cu*p* present or absent 24
24. cu*p* always present and well developed; postocellar bristles convergent; arista long plumose above and below; Sc incomplete .... Drosophilidae
cu*p* present or absent; postocellar bristles parallel or divergent; arista pubescent; proboscis short and stout .... Carnidae

B. SIGNIFICANCE OF FOOD-CONTAMINATING FAMILIES

Table 3 lists the families of flies that are most commonly encountered in foods, as well as the significance of the contamination (disease vector, myiasis agent, filth pest, stored-product pest or incidental pest) associated with each family.

IV. DISCUSSION OF FAMILIES

A. COMMON FOOD-CONTAMINATING FAMILIES

1. Anthomyiidae

The anthomyiids resemble house flies (Muscidae) in general shape and form, ranging in size up to 12 mm. The wings are calyptrate and usually hyaline, occasionally dusky or, rarely, spotted. Vein M_1 is straight, not curved sharply forward. The A_1 anal vein is longer than A_2 and the two veins are positioned parallel to each other or diverging beyond the alular incision. Body color varies but is never metallic. On the head, the compound eyes are bare, lacking hairs between the eye facets. Mouthparts are the sponging type flanked by one or more laterad pairs of robust, bristlelike vibrissae. In the subfamily commonly encountered in food products (Anthomyiini), the scutellum is haired underneath and the wing vein A_1 can be traced to the wing margin as a dorsally convex fold.[27,28] *Anthomyia* spp. are gray with distinctive black spots on the mesonotum. They live in bird nests.[27] *Delia* (= *Paregle*) spp. are grayish Anthomyiini that may breed in manure or decomposed material. *Hylemya* spp. are field pests of the roots of onions, cabbages and other crops, and *Pegomya* spp. are leafminers that may infest crops in the field.[27,29] The wing venation of anthomyiids resembles the venation of dung flies, family Scathophagidae. The dung flies can be recognized by their reduced lower calypter.[17,30]

2. Calliphoridae (Blow Flies)

The blow flies are medium to large, up to 16 mm, metallic-colored musciform filth flies. The metallic body color ranges from blue-black to blue or green to bronze or brassy. A few calliphorids, mostly parasites of invertebrates, are nonmetallic in color. The wing (Figure 4J) is hyaline and calyptrate with vein M_1 bending sharply forward close to the wing margin and intercepting the costa (C) before the wing apex. The d-m cross vein of the discal cell is slightly sigmoid. On the head, the compound eyes of the male are usually closer together at the vertex than the eyes of the female.

The mouthparts are the sponging type flanked by one or more pairs of laterad vibrissae. The palpi are yellow. The arista of the antenna is plumose above and below at least on the basal two-thirds and often completely plumose to the apex.[1,23,31,32] Blow flies are primarily filth and carrion feeders but many are endoparasites.[6,17,18,33] The screw-worm fly, *Cochliomyia hominivorax* (Coquerel), for example, is an obligate parasite that causes myiasis in livestock.[17-19] Blow flies of the genus *Chrysomya* include parasitic flies and filth flies. The Oriental latrine fly, *Chrysomya megacephala* (F.), is a metallic blue filth fly that is spread by commerce.[34-36] Its distinctive males are holoptic with the compounds eyes meeting at the vertex and the upper facets of the eye larger than the lower facets with a demarcation line between the upper and lower facets. It breeds in excrement and carrion but is facultatively parasitic in open wounds of vertebrates.[37] Other synanthropic filth fly genera in this family include the bluebottle flies, *Calliphora* spp.; the greenbottle flies, *Lucilia* spp. and *Phaenicia* spp.; and the black blow fly, *Phormia regina* (Meigen).[3,6,38,39] The latter blow fly, *P. regina*, has distinctive bright orange hairs on the anterior thoracic spiracle.[17] All are attracted to food products. The cluster fly, *Pollenia rudis* (F.), is a nonmetallic blow fly that may accumulate in buildings in large numbers.[40] It is parasitic on earthworms and is easily recognized by a petiolate wing cell R_{4+5} on an otherwise typical calliphorid wing.

3. Carnidae

The carnids are small, 1–3 mm, shiny black acalyptrate filth flies. They are differentiated from similar flies by distinctive combined characteristics of the head. The postocellar bristles are parallel to slightly divergent. The compound eye is ovoid and the gena beneath it is broad from top to bottom. The vibrissae extend in a row across the middle of the gena from the lower corner of the frontal margin to behind the eye. The sponging mouthparts have a stout proboscis with a short, round inconspicuous labellum. The three genera can be differentiated by wing venation. In *Hemeromyia* spp. the wing venation is similar to the milichiid genus *Madiza*. The venation of *Carnus hemapterus* Nitzsch lacks a d-m cross vein and vein M_1 is indistinct beyond the r-m cross vein, which is located basally. In *Meoneura* spp. (Figure 5E) both r-m and d-m cross veins are basal and M_1 is indistinct beyond d-m. Vein A_1 and basal cell cu*p* are absent in *C. hemapterus* and *Meoneura* spp. The carnids live in excrement except for *C. hemapterus*, which lives in bird nests and has the unusual habit (for a fly) of dealating.[1,23,41]

4. Ceratopogonidae (Biting Midges)

The biting midges are small, 1/2–6 mm, gnats with long, 13-segmented plumose antennae. The wing venation is characterized by a compact system of radial veins that are grouped close to the costa (C). The costa and radial veins usually end at the midpoint of the anterior margin of the wing or slightly beyond. Other veins are indistinct or colorless, if present. The head bears relatively large reniform compound eyes that wrap around the antennal sockets. The antennal segments are beadlike. Mouthparts are the biting and sucking type and form a short beak about as long as the head. In some groups, the mouthparts are reduced.[1,23,42]

The genus *Dasyhelea* occurs in garbage. It has reduced mouthparts and the integument of the antennal segments is distinctively sculptured.[6]

5. Chironomidae (Midges)

The midges resemble mosquitoes, except they lack scales on their wings and their mouthparts are reduced. They do not have a long proboscis for biting. The wings have a strong system of radial veins grouped anteriorly near the costa (C) and the other veins are usually uncolored or indistinct. Unlike most of the biting midges, the radial vein system of chironomids extends nearly to the apex of the wing. Midges live in moist and aquatic habitats. The microanalytical entomologist may encounter the occasional adult in food but is more likely to encounter larvae and their fragments attributable to the use of nonpotable water in the manufacture of a product. The midge larva is an elongate, wormlike (vermiform) animal lacking segmented legs. The prothorax bears a pair of fleshy prolegs armed with terminal spines, and the terminal segment of the abdomen usually bears a pair of prolegs and sometimes a third fleshy projection, the procercus. The larval head capsule is sclerotized and bears lateral eyespots. Antennae, labia, mandibles and maxillae are present. The mouthparts are the chewing type. Ventral to the mouthparts lies a central, heavily sclerotized plate usually bearing a row of apical teeth.[23,43]

6. Chloropidae (Eye Gnats)

The small, 1–5 mm, acalyptrate flies can be recognized by their large ocellar triangle, lack of bristles on the anterior and middle portions of the mesonotum and a distinctive vertical ridge dividing the propleuron (sclerite in front of the anterior thoracic spiracle). Mouthparts are the sponging type and the oral vibrissae are thin and hairlike. The hyaline acalyptrate wing (Figure 4D) has one costal break, near the radial R_1 vein. The costa (C) reaches nearly to the apex of the wing or beyond. Subcosta (Sc) is incomplete and other venation is reduced. The only cell below the r-m cross vein is the discal cell, which is usually closed. Basal cells bm and cu*p* are always absent and the anal veins are short or absent.[44] The CuA vein has a thickened node or kink midway, even in the few cases where the discal cell is open.[6] The tawny yellow *Cadrema pallida* (Loew) is a filth fly that breeds in excrement, carrion and decayed material.[6,23] They can be recognized by a long spur on the apex of the tibia that is almost as long as the first tarsal segment and by a flattened, broad scutellum with a single pair of long bristles near the rounded apex that converge, crossing over each other.[6,23] The shiny black *Hippelates* spp. are the pesky eye gnats that are attracted to the secretions of the eyes of cattle and other animals.[17] Two other shiny black chloropids, *Rhodesiella* spp. and *Apotropina* spp. (=*Lasiopleura)* are attracted to carrion and excrement. *Thaumatomyia* spp. are conspicuous yellow and black striped filth flies.[6,45]

7. Drosophilidae (Pomace Flies)

The familiar "genetics" fruit fly, *Drosophila melanogaster* Meigen, is also a filth fly, attracted to garbage and excrement as well as to fruit and other foods. This small, 2–3 mm, fly can be recognized by its bright red eyes, brownish color with spotted mesonotum that lacks strong bristles anterior to the transverse suture and typical wing venation. The hyaline, acalyptrate wing (Figure 4E) has two costal breaks. The subcosta (Sc) is incomplete. The basal cell (bm) is not differentiated from the discal cell (dm), but the anal cell (cu*p*) is distinctly formed. On the head, the antennae are geniculate with a distinctively sparse plumose with the dorsal hairs biordinal (two sizes) and more numerous (2×) long dorsal hairs than ventral ones. The mouthparts are the sponging type.[46] Male *Drosophila* spp. have a tibial comb of densely packed, stout black setae located dorsally near the apex of the protibia only.

8. Ephydridae (Shore Flies)

This is a large and morphologically varied group consisting mostly of aquatic and semiaquatic species. They are medium-sized flies for the most part, but sizes can range from 1–10 mm. They are recognized by their convex faces and a wing venation pattern that resembles *Drosophila* spp. except cell cup is not formed.[47] *Discomyza* spp. filth flies are attracted to carrion and cadavers.[6] They are differentiated from other shore flies by a characteristic metallic green facial area that is sculpted with coarse, transverse rugosities between the eyes and by dark markings on the wings. Larvae and larval fragments of *Discomyza* spp. are occasionally encountered in foods. Ephydrid larvae resemble rat-tailed maggots (Syrphidae) except the anal breathing tube is bifurcated apically. *Discomyza* larvae have branched lateral processes, similar to *Fannia* spp. (Fannidae), and the two genera are sometimes collected together in the field.[48] Other filth flies in this family include the brownish-black *Teichomyza fusca* Macquart and the green *Paralimna* spp., which both breed in excrement.[6,17]

9. Milichiidae

The milichiids are small, 1–5 mm, black filth flies. The wing (Figure 5F) is acalyptrate, hyaline and with strong, distinct veins. Vein C is broken twice, near the humeral cross vein

(h) and the juncture with R_1. In the subfamily Milichiinae (e.g., *Milichia* spp. and *Milichiella* spp.) the second costal break is deeply notched to form a lappet where part of the vein before the break overlaps part of the vein beyond the break. In most cases, vein C continues around the wing apex to join with vein M_1. The subcosta is usually incomplete and closely parallel to R_1. The radius is three branched with R_1 ending in the costa (C) at or before the midpoint of the anterior wing margin. R_{2+3} is long and terminates at the margin just above the wing apex, whereas R_{4+5} are straight and parallel for most of their length. The discal cell (dm) is elongate and prominent. Basal cells bm and cu*p* are closed and distinct, but the anal veins A_1 and A_2 either do not extend beyond cu*p* or extend for only a short distance that is less than the width of cu*p*. The proboscis is geniculate and generally slender and elongate with the sponging labellum usually slender.[49] The milichiids are mostly filth flies but a few species are associated with ants.[3,49] A common filth-inhabiting group of milichiids, *Desmometopa* spp., are recognized by two longitudinal rows of bristles that are located on two raised silvery sclerotized strips between the compound eyes on the face.[6,23] These small, 2 mm, flies breed in excrement and are also attracted indoors to lights at night. Another common filth-inhabiting group, the shiny black *Milichiella* spp., is recognized by a distinct notch on the posterior margin of the compound eye. The synanthropic milichiids include *Desmometopa*, *Leptometopa*, *Madiza*, *Milichia* and *Milichiella*.[3,6]

10. Muscidae (House Flies)

The family Muscidae is a highly variable group whose taxonomy is in a state of flux. This discussion will focus on the common synanthropic food-contaminating members of the group. The muscids are mostly gray to black, bristly flies about the size of the prototypical house fly, 7–8 mm, or slightly smaller. The wing has three costal breaks: one basally midway between the attachment to the thorax and the humeral cross vein (h), one just beyond the humeral and one just before the subcosta (Sc). The subcosta is complete, ending in the costa (C) before the midpoint of the wing's anterior margin. The length of the costa is variable, sometimes extending past the wing apex, sometimes not. The costa usually meets and ends at vein M_1. The radius is three branched with R_1 meeting the costa at about the midpoint of the anterior wing margin. R_{2+3} and R_{4+5} are long and parallel but widely spaced for most or all of their lengths. Vein M_1 is straight or curved forward. The closed discal cell is large and elongate with basal cell (bm) differentiated. Cross vein dm is long and straight to slightly concave in the middle. The anal cell (cu*p*) is closed and relatively large. Veins A_1 and A_2 extend beyond the anal cell but never extend all the way to the wing margin. The muscids are differentiated from similar-appearing anthomyiids by the lack of fine hairs on the underside of the scutellum in muscids and by the A_1 wing vein that never reaches the wing margin in muscids but usually reaches the margin as a vein or distinct fold of the wing membrane in anthomyiids. Muscids are differentiated from calliphorids and sarcophagids by the M_1 wing vein that is straight or gently curved in muscids but always sharply curved in the other two families and by the presence of a vertical series of bristles on the hypopleuron of calliphorids and sarcophagids that is absent on muscids.[1,3,50]

Musca domestica **L.**—The house fly, *M. domestica*, is a synanthropic pest that is familiar to almost everyone as ubiquitous around human settlements. It is distinguished by its typical markings and morphology. The bristled body is up to 9 mm long and black with grayish pollenosity or overtones. The mesonotum has four longitudinal dark stripes and the abdomen has diffuse laterobasal yellow areas on the third and fourth segments. The eyes are dark red and the mouthparts are the sponging type. The pteropleuron, or sclerite, directly beneath the wing attachment (anepimeron of some authors) has fine hairs. Vein M_1 of the wing (Figure 4I) is curved forward, meeting the costa (C) at a point anterior to the wing apex, near the juncture of R_{4+5} with C.[3,11] House flies breed in garbage, are attracted to food indoors and are known passive vectors of food-borne pathogens.[3,9,12,17,51] A related species, *Musca sorbens* Wiedemann, is differentiated by having two, not four, mesonotal stripes.[17,50]

***Fannia* spp.**—The little house flies, *Fannia canicularis* (L.) and *Fannia scalaris* (F.), are similar to *M. domestica* in color and general morphology except smaller (5–7 mm). The protopleuron is bare and vein M_1 is straight, unlike *M. domestica*. A recognition characteristic is the configuration of wing veins A_1 and A_2 (Figure 4H). Although these veins do not extend much beyond the cu*p* cell, the A_2 vein is uniquely curved forward so that an imaginary extension of the vein would intersect a similar extension of vein A_1 before either extension reached the wing margin.(3,17,52) Some authors treat *Fannia* spp. and related species as belonging to a separate family, Fanniidae. The larvae of *Fannia* spp. have branched, fleshy lateral projections on each body segment that provide a ready recognition characteristic. They breed in excrement.(53)

Many other muscids are synanthropic pests.(54) *Hydrotaea* spp. are the shiny blue-black "dump" flies, 4–6 mm long, some of which were previously classified as *Ophyra* spp.(55,56) *Atherigona orientalis* Schiner is a small, 4 mm, yellowish gray filth fly that breeds in excrement and is attracted to food products. In *Atherigona* and *Hydrotaea* vein M_1 is straight or only slightly curved, ending below the apex of the wing. The stable fly, *Stomoxys calcitrans* (L.), and the false stable fly, *Muscina stabulans* (Fallen), are grayish black flies, 6–8 mm, with vein M_1 curved forward gently but sufficient to meet the costa (C) at or above the wing apex. *Stomoxys* spp. have a rigid proboscis equipped for piercing the skin of animals, whereas the mouthparts of *Muscina* spp. are the sponging type.(17) Another large, 10 mm, muscid that is attracted to food products is the gray-colored *Synthesiomyza nudiseta* (van der Wulp). This fly has a striped thorax, tesselated abdomen and sharply curved M_1 vein, giving it the appearance of a flesh fly (Sarcophagidae). The antennae, palpi and tip of the abdomen are bright orange-yellow. A bare arista and lack of hypopleural bristles distinguishes this filth fly from the sarcophagids.(57)

11. Phoridae (Humpbacked Flies)

The humpbacked flies are recognized by their humpbacked appearance, small size (1/2–2 mm in most cases) and distinctive wing venation. The wing (Figure 4B) of these acalyptrate flies is relatively large, with the costa extending only halfway to the apex, along the anterior margin. In *Megaselia* spp., a common filth-inhabiting group, the subcosta is reduced or absent. The radial branches are strongly expressed, with the anterior branch meeting the costa at the latter's midpoint and the posterior branch reaching to the end of the costa. The posterior radial branch forks just before its end forming a triangular cell with the tip of the costa. Phorids exhibit a characteristic darting running behavior when exploring a feeding or oviposition site.(58) They breed in filth and excrement.(6,59) Some phorids are pests in mushroom houses.(60)

12. Piophilidae (Skipper Flies)

The cheese skipper, *Piophila casei* (L.), is a pest in the food industry, infesting cheese and other products.(17,61) It is a black, acalyptrate fly with yellow markings on the lower head, antennae and legs. The wing is hyaline with the costa (C) extending around the apex. The subcosta (Sc) is complete and closely parallels R_1 for nearly its entire length. There is one costal break, at the juncture with Sc. The membrane between Sc and R_1 is darkly colored. The discal cell is closed and elongate with the basal cell (bm) differentiated. The anal cell (cu*p*) is closed and the anal vein A_1 ends abruptly before reaching the wing margin, unlike in most other piophilids.(62)

13. Psychodidae (Sewer Flies)

Also known as moth flies, the psychodids are characterized by their vestiture of fine hairs or scales on the body and wings. The wing design (Figure 4A) is also characteristic and distinctive. The shape is ovoid, often with the apex pointed. All the veins extend straight to the wing margin so that their ends are evenly spaced along the margin. The veins are strategically forked to fill in spaces as the wing widens so that the vein branches are evenly spaced throughout. There are no distal cross veins. On the head, the antennae are filiform with 10–14 beadlike segments, each bearing whorls of hairs and thin-walled projections called sensilla. Some groups have a complete eye bridge. The proboscis is short and often the piercing type.[63] Some psychodids (e.g., *Psychoda* spp.) live in sewers and drains from which they can emerge to infest foods.[29]

14. Platystomatidae (Picture-Winged Flies)

In the tropics, a large (up to 8 mm) platystomatid, *Scholastes* spp., is a synanthropic filth fly that breeds in fallen coconuts but frequents latrines and garbage receptacles. The wing (Figure 5G) has one costal break, just beyond the humeral cross vein, and is transversely banded with diffuse brown markings that are tinged yellow basally. The head and thorax are brown and the abdomen vaguely metallic brown. The wing venation includes a complete subcosta (Sc) and R_1 that is setulose dorsally along its entire length. The costa reaches the wing apex where it meets M_1. The discal cell is closed by a dm cross vein that undulates in the middle. The basal cell (bm) is differentiated and the anal cell (cu*p*) is closed. The CuA_1 anal vein beyond the discal cell is much shorter than the dm cross vein (i.e., the discal cell reaches nearly to the wing margin posteriorly).[23,64] The maggots living inside fallen, rotten coconuts are often bright blue in color.

15. Sarcophagidae (Flesh Flies)

The flesh flies are large, up to 18 mm, calyptrate flies that are typically gray with dark thoracic stripes and an abdomen that is tesselated (checkered), sometimes striped or spotted. The wing (Figure 5H) resembles the calliphorid or muscine wing with three costal breaks (basal, humeral and subcostal), a complete subcosta (Sc) and an angular vein M_1 occurring closer to cross vein d-m than to the wing margin. R_{4+5} is setulose above and below from its branch node toward the r-m cross vein.[3,17,65] The red-tailed flesh fly, *Sarcophaga haemorrhoidalis* (Fallen) (= *Bercaea haemorrhoidalis* of some authors) is a common synanthropic pest that breeds in carrion and readily enters buildings. It has caused intestinal myiasis in humans after being ingested on contaminated food.[17,20] The terminal segment of the abdomen of the male is bright red, giving this fly its common name.[17]

16. Scathophagidae (Dung Flies)

The yellow dung fly, *Scathophaga stercoraria* (L.) and related species are synanthropic filth flies that breed in dung.[3,17] The bristly yellow fly is about the size of a house fly. Scathophagids are calyptrate flies with the lower calypter reduced to a membranous fold. The hypopleuron is bare. The wing (Figure 5D) has three costal breaks and a complete subcosta (Sc). The costa (C) extends around the wing apex. The closed discal cell is elongate, nearly reaching the wing margin at the posterior apical corner. Basal (bm) and anal (cu*p*) cells are closed. The yellow dung fly adult is a predator on house flies, blow flies and other insects. The labellum of the sponging mouthparts is equipped with prestomal teeth that are used to penetrate membranous areas of the bodies of their prey.[30,66]

17. Scatopsidae (Minute Black Scavenger Flies)

These small, 1/2–4 mm, dark-colored flies resemble phorids in appearance and reduced wing venation. They are easily distinguished from phorids by their filiform antennae as opposed to the aristate phorid antennae. *Scatopse notata* (L.) is a synanthropic fly that breeds in dung and readily enters buildings.[3] The wing venation of *Scatopse* spp. is diagnostic. The unbroken costa (C) extends past the midpoint of the anterior wing margin, but only slightly. The subcosta (Sc) is incomplete, ending in the wing membrane before the humerus. The r-m cross vein is located basally below the node where the radius (R) branches, forming a cross-shaped node complex of four veins extending in four directions (anterior R_1, posterior r-m, apical R_5 and basal R stem vein). Vein M_1 has a stump branch projecting forward from a point apical to the midwing branch node of vein M. Between M and Cu, a straight spurious vein extends nearly to the wing margin. There are no cross veins or closed cells below M_1, only independent veins Cu_1, Cu_2 and A.[67]

18. Scenopenidae (Window Flies)

The window flies *Scenopinus fenestralis* (L.) and *S. glabrifrons* Meigen are predators of stored-product insects, especially beetles, in the flies' larval stage. The larvae are elongate, legless, white, wormlike insects with a conical sclerotized head. The abdomen is secondarily segmented giving the appearance of 19 total body segments. They move with a rapid, whipping, serpentine form of locomotion. The adult wing (Figure 5C) has an unbroken costa (C) ending at the wing tip. The subcosta (Sc) is complete and the radius is four branched. Vein M_1 curves gently forward to meet the costa. The discal cell is closed, elongate and the basal cell (bm) is differentiated. The cu*p* is closed by an arcing CuA_2 vein. A_1 and CuA_2 nearly reach the wing margin. A_2 is indistinct beyond the alular incision. The head is almost totally dominated by the large spherical compound eyes. Males are holoptic, and the upper facets are coarser than the lower facets. The antenna is stylate with a peglike apical stylus on the third segment. The body is clothed with fine hairs and no bristles. The abdomen is dorsoventrally flattened and consists of ten segments.[61,68] *Scenopinus fenestralis* is a stored-product pest as a predator of stored-product beetles. It is often associated with infestations of carpet beetles (Dermestidae) and has been found in feed mills infested with stored-product beetles and moths.[29,61,68] We have seen it in wheat that was heavily infested with sawtoothed grain beetles, *Oryzaephilus* spp.

19. Sciaridae (Dark-Winged Fungus Gnats)

This group and the related fungus gnats (Mycetophilidae) are pests of mushrooms and general fungivores.[61] Sciarids are usually dark (but some are yellow) gnats with a characteristic eye bridge above the 14-segmented filiform antennae. The wings have strongly expressed anterior veins and weaker, yet distinct, posterior ones. The costa (C) vein extends to the wing apex but does not meet M_1, ending between R_{4+5} and M_1. The subcosta (Sc) is incomplete. A peculiarity of the sciarid wing is the horizontal position of the r-m cross vein, which makes it appear to be a section of vein R_{4+5}. If one traces the basal origins of the veins R and M, it becomes clear that r-m has shifted to a horizontal position and what appears to be a vertical cross vein is, in fact, a branch of R that is connected to the M stem vein by a horizontal r-m.[3,6,69] Mycetophilids have a more traditional r-m cross vein and most lack the eye bridge that is typical of sciarids.[70] The two families are very similar in most other respects.

20. Sepsidae (Black Scavenger Flies)

These generally small, 2–6 mm, acalyptrate filth flies can be recognized by the antlike appearance of their bodies with the abdomen more or less constricted basally and proportionately long legs. The front of the head is smooth and bare between the compound eyes, and the mesonotum lacks bristles except along the side and posterior margins in *Sepsis* spp. and other groups. The wing (Figure 4F) has an unbroken costa (C) that extends around the wing apex to meet vein M_1. The subcosta is complete and the radius three branched. The discal cell is closed and elongate with the basal cell (bm) differentiated. The anal cell (cu*p*) is closed. Vein A_1 does not reach the wing margin and A_2 is indistinct beyond the alular incision.[71] The common filth-inhabiting group, *Sepsis* spp., is black with hyaline wings that bear a dark stigma at the end of vein R_{4+5}.[3,17] Sepsids breed in excrement and have a distinctive habit of rapidly flipping their wings in a wasplike fashion as they run about.[6]

21. Sphaeroceridae (Small Dung Flies)

The sphaerocerids most frequently encountered contaminating food products are members of the subfamily Limosiniae (e.g., *Coproica*, *Leptocera*, *Poecilosomella*, *Pullimosina* and *Thoracochaeta*).[6] These small (1–5 mm) acalyptrate synanthropic flies are characterized by a reduced wing venation (Figure 4C). The subcosta (Sc) is incomplete and lies very close to the radius. The discal cell is closed and the basal cell (dm) is undifferentiated. Veins M and CuA end at cross vein dm or barely beyond. In some species, the discal cell is open. Vein A_1 never reaches the wing margin and the anal cell (cu*p*) is always open. The costa has three breaks and the radius is three branched. The wings are hyaline and the body color is dark brown to black, sometimes with yellow head or legs. Mouthparts are the sponging type. The first tarsal segment of the hind leg is enlarged and shorter than the second, a general characteristic of the family. Sphaerocerids breed in excrement and are known to infest food processing facilities. They have caused intestinal myiasis in humans and are associated with a number of food-borne pathogens.[3,6,72,73]

22. Stratiomyidae (Soldier Flies)

The soldier fly, *Hermetia illucens* (L.), is a synanthropic filth fly.[6,17] This large (15–20 mm), black fly is easily recognized by its clubbed antennae with the apical segment flattened and longer than the proceeding eight segments. The entire antenna is longer than the head. The abdomen has a lateral pair of large translucent areas at its base. The wing (Figure 5B) is dusky and bears typical stratiomyid venation characterized by a small, compact discal cell and large, elongate basal (bm) and anal (cu*p*) cells.[17,74] The synanthropic *H. illucens* breeds in excrement, carrion, beehives and vinegar vats. The large (up to 25 mm), vigorous larvae cause severe intestinal myiasis in humans.[20,75] They have been found in decayed fruits and vegetables, catsup and honey.[17]

23. Syrphidae (Rat-Tailed Maggots)

The adults of the food-contaminating rat-tailed maggots, *Eristalis tenax* (L.) and *Erstalinus aeneus* (Scopoli), resemble drone honeybees in size, color and flight characteristics. The wing (Figure 4G) has a distinctive closed R_{4+5} cell that is peanut shaped and petiolate. The characteristic syrphid spurious vein is strongly expressed, extending unattached from the basal area across vein r-m.[1,17,76] The rat-tailed maggot, *Eristalis tenax*, lives in sewage, excrement and liquids with high organic content. The large, 30 mm, maggots have seven pairs of prolegs on a grublike body that is secondarily segmented. The long "rat-tail" consists of three segments with the apical segment bearing a pair of spiracles. The apical and subapical segments of this breathing tube can be retracted into the basal segment.[48] The larvae of *Eristalinus aeneus* can be distinguished from *Eristalis tenax* by the presence of sclerotized ventral "denticles" on the anterior head.[21] Adults of *Eristalinus aeneus* have dark spots on the compound eyes, whereas the eyes of *Eristalis tenax* are uniform in color.[77] The maggots of *Eristalinus aeneus* infest pickle vats, where they can be observed freely swimming in the brine with a serpentine undulating motion.

B. OTHER FAMILIES

The families discussed above contain members that are likely to contaminate food held or processed under insanitary conditions. Other families may occur either infrequently, because they are asynanthropic, or frequently, but as incidental contaminants unrelated to gross insanitation unless found in great numbers. The march flies (family Bibionidae) are primarily agricultural pests of forage crops, although some species breed in dung. They are characterized by a radius that first branches close to the r-m cross vein and is slanted posteriorad.[78] The Cecidomyiidae are small midges that include leafminers and gallmakers.[79] Some cecidomyiids may infest commercial mushroom crops as an agricultural pest, and the FDA has a defect action level that delineates tolerable levels of cecidomyiid larvae in mushrooms.[80,81] The family Anisopodidae includes the synanthropic species *Sylvicola fenestralis* (Scopoli), which may cause intestinal myiasis. It has been found in honeycombs and homemade wines and ciders.[17,82] Figure 5A illustrates the diagnostic wing of *Sylvicola fenestralis*.[3] Because they are major pests of humankind, mosquitoes (family Culicidae) are often found as incidental contaminants in foods even though the chief attractant is the people near the food. They are midgelike flies covered with scales on their wings and body.[83]

The true fruit flies, family Tephritidae, include many agricultural pests that are encountered, usually as fragmented contaminants, in foods. On occasion, sterile males are found as a result of controlled releases in urban areas for control of invading pests such as the Mediterranean fruit fly, *Ceratitis capitata* (Wiedemann). The leafminers of the family Agromyzidae are similarly serious agricultural pests whose fragments may be encountered in processed foods made from infested crops. Flies of the families Dryomyzidae and Heleomyzidae may breed in dung and are occasionally found near buildings.[6] The family Canaceidae consists mainly of intertidal shore flies that also occur in seafoods as incidental contaminants.

Finally, certain inquilines and ectoparasites are known to occur in food through insanitary practices relating to their host species. The bee louse, *Braula coeca* Nitsch (Braulidae), may infest honeycombs of commercial beehives.[84] These inquilines feed on the comb as larvae, often causing economic losses. The pigeon louse, *Pseudolynchia canariensis* (Macquart), and other members of the family Hippoboscidae are ectoparasites of synanthropic birds.[85] They have been found in foods and food-handling establishments. The bat parasites of the families Nycteribidae and Streblidae may also occur in products that were handled in buildings infested by bats. The spiderlike *Basilia* spp. (Nycteribiidae) occur on vespertiliontid bats (see chapter 12), which frequently infest buildings and whose hairs are often found in foods.[86-88]

V. REMARKS

Many of the filth flies that occur in foods also command the interest of medical and forensic entomologists.[10,89-91] As a consequence of these overlapping interests, there are many identification keys for fly adults and larvae that have applications in medical entomology, forensic entomology and food sanitation entomology.[92,93] In addition to the general works of McAlpine and Stehr, there are a number of recent works that allow identifications to the species level for many fly groups. Greenberg's text on synanthropic flies covers most of the groups described in this chapter with keys to larvae and adults.[9] Zumpt, Smith and Gorham provide keys for medical entomology, forensic entomology and microanalytical entomology, respectively.[18,33,89] The works of Kano and Shinonaga and of Kurahashi provide coverage of Asian species.[39,94] For rapid determinations, the U.S. Public Health Service pictorial keys often prove helpful.[95-101] Whether interested in medicine, forensics or food sanitation, the entomologist will find applications in these identification keys to carry beyond the scope of this chapter.

REFERENCES

1. Borror, D.J., Triplehorn, C.A., and Johnson, N.F., *An Introduction to the Study of Insects*, 6th ed., Saunders Coll. Publ., Philadelphia, 1-875, 1989.
2. McAlpine J.F., Introduction, in *Manual of Nearctic Diptera*, Vol. 1, McAlpine, J.F., Ed., Agric. Canada Monogr. No. 27, 1-7, 1981.
3. Greenberg, B., *Flies and Disease. Volume I: Ecology, Classification and Biotic Associations*, Princeton Univ. Press, Princeton, 1-856, 1971.

4. Nourteva, P., Synanthropy of blowflies (Dipt. Calliphoridae), *Finland Ann. Entomol. Fenn.*, 29, 1-49, 1963.
5. Norris, K.R., The bionomics of blowflies, *Annu. Rev. Entomol.*, 10, 47-68, 1965.
6. Bohart, G.E. and Gressitt, J.L., Filth-inhabiting flies of Guam, *Bernice P. Bishop Mus. Bull.*, 204, 1-152, 1951.
7. Wilton, D.P., Refuse containers as a source of flies in Honolulu and nearby communities, *Proc. Hawaii. Entomol. Soc.*, 17, 477-481, 1962.
8. Schnack, J.A., Mariluis, J.C., Muzon, J., and Spinelli, G.R., Synanthropy of Calliphoridae: A first approach in Argentina (Insecta, Diptera), *Eos*, 65, 273-280, 1989.
9. Greenberg, B., *Flies and Disease. Volume II: Biology and Disease Transmission*, Princeton Univ. Press, Princeton, 1-447, 1973.
10. Harwood, R.F. and James, M.T., *Entomology in Human and Animal Health*, 7th ed., Macmillan, NY, 1-548, 1979.
11. West, L.S., *The Housefly*, Comstock Publ. Co., Ithaca, 1-584, 1951.
12. Lindsay, D.R. and Scudder, H.I., Non-biting flies and disease, *Annu. Rev. Entomol.*, 1, 323-346, 1956.
13. Gorham, J.R., HACCP and filth in food, *J. Environ. Health*, 52(2), 84-86, 1989.
14. Gorham, J.R., The significance for human health of insects in food, *Annu. Rev. Entomol.*, 24, 209-224, 1979.
15. Muirhead-Thomson, R.C., *Ecology of insect vector populations*, Academic Press, London, 1-174, 1968.
16. Chow, C.Y., The common blue-bottle fly, *Chrysomyia megacephala*, as a carrier of pathogenic bacteria in Peping, China, *Chin. Med. J.*, 57, 145-153, 1940.
17. James, M.T., *The Flies That Cause Myiasis in Man*, U.S. Dept. Agric. Misc. Bull. No. 631, 1-175, 1947.
18. Zumpt, F., *Myiasis in Man and Animals in the Old World*, Butterworths, London, 1-276, 1965.
19. Ferrar, P., A guide to the breeding habits and immature stages of Diptera Cyclorrhapa: 12, Calliphoridae, *Entomographia*, 8, 83-98, 1987.
20. Scott, H.G., Human myiasis in North America (1952-1962 inclusive), *Fla. Entomol.*, 47, 255-261, 1964.
21. Cole, F.R., *The Flies of Western North America*, Univ. Calif. Press, Berkeley, 1-360, 1969.
22. McAlpine, J.F. Morphology and Terminology—Adults, in *Manual of Nearctic Diptera*, Vol. 1, McAlpine, J.F., Ed., Agric. Canada Monogr. No. 27, 65-88, 1981.
23. Colless, D.H. and McAlpine, D.K., Diptera (Flies), in *The Insects of Australia*, CSIRO, Melbourne Univ. Press, Carlton, Victoria, 656-740, 1970.
24. Carson, N.A. and Martinez, E.F., Fly identification by the micromorphology of the head and head appendages, *J. Assoc. Off. Anal. Chem.*, 50, 1146-1193, 1967.
25. Kano, R. and Sato, K., Notes on the flies of medical importance in Japan. V. Classification of female blowflies by the characters of tergites and sternites, *Jpn. J. Exper. Med.*, 21, 229-241, 1951.
26. Sabatino, F.J., Identification of insect setae as an index of contamination in dairy products, *J. Assoc. Off. Agric. Chem.*, 37, 960-966, 1954.
27. Huckett, H.C., Anthomyiidae, in *Manual of Nearctic Diptera*, Vol. 2, McAlpine, J.F., Ed., Agric. Canada Monogr. No. 28, 1099-1114, 1987.
28. Huckett, H.C., The Anthomyiidae of California exclusive of the subfamily Scatophaginae (Diptera), *Bull. Calif. Insect Surv.*, 12, 1-121, 1971.
29. Oldroyd, H., *The Natural History of Flies*, Weidenfield & Nicolson, London, 1-324, 1964.
30. Seguy, E., The family Scatophagidae, *Genera Insect*, 209e, 1-107, 1952.
31. Hall, D.G., *The Blowflies of North America*, Thomas Say Found., College Park, MD, 1-477, 1948.
32. Shewell, G.G., Calliphoridae, in *Manual of Nearctic Diptera*, Vol. 2, McAlpine, J.F., Ed., Agric. Canada Monogr. No. 28, 1133-1145, 1987.
33. Smith, K.G.V., *A Manual of Forensic Entomology*, Brit. Mus. (Natur. Hist.), London, 1-205, 1986.
34. Olsen, A.R. and Sidebottom, T.H., Biological observations on *Chrysomya megacephala* (Fabr.) (Diptera: Calliphoridae) in Los Angeles, California and the Palau Islands, *Pan-Pac. Entomol.*, 66, 126-130, 1990.
35. Olsen, A.R., Angold, S.C., Gross, D.F., and Sidebottom, T.H., New record of the blowfly, *Chrysomya megacephala* (Fabr.) from Ecuador (Diptera: Calliphoridae), *Pan-Pac. Entomol.*, 68, 280-281, 1992.
36. Olsen, A.R., Sidebottom, T.H., and Bennett, S.G., The Oriental latrine fly, *Chrysomya megacephala* (Fabricius 1794) (Diptera: Calliphoridae), as an invading blow fly of public health importance, *J. Soc. Vector Ecol.*, 18, 133-146, 1993.
37. Roy, P. and Dasgupta, B., Behavior of *Chrysomya megacephala* (Fabr.) and *Hemipyrellia ligurriens* (Wied.) as parasites of living animals under experimental conditions, *S. Afr. J. Med. Sci.*, 36, 85-91, 1971.
38. James, M.T., The blowflies of California (Diptera: Calliphoridae), *Bull. Calif. Insect Surv.*, 4(1), 1-34, 1955.
39. Kano, R. and Shinonaga, S., *Illustrated Keys to Adult Filth Flies of Japan,* 406th Med. Lab. Annu. Rep. (Addendum), U.S. Army Medical Command, Japan, 1-19, 1965.
40. Yahnke, W., and George, J.A., Rearing and immature stages of the cluster fly (*Pollenia rudis*)(Diptera: Calliphoridae) in Ontario, *Can. Entomol.*, 104, 567-576, 1972.
41. Sabrosky, C.W., Carnidae, in *Manual of Nearctic Diptera*, Vol. 2, McAlpine, J.F., Ed., Agric. Canada Monogr. No. 28, 909-912, 1987.

42. Downes, J.A. and Willis, W.W., Ceratopogonidae, in *Manual of Nearctic Diptera*, Vol. 1, McAlpine, J.F., Ed., Agric. Canada Monogr. No. 27, 393-421, 1981.
43. Oliver, D.R., Chironomidae, in *Manual of Nearctic Diptera*, Vol. 1, McAlpine, J.F., Ed., Agric. Canada Monogr. No. 27, 423-458, 1981.
44. Sabrosky, C.W., Chloropidae, in *Manual of Nearctic Diptera*, Vol. 2, McAlpine, J.F., Ed., Agric. Canada Monogr. No. 28, 1049-1067, 1987.
45. Sabrosky, C.W., A revised synopsis of Nearctic *Thaumatomyia* (= *Chloropisca*)(Diptera: Chloropidae), *Can. Entomol.*, 75, 109-117, 1943.
46. Wheeler, M.R., Drosophilidae, in *Manual of Nearctic Diptera*, Vol. 2, McAlpine, J.F., Ed., Agric. Canada Monogr. No. 28, 1007-1018, 1987.
47. Wirth, W.W., Moths, W.N., and Vockeroth, J.R., Ephydridae, in *Manual of Nearctic Diptera*, Vol. 2, McAlpine, J.F. Ed., Agric. Canada Monogr. No. 28, 1027-1047, 1987.
48. Foote, B.A., Syrphidae (Syrphoidea), in *Immature Insects*, Vol. 2, Stehr, F.W., Ed., Kendall/Hunt, Dubuque, IA, 792-794, 1991.
49. Sabrosky, C.W., Milichiidae, in *Manual of Nearctic Diptera*, Vol. 2, McAlpine, J.F., Ed., Agric. Canada Monogr. No. 28, 903-908, 1987.
50. Huckett, H.C. and Vockeroth, J.R., Muscidae, in *Manual of Nearctic Diptera*, Vol. 2, McAlpine, J.F., Ed., Agric. Canada Monogr. No. 28, 1115-1131, 1987.
51. Haines, C.P. and Rees, D.P., *A Field Guide to the Types of Insects and Mites Infesting Cured Fish*, FAO Fisheries Tech. Paper No. 303, 1-33, 1989.
52. Huckett, H.C., The Muscidae of California exclusive of the subfamilies Muscinae and Stomoxyinae, *Bull. Calif. Insect Surv.*, 18, 1-148, 1975.
53. Dahlem, G.A., Muscidae (Muscoidea), in *Immature Insects*, Vol. 2, Stehr, F.W., Ed., Kendall/Hunt, Dubuque, IA, 857-862, 1991.
54. Skidmore, P., The biology of the Muscidae of the world, *Ser. Entomol.*, 29, 1-550, 1985.
55. Delfinado, M.D. and Hardy, D.E., Eds., *A Catalog of the Diptera of the Oriental Region, Vol. II, Suborder Brachycera through Division Aschiza, suborder Cyclorrhapha*, Univ. Haw. Press, Honolulu, 1-459, 1975.
56. Sabrosky, C.W., The muscid genus *Ophyra* in the Pacific region (Diptera), *Proc. Hawaii. Entomol. Soc.*, 13, 423-432, 1948.
57. Eldridge, B.F. and James, M.T., The typical muscid flies of California, *Bull. Calif. Insect Surv.*, 6, 1-17, 1957.
58. Peterson, B.V., Phoridae, in *Manual of Nearctic Diptera*, Vol. 2, McAlpine, J.F., Ed., Agric. Canada Monogr. No. 28, 689-712, 1987.
59. Robinson, W.H., *Megaselia* (*M.*) *scalaris* (Diptera: Phoridae) associated with laboratory cockroach colonies, *Proc. Entomol. Soc. Wash.*, 77, 384-390, 1975.
60. Robinson, W.H., Phoridae (Diptera) associated with cultivated mushrooms in eastern North America, *Proc. Entomol. Soc. Wash.*, 79, 452-462, 1977.
61. Kurtz, O.L. and Harris, K.L., *Microanalytical Entomology for Food Sanitation Control*, Assoc. Off. Agric. Chem., Washington, DC, 1-576, 1961.
62. McAlpine, J.F., Piophilidae, in *Manual of Nearctic Diptera*, Vol. 2, McAlpine, J.F., Ed., Agric. Canada Monogr. No. 28, 845-852, 1987.
63. Quate, L.W. and Vockeroth, J.R., Psychodidae, in *Manual of Nearctic Diptera*, Vol. 1, McAlpine, J.F., Ed., Agric. Canada Monogr. No. 27, 293-300, 1981.
64. Steyskal, G.C., Platystomatidae, in *Manual of Nearctic Diptera*, Vol. 2, McAlpine, J.F., Ed., Agric. Canada Monogr. No. 28, 809-812, 1987.
65. Shewell, G.E., Sarcophagidae, in *Manual of Nearctic Diptera*, Vol. 2, McAlpine, J.F., Ed., Agric. Canada Monogr. No. 28, 1159-1186, 1987.
66. Vockeroth, J.R., Scathophagidae, in *Manual of Nearctic Diptera*, Vol. 2, McAlpine, J.F., Ed., Agric. Canada Monogr. No. 28, 1085-1097, 1987.
67. Cook, E.F., Scatopsidae, in *Manual of Nearctic Diptera*, Vol. 1, McAlpine, J.F., Ed., Agric. Canada Monogr. No. 27, 212-219, 1981.
68. Kelsey, L.P., Scenopenidae, in *Manual of Nearctic Diptera*, Vol. 1, McAlpine, J.F., Ed., Agric. Canada Monogr. No. 27, 525-528, 1981.
69. Steffen, W.A., Sciaridae, in *Manual of Nearctic Diptera*, Vol. 1, McAlpine, J.F., Ed., Agric. Canada Monogr. No. 27, 247-255, 1981.
70. Vockeroth, J.R., Mycetophilidae, in *Manual of Nearctic Diptera*, Vol. 1, McAlpine, J.F., Ed., Agric. Canada Monogr. No. 27, 223-246, 1981.
71. Steyskal, G.C., Sepsidae, in *Manual of Nearctic Diptera*, Vol. 2, McAlpine, J.F., Ed., Agric. Canada Monogr. No. 28, 945-950, 1987.
72. Marshall, S.A. and Richards, D.W., Sphaeroceridae, in *Manual of Nearctic Diptera*, Vol. 2, McAlpine, J.F., Ed., Agric. Canada Monogr. No. 28, 993-1006, 1987.

73. Fredeen, H. and Taylor, M.E., Borborids (Diptera: Sphaeroceridae) infesting sewage disposal tanks with notes on the life cycle, behavior and control of *Leptocera (Leptocera) caenosa* (Rondani), *Can. Entomol.*, 96, 801-808, 1964.
74. James, M.T., Stratiomyidae, in *Manual of Nearctic Diptera*, Vol. 1, McAlpine, J.F., Ed., Agric. Canada Monogr. No. 27, 497-511, 1981.
75. James, M.T., The soldier flies or Stratiomyidae of California, *Bull. Calif. Insect Surv.*, 6, 79-122, 1960.
76. Vockeroth, J.R. and Thompson, F.C., Syrphidae, in *Manual of Nearctic Diptera*, Vol. 2, McAlpine, J.F., Ed., Agric. Canada Monogr. No. 28, 713-743, 1987.
77. Telford, H.S., *Eristalis* (Diptera: Syrphidae) from America north of Mexico, *Ann. Entomol. Soc. Amer.*, 63, 1201-1210, 1970.
78. Hardy, D.E., Bibionidae in *Manual of Nearctic Diptera*, Vol. 1, McAlpine, J.F., Ed., Agric. Canada Monogr. No. 27, 217-222, 1981.
79. Gagne, R.J., Cecidomyiidae, in *Manual of Nearctic Diptera*, Vol. 1, McAlpine, J.F., Ed., Agric. Canada Monogr. No. 27, 257-292, 1981.
80. Wyatt, I.J., Cecidomyiidae as pests of cultivated mushrooms, *Proc. Assoc. Appl. Biol.*, 48, 430-432, 1948.
81. Foote, R.H. and Steyskal, G.C., Tephritidae, in *Manual of Nearctic Diptera*, Vol. 2, McAlpine, J.F., Ed., Agric. Canada Monogr. No. 28, 817-831, 1987.
82. Peterson, B.V., Anisopodidae, in *Manual of Nearctic Diptera*, Vol. 1, McAlpine, J.F., Ed., Agric. Canada Monogr. No. 27, 305-312, 1981.
83. Stone, A., Culicidae, in *Manual of Nearctic Diptera*, Vol. 1, McAlpine, J.F., Ed., Agric. Canada Monogr. No. 27, 341-350, 1981.
84. Peterson, B.V., Braulidae, in *Manual of Nearctic Diptera*, Vol. 2, McAlpine, J.F., Ed., Agric. Canada Monogr. No. 28, 913-918, 1987.
85. Maa, T.C. and Peterson, B.V., Hippoboscidae, in *Manual of Nearctic Diptera*, Vol. 2, McAlpine, J.F., Ed., Agric. Canada Monogr. No. 28, 1271-1281, 1987.
86. Peterson, B.V. and Wenzel, R.L., Nycterbiidae, in *Manual of Nearctic Diptera*, Vol. 2, McAlpine, J.F., Ed., Agric. Canada Monogr. No. 28, 1283-1291, 1987.
87. Wenzel, R.L. and Peterson, B.V., Streblidae, in *Manual of Nearctic Diptera*, Vol. 2, McAlpine, J.F., Ed., Agric. Canada Monogr. No. 28, 1293-1301, 1987.
88. Olsen, A.R., Distinguishing common food-contaminating bat hairs from certain feather barbules, *J. Assoc. Off. Anal. Chem.*, 64, 786-791, 1981.
89. Gorham, J.R., Ed., *Insect and Mite Pests in Food: An Illustrated Key*, Vols. 1 & 2, U.S. Dept. Agric., Agric. Handbook No. 655, 1-767, 1987/1991.
90. Keh, B., Scope and applications of forensic entomology, *Annu. Rev. Entomol.*, 30, 137-154, 1985.
91. Catts, E.P. and Goff, M.L., Forensic entomology in criminal investigations, *Annu. Rev. Entomol.*, 37, 253-272, 1992.
92. McAlpine, J.F., Ed., *Manual of Nearctic Diptera*, Vols. 1 & 2, Agric. Canada Monogr. 27 and 28, 1-1332, 1981/1987.
93. Stehr, F.W., Ed., *Immature Insects*, Vol. 2, Kendall/Hunt, Dubuque, IA, 1-975, 1991.
94. Kurahashi, H., Studies on the Calypterate Muscoid flies from Japan. VI. Revision of the tribes Bengaliini and Pollenini of the subfamily Calliphoridae and the subfamilies Chrysomyiinae and Rhiniinae (Diptera, Calliphoridae), *Sci. Rep. Kanazawa Univ.*, 12, 255-302, 1967.
95. Dodge, H.R., Diptera: Pictorial Key to Principle Families of Public Health Importance, in *CDC Pictorial Keys. Arthropods, Reptiles, Birds and Mammals of Public Health Significance*, U.S. Public Health Service, 120, 1967.
96. Dodge, H.R., Domestic Flies: Pictorial Key to Common Species in the U.S., in *CDC Pictorial Keys. Arthropods, Reptiles, Birds and Mammals of Public Health Significance*, U.S. Public Health Service, 121, 1967.
97. Scott, H.G. and Borom, M.R., Domestic Flies: Pictorial Key to Common Species, in *CDC Pictorial Keys. Arthropods, Reptiles, Birds and Mammals of Public Health Significance*, U.S. Public Health Service, 122, 1967.
98. Dodge, H.R., Domestic Flies: Pictorial Key to Common Species in Southern U.S., in *CDC Pictorial Keys. Arthropods, Reptiles, Birds and Mammals of Public Health Significance*, U.S. Public Health Service, 123, 1967.
99. Seago, J.M., Fly Larvae: Pictorial Key to Some Common Species, in *CDC Pictorial Keys. Arthropods, Reptiles, Birds and Mammals of Public Health Significance*, U.S. Public Health Service, 124, 1967.
100. Stojanovich, C.J., Pratt, H.D., and Bennington, E.E., Fly Larvae: Key to Some Species of Public Health Importance, in *CDC Pictorial Keys. Arthropods, Reptiles, Birds and Mammals of Public Health Significance*, U.S. Public Health Service, 125-133, 1967.
101. Dodge, H.R., Identifying common flies, *Publ. Health Rep.*, 68, 345-350, 1953.

102. Gorham, J.R., Filth and extraneous material in food, in *Encyclopedia of Food Science and Technology*, Hui, Y.H., Ed., Wiley-Interscience, NY, 847-868, 1991.
103. Rosef, O. and Kapperud, G., House flies (*Musca domestica*) as possible vectors of *Campylobacter fetus* subsp. *jejuni*, *Appl. Environ. Microbiol.*, 45, 381-383, 1983.
104. Buxton, P.A., The importance of the house fly as a carrier of *E. histolytica*, *Brit. Med. J.*, 1, 142-144, 1920.
105. Echeverria, P., Harrison, A., Tirapat, C., and McFarland, A., Flies as a source of enteric pathogens in a rural village in Thailand, *Appl. Environ. Microbiol.*, 46, 32-36, 1983.
106. Biwald, S.P., Edeson, J.F.B., Ibrahim, J., and Matossian, R.M., The role of non-biting flies in the transmission of enteric pathogens (*Salmonella* species and *Shigella* species) in Beirut, Lebanon, *Ann. Trop. Med. Parasitol.*, 72, 117-121, 1978.
107. Bolanos, R., Frequency of *Salmonella* and *Shigella* in domestic flies collected in the city of San Jose, (in Spanish) *Rev. Biol. Trop.*, 7, 207-210, 1959.
108. Hormaeche, E., Peluffo, C.A., and Aleppo, P.L., Investigations on the occurrence of bacteria of the genera *Salmonella* and *Shigella* in flies, (in Spanish) *An. Inst. Hig. Montevideo*, 4, 75-79, 1950.
109. Levine, O.S. and Levine, M.M., Houseflies (*Musca domestica*) as mechanical vectors of shigellosis, *Rev. Infect. Dis.*, 13, 688-696, 1991.
110. Richards, C.S., Jackson, W.B., DeCapito, R.M., and Maier, P.P., Studies on rates of recovery of *Shigella* from domestic flies and from humans in southwestern United States, *Amer. J. Trop. Med. Hyg.*, 10, 44-48, 1961.
111. Akinboade, O.A., Hassan, J.O., and Adejinmi, A., Public health importance of market meat exposed to refuse flies, *Int. J. Zoonoses*, 11, 111-114, 1989.

Chapter 7

MOTHS (INSECTA: LEPIDOPTERA)

Alan R. Olsen

CONTENTS

I. Introduction 93

II. Morphology and Identification 93

III. Discussion of Families 97
- A. Pyralidae 97
  - 1. Pyralinae 97
  - 2. Galleriinae 97
  - 3. Phycitinae 98
- B. Tineidae 98
- C. Oecophoridae 99
- D. Gelechiidae 99
- E. Other Families 99

IV. Immatures 99
- A. Morphology 100
- B. Identification 100

References 102

I. INTRODUCTION

Stored-product moths are among the most destructive pests of stored food products.[1-3] The principal damage is caused by the larval caterpillars that not only consume product but also contaminate much more than they eat, with their webbing and excrement. The webbing that is continuously produced by the larva adheres to the food product and is nearly impossible to completely remove.[3] The adults are covered with small, flat scales that easily detach and become mingled with a food product.[1,4] Along with the stored-product beetles, these moths and moth larvae account for the most frequently encountered fragments in processed foods.[1,5] Table 1 lists the common food-contaminating moths.

II. MORPHOLOGY AND IDENTIFICATION

The stored-product pests in the order Lepidoptera are all small to medium-sized grayish or brownish moths with a body length of 5–15 mm and wing span of 10–30 mm (Figure 1). As with all Lepidoptera, the adult's body and wings are covered with a layer of thin, flat scales. The mouthparts consist of a coiled tubelike proboscis that is uncoiled for feeding. The proboscis is composed of the fused galea of the maxillae. The maxillary palpi are short or missing. The labial palpi are usually long. The labrum is much reduced and the mandibles are nearly always absent.[6-8] In the stored-product moths, the antennae are filiform in both males and females and are shorter than the length of the wing.[4,9]

TABLE 1
List of Common Food-Contaminating Moths

Scientific Name	Common Name
Cosmopterigidae	
Pyroderces rileyi (Walsingham)	Pink scavenger caterpillar
Oecophoridae	
Endrosis sarcitrella (L.)	Whiteshouldered house moth
Hofmannophila luteolus (Erichson)	
Hofmannophila pseudospretella (Stainton)	Brown house moth
Pyralidae	
Achroia grisella (F.)	Lesser wax moth
Aglossa caprealis (Hubner)	Murky meal caterpillar
Anagasta kuehniella (Zeller)	Mediterranean flour moth
Cadra cautella (Walker)	Almond moth, warehouse moth
Cadra figulilella (Gregson)	Raisin moth
Corcyra cephalonica (Stainton)	Rice moth
Ephestia elutella (Hubner)	Tobacco moth
Galleria mellonella (L.)	Greater wax moth
Paralipsa gularis (Zeller)	Stored nut moth
Plodia interpunctella (Hubner)	Indianmeal moth
Pyralis farinalis L.	Meal moth
Vitula serratilineella Ragonot	Driedfruit moth
Tineidae	
Haplotinea insectella (F.)	
Nemapogon granella (L.)	European grain moth
Setomorpha rutella Zeller	Tropical tobacco moth
Tinea pallescentella (Stainton)	Large pale clothes moth
Tinea pellionella L.	Casemaking clothes moth
Tineola bisselliella (Hummel)	Webmaking clothes moth
Trichophaga tapetzella (L.)	Tapestry moth
Gelechiidae	
Sitotroga cerealella (Olivier)	Angoumois grain moth

Identification is made by wing shape and venation patterns (Figure 2). Wing venation combined with characteristics of the mouthparts, antennae and ocelli are sufficient to identify adult stored-product moths to the family level.[9-11] Identification to genus and species level usually requires examination of the genitalia. The wing venation (Figures 3–5) of the families with stored-product species is complex but relatively uniform within the group of families. The forewing has an anterior unbranched subcostal vein and no humeral cross vein. The radial vein has four or five branches (R_1 through R_5) and the medial vein has three (M_1 through M_3). The discal cell is large and closed by a cross vein that is actually the first fork of the medial vein. The basal portion of the medial vein is lost but a short stem remnant is sometimes noticeable on the apical cross vein of the discal cell. This condition results in branch M_1 being associated with R_5, and M_2 and M_3 with the cubitus. The cubitus is usually two branched (Cu_1 and C_2) and there are one or two unbranched anal veins (1A and 2A).

The hindwing of these families has the subcosta and radius fused or very closely parallel opposite the closed discal cell. Beyond the discal cell, they branch into Sc, R_1 and Rs (radial sector). The arrangement of the discal cross vein and veins M_1–M_3 resembles that of the forewing. The cubitus is two-branched and up to three unbranched anal veins may be present.

FIGURE 1. Adult Mediterranean flour moth, *Anagasta kuehniella* (Phycitinae).

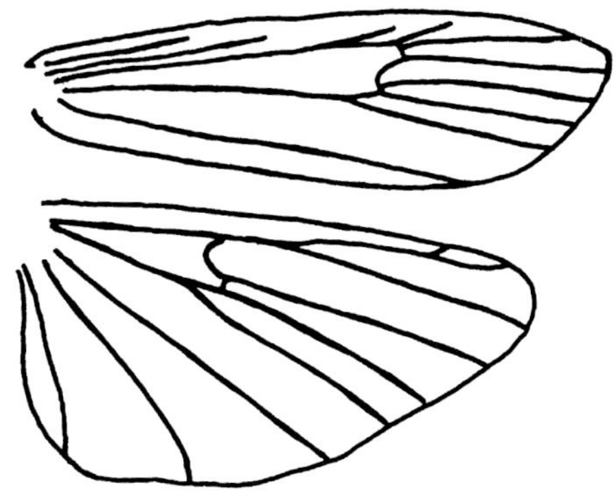

FIGURE 2. Forewing and hindwing of *A. kuehniella*, showing typical venation.

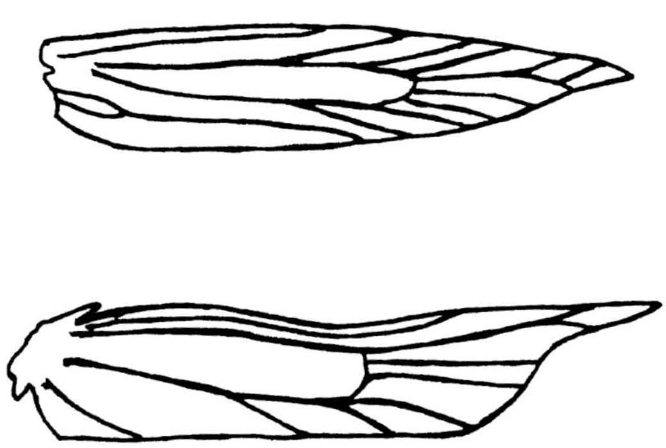

FIGURE 3. Forewing and hindwing of the Angoumois grain moth, *Sitotroga cerealella* (Gelechiidae), showing wing venation.

The base of the cubitus may bear a dorsal comb of dark stiff hairs known as the cubital pecten. The anterior margin of the hindwing bears a basal curved spine, the frenulum. In general, the hindwing is wider (anterior to posterior margin) than the forewing in the families discussed below.

The fragments of stored-product moths that are most often found in processed foods are fragments of the caterpillar larvae. Diagnostic larval fragments include the mouthparts, the frons of the head capsule and the antennae. Moth larval fragments are fully discussed and

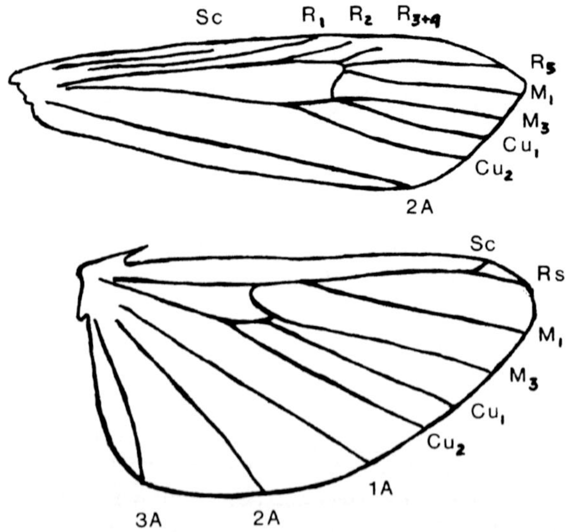

FIGURE 4. Forewing and hindwing of *A. kuehniella* (Phycitinae), showing veins Sc (subcosta), R (radial branches of forewing), Rs (radial sector of hindwing), M (medial branches), Cu (cubito-anal branches) and A (anal branches).

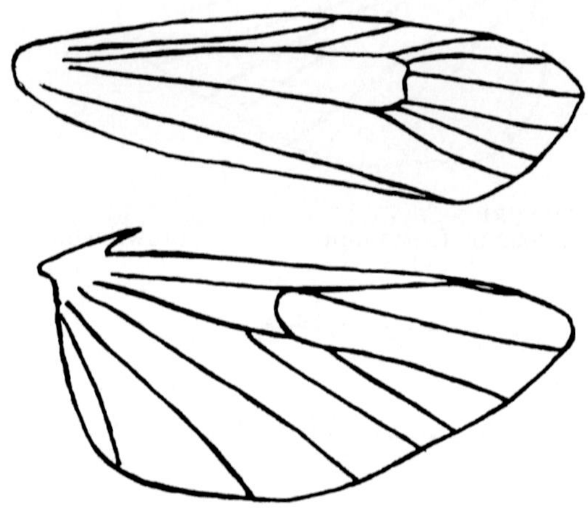

FIGURE 5. Forewing and hindwing of the Indianmeal moth, *Plodia interpunctella* (Phycitinae), showing wing venation.

described in the scientific literature.[1,5,12] Diagnostic adult fragments include the wings and the genitalia. The male genitalia consist of a pair of flat claspers articulated on the terminal abdominal segment. The paired claspers conceal a fleshy uncus, a central aedeagus and sclerotized supporting rods. The shapes of the claspers, uncus and associated structures are diagnostic to the species level.[1,9-11] The female genitalia consist of internal structures and an extrusable fleshy ovipositor the tip of which is often diagnostic in shape and proportions.[1,9,11]

Identification of moth adults involves observing the wing venation and genitalia and, in most cases, comparing these with authentic specimens. With wings, it is often necessary to remove scales to observe the veins. In many cases removal of a few strategic scales will reveal the desired vein structures. An alternative is to bleach the wings to reveal the venation. For many stored-product moths, mounting wings in glycerin jelly and viewing the mount with a transmitted-light microscope allows the identifier to trace the venation beneath the scales. Male genitalia can be observed under a widefield microscope by gently spreading the claspers to reveal the uncus and other structures. Female genitalia can often be forced to extrude from fresh specimens by squeezing the abdomen with a forceps or tweezers.

III. DISCUSSION OF FAMILIES

A. PYRALIDAE (MEAL MOTHS)

The meal moths are small to medium-sized gray, occasionally brown, moths with a wing span ranging from 10–30 mm. As a rule, pyralids can be recognized by the distinctive configuration of the subcostal and radial veins of the hindwing. The subcosta and radius are closely subparallel but separate opposite the discal cell. Shortly past the discal cell, the two veins coalesce, after which they separate again, becoming widely divergent at the wing margin.[13] In addition to this characteristic, pyralid wings are fully clothed with scales and the hindwing is wider (anterior to posterior margin) than the forewing, which is usually elongate in its proportions. The cubitus of the forewing appears four branched (M_2, M_3, Cu_1, Cu_2) and there is one anal vein. The discal cell is large and there is no accessory cell. On the hindwing there are three anal veins and the frenulum is well developed. The adult antennae are filiform in both male and female and ocelli are present. The proboscis is a typical coiled tube in stored-product species. The three-segmented labial palpi extend forward giving the appearance of a beak, hence the common name "snout moths" that is sometimes applied to this group. The principal stored-product pest species belong to the subfamilies Galleriinae, Pyralinae and Phycitinae.

1. Pyralinae

The subfamily Pyralinae is differentiated from other pyralids by a combination of vein Cu present on the hindwing but lacking a cubital pecten, R_5 present on the forewing and labial palpi forming a distinct beak. The meal moth, *Pyralis farinalis* L., is a medium-sized moth (20–30 mm) with a wing span of 15–25 mm.[8,14,15] It is brown with three color bands on the forewing. The color zones are divided by two longitudinal, wavy white lines, each line with a large chocolate-brown spot at either end.[9,16] The lines delineate a lighter colored middle zone. The hindwing has a similar pair of wavy white lines although the differences in coloration of the zones defined by the lines is less marked than in the forewing.[10] Another pyraline stored-product pest is the murky meal caterpillar, *Aglossa caprealis* (Hubner). The forewing of this species is mottled brown and the hindwing is white with a gold bordered margin.[16]

2. Galleriinae

The subfamily Galleriinae includes the rice moth and the wax moths. The latter infest beehives as well as occurring in stored-products. They are distinguished from other pyralids by the absence of ocelli and a vestigial or absent proboscis. Vein R_5 is present on the forewing and Cu is present on the hindwing with a distinct cubital pecten. The labial palpi are present in the female and are short or absent in the male. The front of the head has a conical projection of scales.[3] The female has a multiple frenulum. The rice moth, *Corcyra cephalonica* (Stainton), is a smaller (15–18 mm), grayish-brown to yellow moth with the hindwings forming a point apically and the posterior margin moderately fringed. The antennae are whitish distally and the head and thorax are paler than the forewings. The rice moth lacks a proboscis and the labial palpi are inconspicuous.[10] The rice moth infests a wide range of products. The greater wax moth, *Galleria mellonella* (L.), is one of the larger pyralids with a 25–30 mm wing span.[7,17] It is brown-gray in color and the hindwings are rounded apically. In the female, the apical margin of the forewing is concave and in the male it is notched.[8,10,17] A similar but smaller Old World species, the lesser wax moth, *Achroia grisella* (F.), is now worldwide in distribution, as are the rice moth and the greater wax moth. The lesser wax moth lacks a proboscis although the labial palpi are conspicuous. Another distinctive species is the stored nut moth, *Paralipsa gularis* (Zeller), with buff-colored forewings, each bearing a yellow patch and single black dot in the middle color zone.[10] The dot is small in the male and large in the female.[3]

3. Phycitinae

The subfamily Phycitinae contains the majority of stored-product pest species. Most are cosmopolitan. They are distinguished by the absence of vein R_5 on the forewing and the cubital pecten.[7–9] The discal cell of the hindwing is closed with a faint but distinct cross vein.[1] In stored-product species, the cubitus of the hindwing has only three apparent branches.[10] The proboscis and ocelli are present and the labial palpi are well developed and porrect. The Indianmeal moth, *Plodia interpunctella* (Hubner), is a striking member of this group, with the forewings colored light gray basally and metallic copper on the distal two-thirds.[14,15,18] The Mediterranean flour moth, *Anagasta* (= *Ephestia*) *kuehniella* (Zeller),* is a dark mottled gray moth with a pattern of dark zigzag lines on the forewing. Similar-appearing pests include the almond moth, *Cadra cautella* (Walker) (also called the tropical warehouse moth); raisin moth, *Cadra figulilella* (Gregson); carob moth, *C. calidella* (Guenee); and tobacco moth, *Ephestia elutella* (Hubner).[7-11,15] The Mediterranean flour moth is distinguished from these other gray moths by a lack of costal fold on the forewing of the male.[9,10] The costal fold is a basal flap on the front margin of the wing that is usually folded under the wing.

In western North America the driedfruit moth, *Vitula serratilineella* Ragonot, is a pest of stored raisins, prunes, dried apple and other dried fruit product.[19] It is distinguished from previously mentioned phycitine moths by relatively lighter colored forewings and nearly white (not brownish) hindwings. In eastern North America a closely related species, *Vitula edmandsii* (Packard), has brownish hindwings. It invades beehives but is not known to be a true pest of stored-products. Some authorities consider *V. edmandsii* and *V. serratilineella* a single species, with a western subspecies *V. edmandsii serratilineella*.[10] Others retain the division between eastern and western species.[9,20] All phycitine moth adults should be identified by examination of the genitalia, which are superbly illustrated by Neunzig, Heinrich, and Richards and Thomson.[9,11,21]

B. TINEIDAE (CLOTHES MOTHS)

The tineids are small gray moths with complete wing venation. The head has a rough appearance due to tufts of scales on the vertex. Antennae are shorter than the wings. The proboscis is usually present and the maxillary palpi are long, five segmented and held folded.[22] The labial palpi are short. The casemaking clothes moth, *Tinea pellionella* L., is a small grayish yellow to brown moth with three black spots on each forewing.[3,7,8,23] The webmaking clothes moth *Tineola bisselliella* (Hummel) has golden-colored scales on its body and golden-red scale tufts on the head. The wings are uniform in color, with no dark spots or other markings.[3,7,8,10,23] The European grain moth, *Nemapogon granella* (L.), is silvery gray with dark brown transverse strips on the forewing. The hindwing is gray with long hairs along the posterior margin.[3,10,15] Other *Tinea* spp. are stored-product pests including *T. pallescentella* (Stainton), the large pale clothes moth, which has also been found on mummified corpses; *T.*

* Previously placed in the genus *Ephestia* Guenee, the Mediterranean flour moth was moved to the genus *Anagasta* Heinrich because of the lack of a costal fold on the forewing of the males, a characteristic that is present in other *Ephestia* spp., and other anomalies in the male morphology. Recently, Neunzig moved the species *A. kuehniella* (Zeller) back into *Ephestia* because of analogies to variable conditions of the costal fold in males of the related genus *Acrobasis* Zeller. *Acrobasis*, however, is a large, unstable group that is likely to have future revisions that could well reevaluate the taxonomic significance of variability of the costal fold characteristic in males, once again affecting the status of *kuehniella*. For now, we are adopting the relatively conservative position of retaining *Anagasta kuehniella* (Zeller) as the scientific name for the Mediterranean flour moth.

cloacella Haworth; *T. ruricolella* Stainton; *T. infimella* Herrich-Schoffer, *T. salella* Zacher; and the tineids *Trichophaga tapetzella* (L.), *Setomorpha rutella* Zeller (tropical tobacco moth), *Haplotinea insectella* (F.) and *H. ditella* Pierce and Metcalfe.[15,24-26] *T. tapetzella*, the tapestry moth, has a distinctive color pattern, with the wings very dark on the basal third and becoming nearly white beyond.[8] As with the pyralids, these moths are best identified by examining the genitalia.[22]

C. OECOPHORIDAE

The oecophorids are small, flat, brown moths with complete wing venation. The antennae have a comb of bristles on the scape and the labial palpi are long, curved upward with an acutely pointed apical tip. The maxillary palpi are vestigial. In the forewing, veins R_4 and R_5 coalesce for at least the basal half, sometimes coalescing entirely. Veins R_2 and Cu_2 arise before the distal margin of the discal cell, and vein A_1 is present. In the hindwing, R_5 and M_1 are separate at their bases and parallel throughout their lengths. The brown house moth, *Hofmannophila pseudospretella* (Stainton), and the related *H. luteolus* (Erichson) are stored-product pests. They are medium to dark brown moths with three dark spots on the forewing and light-colored hindwings.[3,23] Both the adults and larvae of *H. pseudospretella* have also been found on mummified corpses.[25] Another pest, the whiteshouldered house moth, *Endrosis sarcitrella* (L.) infests stored grain, often in association with *H. pseudospretella*.[27] It is a light brown moth with a white head and pronotum.[3]

D. GELECHIIDAE

The gelechiids include a large number of species, one of which, the Angoumois grain moth, *Sitotroga cerealella* (Olivier), is a pest of stored-products.[1-3,6,14] Gelechiids are small moths with veins R_4 and R_5 coalescent basally and diverging apically on the forewing (Figure 3). On the hindwing, vein 2A is forked basally, with two short basal stem veins joining to form a single vein that proceeds to the wing margin. The labial palpi are long and upturned with a long, curved pointed terminal segment. The posterior margin of the hindwing is concavely emarginate, curving sharply forward apically.[3,6,8] The Angoumois grain moth is light brown to gray and may resemble stored-product tineids. It is distinguished from the tineids by the fingerlike projection of the hindwing. The posterior margin of the hindwing is fringed with hairs that are longer than the width of the membrane.

E. OTHER FAMILIES

Some moths that infest field crops or are scavengers in agricultural areas also occasionally infest stored-products. The pink scavenger caterpillar, *Pyroderces rileyi* (Walsingham) (Cosmopterigidae), is a scavenger in cotton fields that sometimes infests warehouses in cotton-growing areas.[17] It is also found in grains and grain residues in the tropics.[28] In the family Tortricidae, the codling moth, *Cydia pomonella* (L.), and the filbertworm, *C. latiferreana* (Walsingham), are sometimes found in dried fruit or nuts.[1] Another olethreutid, *Laspeyresia saltitans* (Westwood), is the Mexican jumping-bean moth, the larva of which is famous for causing the jumping movements of the seeds in which it develops.[2]

IV. IMMATURES

Among stored-product Lepidoptera, the larval feeding stage is the stage that accounts for the majority of adulterants encountered in processed food products. For this reason, special attention to the morphology and identification of larval forms is necessary. Additional illustrations, technical descriptions and keys for the identification of the larvae of specific groups or areas of interest are provided by Stehr, Aitken, Hinton and Capps.[24,29-33] Peterson and Weismann also provide illustrations of the larvae of many stored-product pest species.[34,35]

FIGURE 6. (Left) moth larval mandible.
FIGURE 7. (Right) moth larva head, lateral view, showing stemmata.

A. MORPHOLOGY

The general morphology of the larvae[32] of stored-product moths is the familiar morphology of the caterpillar. The body is soft and wormlike with a distinct sclerotized head. A pair of lateral spiracles occurs on the prothorax and each abdominal segment. The head bears a pair of short antennae located near the ventral mouthparts. The mouthparts are the chewing type consisting of labrum, mandibles and maxillae that are relatively unmodified. The labium is modified with a median distal spinneret for the production of silk. The morphology of the mandible (Figure 6) is diagnostic for moth larvae. The eyes are simple and consist of from 0 to 6 pairs of stemmata, or lenslike structures, that resemble ocelli, located ventrolaterally near the antennae and the bases of the mouthparts (Figure 7).

Each segment of the three thoracic body segments bears a pair of jointed legs. Usually, the ten-segmented abdomen bears a pair of fleshy ventral prolegs on segments 3, 4, 5, 6 and 10. A proleg is not jointed and terminates in a fleshy planta that has a distal series of crochets embedded in it. Crochets resemble small sclerotized hooks on the "sole" of the proleg and are arranged in different ways that are taxonomically important. In stored-product species, as well as other species, the crochets may be arranged in a circle, in ellipsoidal configurations or in a straight line. The individual crochets on a proleg may all be the same approximate size (uniordinal), two alternating sizes (biordinal) or three distinct sizes (triordinal). In a given species, the configuration of the crochets is the same for abdominal prolegs on segments 3–6.

The chaetotaxy of the thoracic segments and the eighth abdominal segment is important in identifying moth larvae. For the following identification key, the system adopted by Stehr is used. The schematic setal maps in Figure 8 show the generalized setal groups for the prothorax, mesothorax and eighth abdominal segment of moth larvae. In some stored-product pests, the number of setae in a group may be reduced. For each species, the relative lengths and the distances and directions of separation among setae are important.

B. IDENTIFICATION

The following key includes only the moth larvae most commonly encountered as pests of stored-products. Identifications should be confirmed by comparison with authentic specimens, technical descriptions or morphological illustrations. In general, the stored-product pest species lack secondary segmentation and secondary setation; are uniformly colored, usually white or pale cream, sometimes with a greenish or pinkish tint; and smooth in appearance, without protuberances, specialized setae or other morphological modifications.

Rapid Key for the Identification of Common Food-Contaminating Moth Larvae

1. Prolegs weakly developed; number of crochets reduced to only a few on each proleg (Gelechiidae) .... *Sitotroga cerealella*
   Prolegs well developed, each with numerous crochets 2

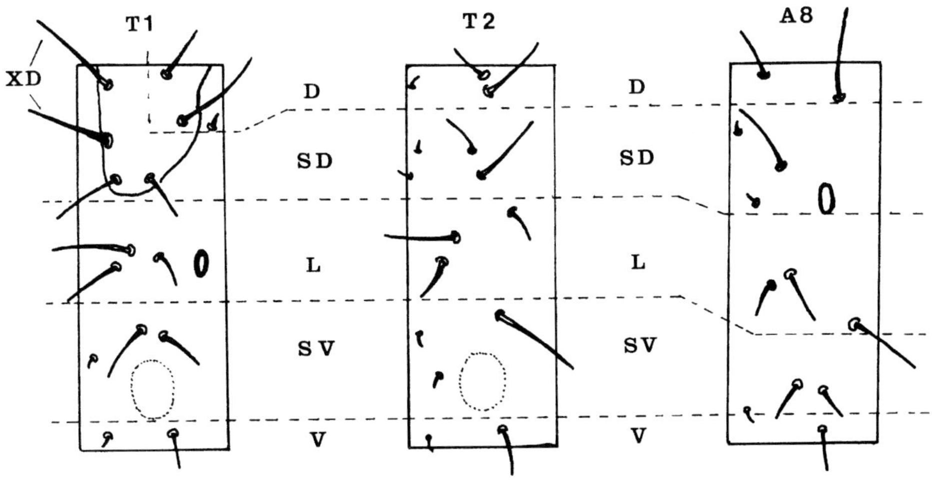

FIGURE 8. Chaetotaxy maps of the prothorax (T1), mesothorax (T2) and eighth abdominal segment (A8) of moth larvae showing setal groups D (dorsal), SD (subdorsal), L (lateral), SV (subventral), V (ventral) and XD (anterior dorsal).

2. Prespiracular seta group (group L) on prothorax bisetose (Pyralidae) 3
   Prespiracular seta group trisetose 13
3. Group L setae on prothorax aligned approximately horizontally (Galleriinae) 4
   Group L setae on prothorax aligned vertically 7
4. Body exceptionally white; mandible with two apical teeth and a smaller, ventral subapical tooth the outer margin of which is mesal to the ventrolateral margin of the mandible .... *Corcyra cephalonica*
   Body yellowish to brownish, sometimes with a pink tint; mandible with three or more apical teeth and no subapical teeth 5
5. Stemmata absent or very rudimentary; pinacula indistinct or not visible .... *Achroia grisella*
   Stemmata present and distinct; pinacula weakly to strongly pigmented 6
6. Some stemmata fused, often fewer than six on each side of head; abdomen with faint to pronounced brown pigment granules dorsally and laterally; head and prothoracic shield lightly pigmented .... *Galleria mellonella*
   Head with six distinctly separate stemmata on each side; head and thoracic shield darkly pigmented and head with reddish brown posterolateral spot on each side .... *Paralipsa gularis*
7. Mesothorax without a broad, sclerotized ring around the dorsolateral seta (SD1) (Pyralinae) 8
   Mesothorax with a broad, darkly sclerotized ring around the dorsolateral seta (SD1) (Phycitinae) 9
8. Body pale; three to four stemmata on each side of head .... *Pyralis farinalis*
   Body dark; five to six stemmata on each side of head .... *Aglossa caprealis*
9. Setae of abdominal segments not arising from sclerotized pinacula .... *Plodia interpunctella*
   Setae of abdominal segments arising from pinacula 10
10. Eighth abdominal segment with seta SD2 separated from the spiracle by a distance equal to or greater than two times the diameter of the spiracle 11
    Eighth abdominal segment with seta SD2 separated from the spiracle by a distance equal to or slightly less than the diameter of the spiracle 12

11. Diameter of the eighth abdominal spiracle as large as or larger than the area enclosed by the sclerotized ring around the seta above the spiracle .... *Anagasta kuehniella*
    Diameter of the eighth abdominal spiracle distinctly smaller than the area enclosed by the sclerotized ring around the seta above the spiracle .... *Ephestia elutella*
12. Eighth abdominal segment with posterodorsal seta (D2) 2–2.5 times as long as anterodorsal seta (D1) (i.e., D1 is about half as long as D2) .... *Cadra cautella*
    Eighth abdominal segment with posterodorsal seta (D2) three to five times or more as long as anterodorsal seta (D1) (i.e., D1 is very short compared with D2) .... *Cadra figulilella*
13. Head with two stemmata, or with four to five stemmata, on each side; crochets biordinal; prespiracular seta group (L) on prothorax aligned approximately horizontally (Oecophoridae) 14
    Head with six, one or no stemmata on each side; crochets uniordinal; prespiracular seta group (L) with seta not aligned in a horizontal row (Tineidae) 15
14. Head with two stemmata on each side .... *Endrosis sarcitrella*
    Head with four to five stemmata on each side .... *Hofmannophila* spp.
15. Planta of prolegs with enlarged, curved cuticular spines in addition to crochets .... *Setomorpha rutella*
    Planta without cuticular spines or only partially developed spines 16
16. Meso- and metathorax with one SV seta on each side 17
    Meso- and metathorax with two SV setae on each side 18
17. Head with six stemmata on each side .... *Nemapogon granella*
    Head with one or two stemmata on each side .... *Haplotinea* spp.
18. SV setae on meso- and metathorax aligned horizontally; seta group L on ninth abdominal segment bisetose; head with no stemmata ... *Tineola bisselliella*
    SV setae aligned vertically on meso- and metathorax; seta group L on ninth abdominal segment trisetose 19
19. Setae L1 and L2 on abdominal segments one to seven aligned vertically, with L1 behind the spiracle .... *Tinea* spp.
    Setae L1 and L2 on abdominal segments one to seven aligned horizontally, with L1 below the spiracle .... *Trichophaga tapetzella*

REFERENCES

1. Kurtz, O.L. and Harris, K.L., *Microanalytical Entomology for Food Sanitation Control*, Assoc. Off. Agric. Chem., Washington, DC, 1-576, 1961.
2. Metcalf, C.L., Flint, W.P., and Metcalf, R.L., *Destructive and Useful Insects: Their Habits and Control*, 4th ed., McGraw Hill, NY, 1-1087, 1962.
3. Munro, J.W., *Pests of Stored Products*, Hutchinson, London, 1-234, 1966.
4. Harris, K.L., Identification of stored product insects by the micromorphology of the exoskeleton. X. Common fragments of mill and grain moths, *J. Assoc. Off. Agric. Chem.*, 43, 444-458, 1960.
5. Ensminger, L.G., Identification of stored product insects by the micromorphology of the exoskeleton. VIII. Moth larvae, *J. Assoc. Off. Agric. Chem.*, 41, 704-738, 1958.
6. Borror, D.J., Triplehorn, C.A., and Johnson, N.F., *An Introduction to the Study of Insects*, 6th ed., Saunders Coll. Publ., Philadelphia, 1-875, 1989.
7. Essig, E.O., *College Entomology*, Macmillan, NY, 1-900, 1942.
8. Comstock, J.H., *An Introduction to Entomology*, 9th ed., Comstock Publ. Co., Ithaca, NY, 1-1064, 1940.

9. Neunzig, H.H., Pyraloidea Pyralidae (part), in *The Moths of America North of Mexico, Fascicle 15.3*, Dominick, R.B., Ed., Wedge Entomol. Res. Found., Natl. Mus. Natur. Hist. MRC-127, Washington, DC, 1-165, 1990.
10. Ferguson, D.C., Adult Moths (Lepidoptera), in *Insect and Mite Pests in Food*, Vol. 1, Gorham, J.R., Ed., U.S. Dept. Agric., Agric. Handbook 655, 231-244, 1987.
11. Heinrich, C., *American Moths of the Subfamily Phycitinae*, U.S. Natl. Mus. Bull. 207, 1-221, 1956.
12. Peace, D.M., *Key for Identification of Mandibles of Stored-Food Insects*, Assoc. Off. Anal. Chem., Arlington, 1-169, 1985.
13. Freeman, P., *Common Pests of Stored Food Products: A Guide to Their Identification*, 6th ed., Brit. Mus. (Natur. Hist.) Econ. Ser. No. 15, London, 1-69, 1980.
14. Schoenherr, W.H. and Rutledge, J.H., *Insect Pests of the Food Industry*, Lauhoff Grain Co., Danville, IL, 159, 1967.
15. Sinha, R.N. and Watters, F.L., *Insect Pests of Flour Mills, Grain Elevators, and Feed Mills and Their Control*, Agric. Can. Publ. 1776, Ottawa, 1-290, 1985.
16. Goater, B., *British Pyralid Moths. A Guide to Their Identification*, Harley Books, Essex, UK, 1-175, 1986.
17. Essig, E.O., *Insects of Western North America*, Macmillan, NY, 1-1033, 1926.
18. Donohue, H., Indian meal moth in California, *J. Econ. Entomol.*, 30, 680-681, 1937.
19. Okumura, G.T., The driedfruit moth (*Vitula edmandsae serratilineella* Ragonot), pest of dried fruits and honeycombs, *Calif. Dept. Agric. Bull.*, 55, 180-186, 1966.
20. Wood, A.M., *Insects of Economic Importance: A Checklist of Preferred Names*, CAB Internat., Wallingford, UK, 1-150, 1989.
21. Richards, O.W. and Thomson, W.S., A contribution to the study of the genera *Ephestia*, GN. (including *Strymax* Dyar) and *Plodia* GN. (Lepidoptera, Phycitidae), with notes on parasites of the larvae, *Trans. Entomol. Soc. London*, 80, 169-250, 1932.
22. Zagulev A.K., True moths (Tineidae), part 3 subfamily Tineinae, *Fauna USSR Lepidoptera*, 4(3), 1-443, 1960 (trans. U.S. Dept. Commerce, 1975).
23. Cornwell, P.B., *Pest Control in Buildings*, Hutchinson, London, 1-192, 1973.
24. Hinton, H.E., The larvae of the species of Tineidae of economic importance, *Bull. Entomol. Res.*, 47, 251-346, 1956.
25. Keh, B., Scope and applications of forensic entomology, *Ann. Rev. Entomol.*, 30, 137-154, 1985.
26. Madrid, F.J. and Sinha, R.N., *Haplotinea insectella* (Fabricius)—a potential pest of stored cereals in North America, *J. Stored Prod. Res.*, 20, 107-110, 1984.
27. Cox, P.D., A survey of stored product Lepidoptera in New Zealand, *N.Z. J. Exp. Agric.*, 14, 71-76, 1986.
28. Dobie, P., Haines, C.P., Hodges, R.J., and Prevett, P.F., *Insects and Arachnids of Tropical Stored Products: Their Biology and Identification*, Trop. Dev. Res. Inst., Slough, Berkshire, UK, 10273, 1984.
29. Hinton, H.E., Notes on the larvae of three injurious species of *Ephestia*, *Bull. Entomol. Res.*, 33, 21-25 1942.
30. Hinton, H.E., The larvae of Lepidoptera associated with stored products, *Bull. Entomol. Res.*, 34, 163-212, 1943.
31. Capps, H.W., *Key for the Identification of Some Lepidopterous Larvae Frequently Intercepted at Quarantine*, U.S. Dept. Agric. Publ. E 475, 1-8, 1963.
32. Stehr, F.W., Ed., *Immature Insects*, Vol. 1, Kendall/Hunt, Dubuque, IA, 1-754, 1987.
33. Aitken, A.D., A key to larvae of some species of Phycitinae associated with stored products and some related species, *Bull. Entomol. Res.*, 54, 175-188, 1963.
34. Peterson, A., *Larvae of Insects. Part 1. Lepidoptera and Hymenoptera*, Edwards Bros., Ann Arbor, MI, 1-315, 1948.
35. Weismann, D.M., Larval Moths (Lepidoptera), in *Insect and Mite Pests in Food*, Vol. 1, Gorham, J.R., Ed., U.S. Dept. Agric., Agric. Handbook 655, 245-268, 1987.
36. Neunzig, H.H., Pyraloidea Pyralidae (part), in *The Moths of America North of Mexico, Fascicle 15.2*, Dominick, R.B., Ed., Wedge Entomol. Res. Found., Natl. Mus. Natur. Hist. MRC-127, Washington, DC, 1-112, 1986.

Chapter 8

ANTS AND PARASITOIDS (INSECTA: HYMENOPTERA)

Vernon K. Avaritt III and Nicholas J. Richter

CONTENTS

I. Introduction ... 106

II. Morphology ... 106

III. Ants (Hymenoptera: Formicidae) ... 107
- A. Significance ... 107
- B. Identification ... 108
- C. Discussion of Subfamilies and Genera ... 109
  1. Aneuretinae ... 110
  2. Cerapachinae ... 110
  3. Dolichoderinae ... 110
  4. Dorylinae ... 111
  5. Ecitoninae ... 111
  6. Formicinae ... 112
  7. Leptanillinae ... 113
  8. Myrmeciinae ... 113
  9. Myrmicinae ... 113
  10. Nothomyrmeciinae ... 115
  11. Ponerinae ... 115
  12. Pseudomyrmecinae ... 115

IV. Parasitoids ... 115
- A. Importance ... 115
- B. Identification ... 116
- C. Discussion of Families ... 119
  1. Megaspilidae ... 119
  2. Ceraphronidae ... 119
  3. Braconidae ... 119
  4. Mymaridae ... 119
  5. Trichogrammatidae ... 119
  6. Eulophidae ... 119
  7. Encyrtidae ... 119
  8. Agaonidae ... 120
  9. Pteromalidae ... 120
  10. Eurytomidae ... 120
  11. Chalcididae ... 120
  12. Eucoilidae ... 120
  13. Diapriidae ... 120
  14. Scelionidae ... 120
  15. Platygasteridae ... 120
  16. Bethylidae ... 120

V. Other Hymenopterans 121

References 121

I. INTRODUCTION

The order Hymenoptera includes the bees, ants and wasps. Honeybees as food-producing insects and the food uses of bees and ants are discussed in chapter 13. Many ants and wasps play other roles in relation to the world's food supply, as pests and contaminants. The ants (Figure 1), family Formicidae, include a number of synanthropic species that are pests in establishments where foods are prepared, processed or packed. The wasps include a number of families of parasitoids that are attracted to insect-infested foods in search of hosts for their larvae. Many of these parasitoid species are more or less specialists in using different groups of stored-product insects as hosts.

II. MORPHOLOGY

The Hymenoptera are characterized by their two pairs of clear, membranous wings with the forewing larger than the hindwing. The hindwings have a series of minute hooks, the hamuli, on the front margin which, in flight, grasp a fold near the posterior margin of the forewing. This allows the forewings and hindwings to move in unison and present a single aerodynamic plane. The mouthparts are the chewing or chewing-lapping type. In the groups discussed in this chapter, the abdomen is petiolate (i.e., bulbous with the base constricted), attaching to the thorax by a stemlike pedicel, or waist, composed of one or two thin segments. Ant workers and soldiers are naturally apterous but can be recognized by their characteristic petiolate abdomen, geniculate antennae, chewing mouthparts, usually small compound eyes and lack of bands of densely packed fine hairs on the abdomen. This latter characteristic differentiates ant workers and soldiers from the similar-appearing females of the wasp family Mutillidae.

The head of Hymenoptera is round and somewhat flattened from front to back. The antennae are usually geniculate, with the first segment (the scape) elongate and the bend, or elbow, occurring between the distal end of the first segment and the second segment (the pedicel). The remainder of the antenna is termed the flagellum. The flagellum takes various forms. In the most common form, it is clubbed distally. The part of the flagellum between the scape and the club is the funiculus. It may consist of the pedicel and one or more ringlike segments or it may be undivided. The club likewise may be segmented or not. The antennae of male ant alates, bee drones and certain other infrequently seen hymenopterans are not geniculate. The antennae arise from the center of the face and their bases are usually close together. Often there are grooves in the head capsule, antennal fossae, that receive the antennae at rest.

The mouthparts of ants and most wasps consist of unmodified labrum, mandibles, maxillae and labium. Above the labrum is a conspicuous sclerite, the clypeus, that is characteristically shaped for many families and genera. A pair of subparallel sutures often connect the bases of each antenna to the clypeus. In ants, these sutures are raised carinae that form distinctive shelves (frontal lobes) over the bases of the antennae.

When viewed dorsally, the thorax has an anterior collarlike notum that may be wide or narrow from front to back but usually reaches across the entire thorax transversely. Behind the notum are the scutum (mesoscutum) and scutellum (mesoscutellum), respectively. The scutum is usually quadrate, sometimes somewhat triangular. Laterally, the scutum is flanked by a pair of roughly triangular scapulae (parapsides), joined to the scutum by the longitudinal parapsidal sutures. These lateral scapulae, or shoulders, have a smaller attached sclerite, the tegula, that projects laterally from the lower outside corner. The posterior scutellum also has a lateral pair of attached sclerites, the axillae. The axillae are displaced forward to occupy a position near the juncture of the scutellum and the scutum. Behind the scutellum, one finds an additional dorsal sclerite, the propodeum, that is often interpreted as being abdominal in origin. The pedicel to the abdomen (not to be confused with the antennal pedicel) is attached to the propodeum dorsally. Laterally, the thorax can be seen to have a "swept back" form, with the lateral sclerites and meso- and metacoxae angled posteroventrally. The wing-bearing portion of the thorax, the alitrunk (the thorax fused with the slender first abdominal segment), is a common contaminant in foods.

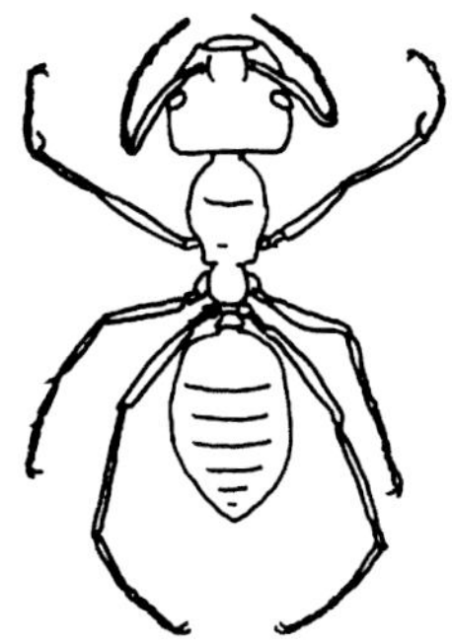

FIGURE 1. (Left) ant worker (Formicidae).
FIGURE 2. (Right) ant protibia and tarsus, showing strigulus.

The abdomen consists of a bulbous gaster connected to the thorax by a thin, stemlike pedicel. The sting of venomous hymenopterans is located in the gaster, extruding from the posterior tip. In some stingless ants, the gaster bears a conspicuous opening, the acidopore, that exudes defensive chemicals when the animal is alarmed or aggressive. The pedicel of ants is composed of one or two distinctively beadlike segments, the petiole and postpetiole. These segments may appear spherical or may have a dorsal projection, or node, that is noticeable in lateral profile.

The legs of ants and wasps are slender and consist of the coxa, one or two trochanters, femur, tibia and tarsus. In ants and many wasps, the ventral apex of the protibia is armed with a comblike organ, the strigulus (Figure 2), that is used in combination with a patch of stiff hairs on the first tarsal segment to clean the antennae of debris. The wing venation of Hymenoptera is frequently reduced. In the parasitoids discussed in this chapter, the reduced venation of the forewing consists of, at most, a costal vein along the anterior wing margin; a subcostal vein; and from one to six closed cells below the subcosta. In most cases, the costa and subcosta join to form an enlarged sclerotized area, the stigma, midway along the anterior wing margin. There may be a closed marginal cell distal to the stigma, the marginal cell. This venation is greatly reduced in many parasitoids, with one to all of the closed cells absent. Extreme reduction of the venation occurs where there is one anterior marginal vein only or a marginal vein with an apical branch that terminates in a stigma on the wing membrane (rather than on the margin), or no veins at all. The wing venation of ants is of marginal use in food sanitation work because most of the contaminants are apterous workers.

Many Hymenoptera, including ants, are social insects that live in large numbers together in a single nest. These social units are divided by castes, where many of the individual insects are nonreproductive workers. In ants, the workers are apterous and frequently exhibit subcastes of different sizes. These subdivisions are called major workers and minor workers, depending on their relative size.

For a definitive analysis of ant morphology, refer to Hölldobler and Wilson.[1]

III. ANTS (HYMENOPTERA: FORMICIDAE)

A. SIGNIFICANCE

Ants in foods pose a significant risk to human health. An ant, *Monomorium latinode* Mayr, was identified as a possible vector of cholera.[2] Two incidents are recorded of the pharaoh's ant, *Monomorium pharaonis* L., as a carrier of *Pseudomonas aeruginosa*, *Staphylococcus* spp., *Salmonella* spp., *Streptococcus* spp. and *Clostridium* spp.[3,4] This ant species must be considered a health hazard and its occurrence should not be dismissed as inconsequential.[5] *Iridomyrmex humilis* Mayr is a vector of *Listeria monocytogenes*; Griffitts suggests *Solenopsis geminata* (F.) as a possible vector of *Shigella flexner;* and Singh et al. report isolating *Salmonella* from ants.[6,7]

Judging from analyses performed by the FDA, more than 98% of the ants found contaminating foods belong to three subfamilies: Myrmicinae, Dolichoderinae and Formicinae. A few ants from the subfamilies Ponerinae and Pseudomyrmecinae have been found in foods, and the subfamilies Dorylinae, Ecitoninae, Cerapachinae, Aneuretinae, Leptanillinae, Myrmeciinae and Nothomyrmeciinae are infrequently, if at all, represented as contaminants in food samples. The following keys are provided to differentiate the ants that are associated with insanitary conditions in and around food storage and manufacturing facilities. The first general key differentiates the various subfamilies of the family Formicidae, to which all ants belong. Following the general key are a series of keys for separating the genera of the three major subfamilies (Dolichoderinae, Formicinae and Myrmicinae) that contain food-contaminating pests.

B. IDENTIFICATION

Rapid Key to Ant Subfamilies

1. Frontal lobes vestigial and do not conceal the antennal sockets 2
   Frontal lobes present and covering antennal sockets 3
2. Eyes absent .... Dorylinae
   Eyes consist of a single ommatidium (very small) .... Ecitoninae
3. Waist consisting of petiole only 4
   Waist consisting of petiole and postpetiole 6
4. Gaster with semicircular to circular acidopore, often looking like a nozzle fringed with setae .... Formicinae
   Acidopore not present 5
5. Sting vestigial or absent .... Dolichoderinae
   Sting present; first segment of gaster slightly differentiated from the following segments of the gaster .... Ponerinae
6. Clypeus straight or weakly arched, level with antennal sockets..... Pseudomyrmecinae
   Center of clypeus triangular, separating frontal lobes .... Myrmicinae

Rapid Key to the Genera of the Subfamily Myrmicinae

1. Postpetiole attached to the dorsal surface of the first gastral segment; gaster roughly heart shaped .... *Crematogaster*
   Postpetiole attached to the apical end of the gaster; gaster not usually heart shaped 2
2. Antennae with two-segmented club 3
   Antennae with three- or more segmented club 5
3. Antennae with ten segments (*Solenopsis*) 4
   Antennae with 11 segments; major workers have large heads and resemble *Pheidole* .... *Pheidologeton*
4. Eye is large and multifaceted .... Fire ants (*Solenopsis*)
   Eye contains only four to six facets .... other *Solenopsis* spp.
5. Propodeum unarmed; antennae containing 10–11 segments (*Monomorium*) 6
   Propodeum armed; antennae variable 9
6. Head and alitrunk rugulose and opaque... *Monomorium pharaonis* (L.), pharaoh's ant
   Head and alitrunk smooth and shiny 7
7. Scape of antennae extending beyond occiput; thorax depressed and flattened laterally and submargined .... *Monomorium destructor* Jerd.
   Scape not extending beyond the occiput 8
8. Head and alitrunk dark brown .... *Monomorium minimum* (Buckley)

Head and alitrunk reddish yellow .... *Monomorium gracillimum* Smith
9. Posterior margin of clypeus raised to form a ridge extending under the antennal insertions (*Tetramorium*) 10
Posterior margin of clypeus not as above; on major workers the head is very large .... *Pheidole*
10. Head without an antennal sulcus; head and alitrunk with longitudinal striae; color brownish to black .... *Tetramorium caespitum* (L.), pavement ant
Head without antennal sulcus, head and alitrunk rugulose; color yellowish to brownish .... other *Tetramorium* spp.

Rapid Key to the Genera of the Subfamily Formicinae

1. Antennae situated far behind the clypeal margin, propodeum uniformly convex .... *Camponotus*
Antennae situated close to clypeal margin 2
2. Orifice of the propodeal spiracle some shape other than circular, and well in front of the side where the alitrunk rounds into declivity 3
Orifice of the propodeal spiracle at least somewhat circular, and bordering the area where the propodeum rounds into declivity 4
3. Masticatory margin of the mandible consists of eight or more teeth; third tooth smaller than the fourth, which is larger than the rest of the teeth ..... *Formica*
Masticatory margin of mandible consisting of five to seven teeth; third tooth larger than the fourth tooth; teeth after the fourth decreasing in basal size .... *Cataglyphus*
4. Head and alitrunk covered with pairs of stout bristles (*Paratrechina*) 5
Head and alitrunk without pairs of stout bristles 6
5. Body slender, with antennae and legs very long .... *Paratrechina longicornis* (L.), crazy ant
Body more or less robust; legs not disproportionately long ..... other *Paratrechina* spp.
6a. Mandibles containing six teeth, sometimes seven; scapes with at least half their length projecting past occiput .... Prenolepis
6b. Mandibles with at least seven teeth, but usually more than seven teeth; antennal scape does not project over half of length past occiput .... Lasius

Rapid Key to the Genera of the Subfamily Dolichoderinae

1. Petiole without a visible node .... *Tapinoma*
Petiole with a node 2
2. Head and alitrunk slender; masticatory margin of mandible with 17 or more teeth or denticles .... *Leptomyrmex*
Head and alitrunk more stocky; masticatory margin usually with 12 or fewer teeth (rarely 15) 3
3. Body heavily sclerotized; thick and armored .... *Hypoclinea*
Body less sclerotized; thin and flexible 4
4. Alitrunk even in profile without an impressed metanotal region .... *Liometopum*
Alitrunk looking concave when viewed in profile; metanotal region lower than the rest of the alitrunk .... *Iridomyrmex*

C. DISCUSSION OF SUBFAMILIES AND GENERA

The following is a discussion of all recognized subfamilies of Formicidae, whether they are found in foodstuff or elsewhere. Not all of the subfamilies appear in the identification keys. The omitted groups are rarely encountered in foods either because of limited geographic distribution or specialized habitat adaptations that isolate them from synanthropic habitats. For the subfamilies that are likely to be encountered in or near food products (Dolichoderinae, Formicinae and Myrmicinae), brief technical descriptions are provided for the most important genera.

1. Aneuretinae

This small subfamily is found only in Sri Lanka. It is a rare subfamily of small, yellowish brown ants characterized by one pedicel. The major tibial spur is simple or with a few tiny barbules. The propodeum is armed, which is unusual for one-pediceled ants; the petiole is pediculate and there is no antennal club. It resembles a myrmicine ant except that it has only one pedicel. There is only one genus in the subfamily. This subfamily is included only for completeness; one would not expect to find it in foodstuffs.

2. Cerapachinae

Hölldobler and Wilson include this group under the subfamily Ponerinae, and it is similar to that subfamily.[1] The pygidium bears rows of erect teeth along both sides that converge toward the rear, and the antennal sockets are nearly touching. Cerapachinae is a rare subfamily found worldwide but is unlikely to be found in food products.

3. Dolichoderinae

This large, cosmopolitan subfamily is one of the major subfamilies found in stored-products. It includes the Argentine ant, the odorous house ant and the velvety tree ant. This subfamily has only one pedicel in its waist and no stinger or acidopore. It has a large Dufour's gland, which excretes butyric acid as a defense.

***Genus* Conomyrma**

Head—oval with occiput emarginate; Mandibles—triangular, masticatory margin with approximately five teeth; Labial palps—three jointed; Maxillary palps—six jointed, with third joint longer than the three succeeding joints combined; Clypeus—fairly straight with no invagination between the antennal sockets; Antennae—12 segmented with no club; Eyes—large, placed laterally on head near middle of head, no ocelli; Thorax—propodeum set off from the rest of the alitrunk by distinct suture, propodeum armed with conelike spine, which projects upward (not backward as in most species); Legs—long and slender; Waist—petiole scalelike, well developed and projecting forward; Abdomen—oval.

The pyramid ant, *Conomyrma insana* (Buckley), occasionally infests houses.[8]

***Genus* Hypoclinea**

Head—triangular; Mandibles—triangular, hypostomal toothlike projection on the ventral surface of the mandibular insertions; Clypeus—triangular, extending between antennal sockets; Antennae—12 segments; Eyes—moderate size; Thorax—declivitous face of propodeum very concave, spine or projections extending outward are simple; Waist—petiole with well-developed erect scale; Abdomen—broadly oval.

These heavily armored ants have been occasionally found as contaminants in food products.

***Genus* Iridomyrmex**

Head—triangular; Mandibles—triangular, very broad at masticatory margin; masticatory margin usually bearing five teeth; third, fourth and basal tooth separated by numerous denticles, and are often a little larger than the denticle; Clypeus—triangular extending between the frontal carinae; Antennae—12 segments; Eyes—moderate size; Thorax—metanotum noticeably impressed, propodeum very distinct from the rest of the alitrunk, propodeum unarmed and rounded; Legs—long and robust, tarsal claws simple; Waist—petiole with well-developed erect scale; Abdomen short, broadly oval.

The Argentine ant, *Iridomyrmex humilis* (Mayr), is an important synanthropic pest species. It is a brown ant, 2–2.5 mm long, that travels in trails.[9,10] It is attracted to almost any type of stored food product.[9-12] *I. humilis* is a feeble biter, but other species of *Iridomyrmex* can inflict painful, but not venomous, bites.[11,13]

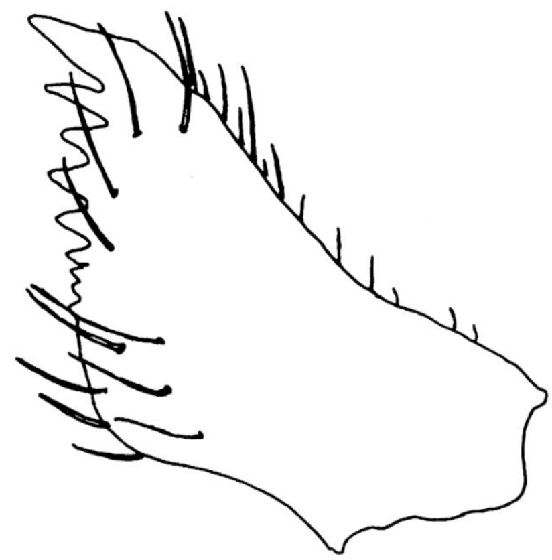

FIGURE 3. Ant mandible, *Tapinoma* sp.

***Genus* Leptomyrmex**

Head—elliptic, with the occiput emarginate, head extremely elongate; Mandibles—triangular, slender, outer margin slightly concave, masticatory margin of mandible with 17 or more teeth or denticle, usually about 20 teeth; Clypeus—triangular emarginate where frontal lobes arise, not extending past frontal lobes; Antennae—12 segments, very long extending past the occiput; Eyes—moderately large; Thorax—extremely elongate and slender, largest body section, propodium unarmed; Legs—very long and narrow, tarsal claws simple; Waist—petiole nodiform and well developed; Abdomen—oval.

***Genus* Liometopum**

Head—triangular; Mandibles—broadly triangular, masticatory margin contains eight to ten teeth, the apical tooth is very long and curved; Clypeus—triangular extending between the frontal lobes, laterally extending over the basal portion of the mandibles; Antennae—12 segments, relatively short scape not extending past the occiput; Eyes—moderate, laterally placed; Thorax—entire alitrunk convex in profile, no noticeable sutures, propodeum unarmed and rounded; Legs—fairly long and slender, tarsal claws simple; Waist—petiole scalelike, erect; Abdomen—broadly oval.

***Genus* Tapinoma**

Head—oval with occiput emarginate; Mandibles—triangular dentate (Figure 3); Labial palps—four segmented; Maxillary palps—six segmented; Antennae—12 segments; Eyes—comparatively large; Thorax—convex, with the metanotal region slightly depressed, propodeum unarmed and rounded, the declivitous face longer than the basal face; Legs—moderately long and slender; Waist—petiole without a well-developed node; Abdomen—oval with the basal segment overhanging the petiole.

A well-known species is *Tapinoma sessile* (Say), the odorous house ant. This ant and the related species *T. melanocephalum* (F.) resemble *I. humilis* except they are darker in color. Both are found in cities, where they forage in buildings for food.[1,9,14]

4. Dorylinae

This subfamily is found only in the Ethiopian region. It consists of the Old World army ants. These ants feed on large arthropods and whatever else they can get, but they do not feed on stored-products. Dorylinae do not have eyes at all, and they do not have frontal lobes. They may have a one- or two-pedicel waist. They are not likely to be found in any food products.

5. Ecitoninae

This subfamily is found in the Nearctic and Neotropical regions, so they are called New World army ants. They are very similar to Dorylinae except they have eyes consisting of one ommatidia only. The waist may consist of one or two pedicles. The genera *Eciton* and *Neviamyrmex* are found in the California desert. *Neviamyrmex* spp. are predators of the Argentine ant, *I. humilis*, and often enter buildings in search of their prey.[11]

6. Formicinae

This is one of the largest subfamilies of ants. They are quickly identified by the acidopore found at the tip of the gaster. They also have a waist with one pedicel. The acidopore is connected to the poison gland, which sprays formic acid. The genera *Paratrechina* and *Lasius* are common contaminants of food.

Genus Acanthomyops

Head—ovaline; Mandible—triangular, masticatory margin with 5–12 teeth; Maxillary palps—three segments, short; Clypeus—triangular, not extending past the frontal lobes; Antennae—12 segments, scape just reaching the occiput; Eyes—small, lateral; Thorax—metanotal region noticeably depressed, propodeum separated from rest of alitrunk, robust ant; Legs—stocky and moderate; Waist—petiole scalelike and erect; Abdomen—broadly oval.

These yellowish ants are known to infest buildings.[8]

Genus Camponotus

Head—oval, occiput emarginate; Mandible—triangular or subtriangular; Labial palp—four segmented; Maxillary palps—six segmented; Clypeus—trapeziform, not reaching frontal lobes or extending past genae; Antennae—12 segmented, not arising near clypeus.

Camponotus spp. are wood-dwelling carpenter ants. Because they are occasionally structural pests, they may be encountered around food-handling operations that are housed in infested wooden structures.

Genus Formica

Head—oval, occiput emarginate; Mandible—triangular, masticatory margin dentate, apical tooth large and curved; Clypeus—triangular, broad and high; Labial palps—four segmented; Maxillary palps—six segments, fourth segment longer than fifth; Antennae—12 segmented, scape and flagellum filiform; Eyes—moderate, near the top of head, frontally placed, ocelli present; Thorax—metanotal region depressed, propodeum rounded and unarmed; Legs—long and slender, tarsal claws simple; Waist—petiole scalelike and erect; Abdomen—globose.

Fragments of *Formica* spp. have been found in food products.

Genus Lasius

Head—triangular; Mandibles—triangular, short and elongate, masticatory margin with seven or more teeth; Clypeus—trapeziform; Antennae—12 segmented, fairly long but not as long as in *Paratrechina*; Eyes—moderate, ocelli very small but present; Thorax—mesonotal region depressed but forms a small bump between metathoracic area and propodeum, pronotum region distinct, propodeum rounded and unarmed, propodeal spiracle orifice circular to subcircular in shape, spiracle bordering on curvature where basal side and diclivious side meet; Legs—somewhat short and fairly robust, tarsal claws simple; Waist—petiole scalelike and erect; Abdomen—broadly oval.

The cornfield ant, *Lasius alienus* (Foerster), and related agricultural pest species may wind up in food products as fragment contaminants.

Genus Oecophylla

Head—squarish; Mandibles—triangular, long, masticatory margin with ten or more teeth or denticles, apical tooth acute and curved, fourth tooth larger than third and fifth teeth; Labial palps—four segmented; Maxillary palps—five segmented; Clypeus—resembling trapeziform, apical region convex, twice as broad as long; Antennae—12 segmented; Eyes—large, frontally placed, no ocelli; Thorax—elongate, slender, mesonotum elongated, propodeum unarmed; Legs—long and slender; Waist—petiole without node, elongate; Abdomen—oval, relatively short, can be bent over the alitrunk.

One species in this group, *Oecophylla smaragdina*, is harvested in Asia to be sold as food (see chapter 13). It has been exported to many countries, including the United States.

FIGURE 4. Ant mandible, *Paratrechina* sp.

Genus Paratrechina

Head—oval, occiput emarginate; Mandibles—triangular, masticatory margin with five or six teeth (Figure 4); Maxillary palps—six segmented; Antennae—12 segmented, as long as head length; Eyes—moderate sized, lateral; Thorax—mesonotum slightly depressed, propodeum separated from rest of the alitrunk, propodeum unarmed; Legs—short and robust to long and slender; Waist—petiole triangular sloping forward; Abdomen—broadly oval, anterior portion overhangs petiole.

The crazy ant, *Paratrechina longicornis* (Latreille), is a common synanthropic pest species. It is spread by commerce.[1,14]

7. Leptanillinae

This small subfamily consists of a type of army ant. They differ from both Ecitoninae and Dorylinae in that they have 12-segmented antennae; they also have no eyes and a two-pedicel waist. They are very rare and are not encountered in food products.

8. Myrmeciinae

This subfamily looks very similar to the subfamily Myrmicinae. The difference is that the Myrmeciinae have a forked tarsal claw and always have ocelli. The Myrmeciinae are found only in Southern Australia. These ants are a very large and primitive subfamily of ants, and they have a pair of mandibles that look like lucanid mandibles.

9. Myrmicinae

This is the largest and most important subfamily. It is cosmopolitan in range and contains most of the ants found in food samples. They include the genera *Monomorium*, *Pheidole*, *Crematogaster*, *Pheidologeton*, *Solenopsis* and *Atta*. This subfamily has a two-pedicle waist, a well-developed stinger and usually no ocelli.

Genus Aphaenogaster

Head—longer than broad, usually posterior portion elongated into a neck; Mandibles—subtriangular, masticatory margin possessing seven or more teeth or denticles, outer margin not strongly curved; Clypeus—flattened, anterior margin transverse; Antennae—12 segmented, four-segmented club not very pronounced; Eyes—moderate sized, lateral, and found near the middle of the head, ocelli absent; Thorax—very slender and elongate, propodeum strongly armed; Legs—very long and slender, tarsal claws simple; Waist—petiole very pedunculate, node subconical, postpetiole longer than broad; Abdomen—oval and somewhat flattened.

Genus Crematogaster

Head—nearly square when viewed from the front with the occipital corners rounded; Mandibles—triangular, strong and thick, but relatively narrow, masticatory margin with four teeth; Labial palps—three segmented; Maxillary palps—five segmented; Antennae—11 segments ending in a club of two, three or four segments; Eyes—moderate sized, lateral, ocelli usually absent; Thorax—propodeum bispinously armed; Waist—petiole more or less broad-

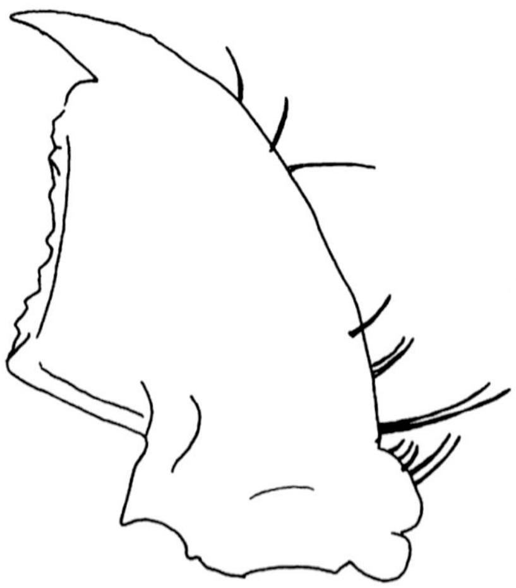

FIGURE 5. Ant mandible, *Pheidole* sp.

ened, and anteriorly concave or flat above, postpetiole has a rounded slightly raised node, sometimes with a longitudinal groove; Abdomen—cordate or subcordate.

Genus Monomorium

Head—rectangular; Mandibles—triangular, narrow; Labial palps—two jointed; Maxillary palps—two jointed; Clypeus—subtriangular, arched anteriorly; Antennae—11 to 12 segments with a three-segmented club; Eyes—lateral, oval; Thorax—relatively long and narrow, propodeum unarmed; Waist—petiole pedunculate, usually higher than the postpetiole; Abdomen—oval, often truncated or emarginate anteriorly.

The pharaoh's ant, *M. pharaonis* (L.), is a small, 1–1.5 mm, yellow ant that infests buildings. It has been found in food manufacturing facilities and in hospitals, where it tracks pathogens onto food, surgical instruments and equipment.[1,3,5,9,15,16] *M. minimum* (Buckley) nests outdoors but forages in buildings for sweet and proteinaceous foods.

Genus Pheidole

Workers of this genus may occur in both major and minor forms.

Major: Head—proportionately large when compared with the body, occiput of cranium appears bilobate; Mandibles—triangular, broad and strong, masticatory margin sometimes with teeth or denticles (Figure 5); Clypeus—narrow, produced immediately behind the base of the antennae, antennal carinae often as long as the scape, antennal scrobes present; Antennae—12 segmented, usually with a three-segmented club (a few species have a four-segmented club); Eyes—moderate to small size, never ocelli; Thorax—propodeum armed; Legs—moderately long and stout, tarsal claws simple; Waist—petiole with an appendix beneath; Abdomen—relatively ovaline.

Minor: In all aspects similar to major except that the minor's head is proportionately smaller. Minors cannot be identified to the species level.

Pheidole megacephala (F.), the bigheaded ant, is a cosmopolitan pest of stored-products.[14] It is recognizable by the large heads of the major workers and a spined alitrunk.

Genus Solenopsis

Head—nearly square, with a medial distinct groove down the front, antennal scrobes and frontal carinae absent; Mandibles—triangular, curved and without teeth; Labial palps—two segmented; Maxillary palps—two segmented; Clypeus—narrow with a U-shaped carina; Antennae—with ten segments ending in a two-segmented club; Eyes—small, round and rather flat; Thorax—slender, propodeum unarmed and rounded; Legs—slender and long; Waist—petiole pedunculate, squamiform, postpetiole varies slightly, usually broader than long; Abdomen—oval, may be slightly truncate.

The thief ant, *S. molesta* (Say), is a pest of food products. This yellowish ant has also been implicated as a vector of dysentery pathogens.[5]

Genus Tetramorium

Head—somewhat rectangular, with a distinct posterior margin; Mandibles—triangular, broad, masticatory margin with five to seven teeth (Figure 6); Labial palps—three jointed; Maxillary palps—four jointed; Clypeus—triangular with the apex curving under the antennal sockets; Antennal carinae—short and wide apart; Antennae—11 to 12 segments with a three-segmented club; Eyes—relatively small, lateral and round; Thorax—propodeum armed; Waist—petiole pedunculate anteriorly, the node curved inward, postpetiole broader than petiole; Abdomen—oval.

T. caespitum (L.), the pavement ant, is black with lighter legs. Pavement ants and related "tramp" species forage in buildings but nest outdoors.[9,14]

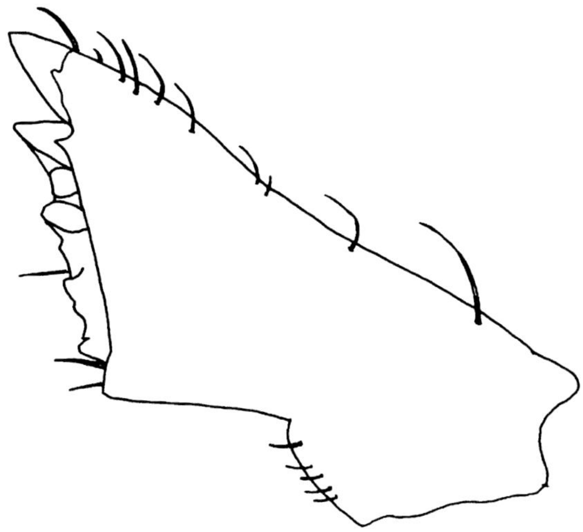

FIGURE 6. Ant mandible, *Tetramorium* sp.

10. Nothomyrmeciinae

This rare and primitive subfamily is found only in Australia. The only genus in this subfamily is *Nothomyrma*. They resemble the Myrmeciinae except that they have only one pedicel in their waist and are much smaller. They are not found in stored-products.

11. Ponerinae

This large cosmopolitan subfamily is not too readily found in food products. Ponerinae have waists with only one pedicel and a well-developed stinger. They are distinguished by the first gastral segment separated from the rest of the gaster by a suture. This subfamily contains the world's largest ant.

12. Pseudomyrmecinae

This subfamily contains only two genera, *Tetraponera* and *Pseudomyrmex*. It is similar to the Myrmicinae in most aspects, but the clypeus is straight and does not separate the two frontal lobes. They are sleek, with short antennae and very large eyes for their size. They are small arbicolus ants noted for their complex relationships with plants. The acacia ant is in this subfamily.

IV. PARASITOIDS

A. IMPORTANCE

The importance of parasitic wasps to microanalytical entomology increased markedly in April 1992, when parasitic wasps and predators were accepted by the EPA as a control measure for stored-product insects in grain silos. Before then, there was a zero tolerance level for any control agent of stored-product pests; indeed, control agents were treated as if they were another pest insect. Since April 1992, a number of parasites and predators have been used to effectively control stored-product insects. The parasitoid *Venturia canescens* (Gravenhorst) is used to control many stored-product moth larvae in the pupal and prepupal stages. The stored-product pests it controls include *Cadra cautella* (Walker) and *Plodia interpunctella* (Hubner). *Bracon hebetor* (Say) and the predator *Xylocoris flavipes* (Reuter) have also been used to control *C. cautella*. Several *Holepyrus* spp. attack various stored-product beetles, including *Sitophilus oryzae* (L.), *Tribolium confusum* Jacquelin Du Val and *Oryzaephilus surinamensis* (L.). *Laelius trogodermatis* (Ashmead) and four other species attack the larvae of Dermestidae. *Cephalonomia tarsalis* (Ashmead) attacks various species of Coleoptera. These examples are only illustrative and do not necessarily represent current EPA permitted uses of specific wasps for biological control applications.

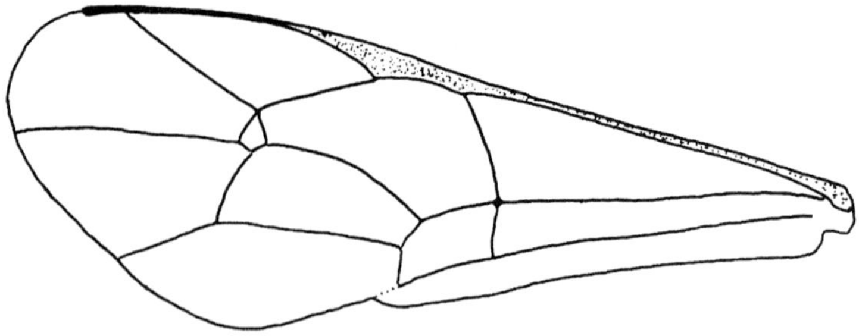

FIGURE 7. Parasitoid forewing, family Ichneumonidae.

B. IDENTIFICATION

Rapid Key for Identifying Wasp Parasites of Stored-Product Pests

1. Wings present 2
   Wings absent 7
2. Forewing with more than two closed cells 3
   Forewing with one, two or no closed cells 4
3. Metasomal tergites 2 and 3 rigidly joined; forewing with a single m-cu cross vein; second segment of the radius about equal to or slightly longer than the first segment and about equal to or shorter than first intercubitus; m-cu vein enters the first cubital cell; second intercubitus present; radius straight, reaching wing margin; basal vein straight at anterior end; 1.5–2.0 mm in length; occipital and prepectal carinae absent; circular–oval opening above mandibles when closed, labrum concave; mandible tips touching and usually overlapping when closed .... *Habrobracon* (Johnson)
   Metasomal tergites 2 and 3 not rigidly joined; forewing with two m-cu veins (Figure 7) .... Ichneumonidae
4. Forewing with one or two closed cells 5
   Forewing with no closed cells 8
5. Scutellum with raised cup or plate .... Eucoilidae
   Scutellum without raised cup or plate, smooth 6
6. Radius does not reach forewing margin, wing with a well-formed medial cell and at most one other cell; antennae filiform, with more than 12 segments; scape and pedicel subequal; mesopleuron bulging; trochanters two segmented .... Braconidae: Aphidiinae
   Radius absent, wing with a well-formed medial cell only; antennae geniculate, 12 segmented; scape about two times as long as pedicel .... Bethylidae: *Cephalonomia tarsalis* (Ashmead)
7. Minute, less than 0.5 mm in length; short, stout, dark brown; antennae seven segmented; scutellum transverse, straplike; metasoma sessile; females .... Scelionidae: Scelioninae: *Baeus* Haliday
   2.0 mm in length, yellowish brown; antlike, but without nodes on the petiole; antennae 12 segmented; lateral margin of propodeum with spiracle on a raised dentiform process and a larger dentiform process posterolaterally; scutum and scutellum fused; scutellum in contact with propodeum; females .... *Cephalonomia gallicola* (Ashmead)
8. Pronotum triangular laterally and reaching the tegula 9
   Pronotum not triangular laterally and does not reach the tegula, or specimen minute and this character difficult to see .... Chalcidoidea 15

9. Antennae inserted well above the mandibles, at a level midway along the length of the compound eyes, usually on a frontal shelf; first gastral tergite slightly overlapping base of the petiole; female with 12-segmented antennae and more or less abrupt club; male with 14-segmented antennae without club .... Diapriidae: Diapriinae
   Antennae inserted near the mandible bases, at or below the lower ocular line 10
10. Head prognathous; wing without closed cells, medial cell at most indicated by a pigmented streak; yellowish brown, males .... Bethylidae: *Cephalonomia gallicola* (Ashmead)
   Head hypognathous 11
11. Forewing with stigmal and usually postmarginal veins, stigmal vein straight; antennae usually with 11 or more segments........................ Scelionidae 13
   Forewing without postmarginal vein 12
12. Forewing with stigmal vein curved.. Ceraphronoidea 14
   Forewing without stigmal and postmarginal veins, often with a single stub vein or veinless; antennae with ten or fewer segments .... Platygasteridae
13. Metasomal tergites wide laterally and loosely attached; second metasomal tergite the longest .... Telenominae
   Metasomal tergites closely attached forming an impressed submarginal ridge; second metasomal tergite not the longest; lateral ocelli usually closer to inner orbits than to median ocellus; marginal vein shorter than stigmal vein .... Scelioninae
14. Forewing with large stigma .... Megaspilidae
   Forewing without stigma, marginal vein with a single break .... Ceraphronidae
15. Minute, usually less than 1 mm; head dorsally with an H-shaped pattern of darkly sclerotized carinae; antennal sockets widely separated, nearer to the compound eyes than to each other; hindwing stalked basally .... Mymaridae 16
   Head without H-shaped carinae; hindwing broader basally 19
16. Tarsi five segmented 17
   Tarsi four segmented 18
17. Abdomen sessile; forewing narrow, oarlike; marginal vein short, less than one half the length of the wing, venation with one large seta and several small setae; hind margin excised beneath venation; mandibles sickle shaped, overlapping; female funicle with five segments .... *Alaptus* Westwood
   Abdomen petiolate; forewing broadening beyond venation; hind margin not excised; mandibles palmate with four teeth; female funicle with eight segments .... *Gonatocerus* Nees
18. Abdomen sessile; forewing narrow, oarlike, recurved, broadened only slightly at the apex; marginal vein short, less than one-half the length of the wing; hypopygium present in females; mandibles do not overlap, three toothed .... *Erythmelus* Enock
   Abdomen petiolate; forewing stalked basally for two-thirds of its length; hindwing hairlike, membrane absent; funicle segment 2 elongate, about ten times as long as wide; scape elongate, at least five times as long as wide, exceeds vertex, constricted medially .... *Mymar* Curtis
19. 1.0–1.5 mm in length; head prognathous, dorsoventrally flattened with antennae inserted in a broad median channel, scape triangular in lateral view; mandibles ventrally with two rasplike appendages; first and third femora swollen and longer than tibiae; females .... Agaonidae: Agaoninae
   Head not prognathous, not dorsoventrally flattened; other characteristics absent..... 20
20. 0.5–0.8 mm in length; tarsi three segmented............................ Trichogrammatidae 21
   Tarsi four or five segmented 24

21. Forewing narrow, oarlike with long marginal fringe resembling Mymaridae; antennae with one funicular segment; females .... *Megaphragma* Timberlake
    Forewing broad, setae in distinct rows 22
22. Antennae with two funicular segments, club unsegmented, one annellus; females .... *Trichogramma* Westwood
    Antennae without funicular segments 23
23. Club unsegmented, with long bristles; males .... *Trichogramma* Westwood
    Club five segmented, fusiform, terminal segment with strong upcurved subapical spine; females .... *Lathromeroidea* Girault
24. Hind femur enlarged, ventral margin finely serrate; head produced anteriorly into two sharp-edged hornlike projections; gaster petiolate, first tergum longitudinally striated; head, thorax, propodeum strongly punctate; dark black .... Chalcididae: Dirhininae
    Head without horns, other characteristics different 25
25. Pronotum rectangular in dorsal view, parallel sided; thorax black with fine reticulation; female antennae with five funicular segments, one annelus and a three-segmented fusiform club; first funicular segment obconical, without linear sensilla; male antenna with four funicular segments, one annelus and a three-segmented fusiform club; segments with long erect setae arranged in whorls .... Eurytomidae: *Systole albipennis* Walker
    Pronotum not rectangular in dorsal view; apex narrower than base 26
26. Mesopleuron oblong, convex, smooth, without a groove for reception of the femur; axillae triangular, transverse, nearly meeting medially; marginal vein very short, at most as long as the stigmal vein .... Encyrtidae
    Mesopleuron concave, impressed with a groove for the reception of the femur 27
27. Axillae strongly and angularly advanced forward of the trans-scutal suture along the hind margin of the notauli; scutellum with a pair of dorsal longitudinal grooves and two pairs of posterolateral setae; forewing usually without postmarginal vein .... Eulophidae: Tetrastichinae
    Axillae not advanced forward of the scutellum; other characteristics absent 28
28. Head prognathous; antennae inserted low on face, below lower ocular line, further than one torulus diameter apart; prothorax bell shaped; abdomen petiolate 29
    Head hypognathous; antennae inserted higher on face, at or above lower ocular line, close together, less than one torulus diameter apart; prothorax narrow, transverse; notauli incomplete; abdomen subsessile; females with hind margin of first metasomal tergite distinctly subangularly produced; female antennae with three annelli, five funicular segments and a three-segmented fusiform club; male antennae with two annelli, six funicular segments and a three-segmented fusiform club .... Pteromalidae: *Anisopteromalus calandrae* (Howard)
29. Head with interantennal projection; body mostly dark glossy brown, with antennal bases, legs and propodeum translucent yellowish brown; notauli and axillae distinctly grooved; forewing with a tuft of hairs at the juncture of the submarginal and marginal veins, infuscated below stigmal vein; antennae inserted in broad median channel, five to six funicular segments; smaller wingless forms occur .... *Choetospila* Westwood
    Head without interantennal projection; head and mesosoma mostly glossy dark black with a few large punctures; notauli and axillae punctate; abdominal terga with bluish sheen and finely reticulated; forewing without a tuft of hairs, hyaline; antennae with seven funicular segments .... *Spalangia* Latreille

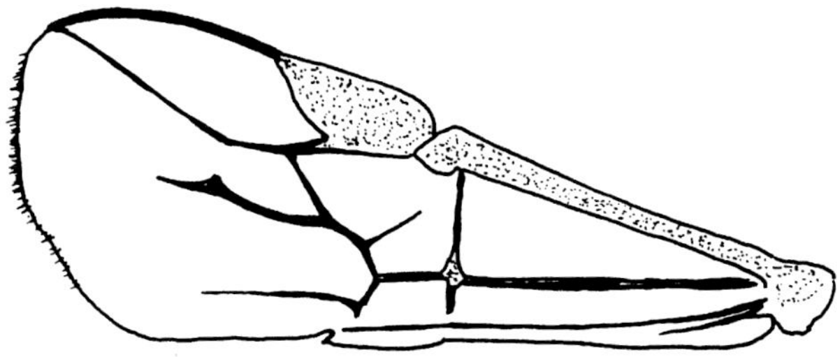

FIGURE 8. Parasitoid forewing, *Habrobracon* sp.

C. DISCUSSION OF FAMILIES

1. Megaspilidae

Megaspilidae are parasitoids of various Diptera, including tachinids, syrphids and agromyzids, and are hyperparasitic on parasitic Hymenoptera.[17]

2. Ceraphronidae

Ceraphronidae are parasitoids of various insect orders, including Diptera, Hymenoptera, Hemiptera and Neuroptera.[17]

3. Braconidae

Habrobracon spp. (Figure 8) are external larval parasitoids of mostly Lepidoptera and Coleoptera, some Diptera and Symphyta. The FDA has recorded *Habrobracon* sp. from India, Mexico and Thailand. All aphidiinae are aphid parasitoids.[18]

4. Mymaridae

Mymaridae are parasitoids primarily of the eggs of Hemiptera and Homoptera. *Mymar* spp. are parasitoids of delphacids; *Alaptus* spp. of psocids; *Erythmelus* spp. of cicadellids, tingids and mirids; and *Gonatocerus* spp. of cicadellids and membracids.[19] The FDA has recorded *Mymar* sp. from Taiwan and China; *Alaptus* sp. from dried herring from the Phillippines and dried fish from Malaysia; *Erythmelus* sp. from China; and *Gonatocerus* sp. from Taiwan.

5. Trichogrammatidae

Trichogrammatids are parasitoids of the eggs of insects. Hosts of *Trichogramma* spp. include economically important noctuids, tortricids and pyralids.[17] It is reared commercially and used widely in biological control. *Lathromeroidea* spp. are known from the eggs of dragonflies and waterstriders. *Megaphragma* spp. are known from the eggs of thrips (D.L. Vincent, pers. commun.). The FDA has recorded *Lathromeroidea* sp. from the Philippines and *Megaphragma* sp. from the Philippines.

6. Eulophidae

Eulophidae is a large and biologically diverse group, including parasitoids of mites and ichneumonids, a few families of Homoptera and Thysanoptera, numerous families of Coleoptera, Lepidoptera, Diptera, and Hymenoptera; spider and mantid eggs, and a few are phytophagous.[20]

7. Encyrtidae

Encyrtidae are parasitoids of various insects and arachnids, mostly Homoptera, but also the eggs or larvae of Coleoptera, synanthropic Diptera, Lepidoptera and Neuroptera, and are hyperparasitic on Hymenoptera, eggs of cockroaches, eggs, nymphs and adults of Hemiptera, nymphs of ixodids and eggs of spiders.[17,20]

8. Agaonidae

All agaonids develop within the figs of various species of *Ficus*. Association with food products other than figs is incidental (E.E. Grissell, pers. commun.).

9. Pteromalidae

Anisopteromalus calandrae (Howard) is a parasitoid of beetles associated with stored grain, especially *Stegobium paniceum* (L.), *Sitophilus granarius* (L.) and *Sitophilus oryzae* (L.).[21] It is the subject of numerous biological studies and is commercially reared for use in biological control. The FDA has recorded this species from India, China and Thailand. Some species of *Spalangia* are parasitoids of synanthropic flies, including *Stomoxys calcitrans*, *Musca domestica*, *Sarcophaga* sp., *Calliphora* sp., *Pycnosoma* sp. and *Chrysomya* sp., and have been used in biological control.[21] They are external pupal parasites.[22] The FDA has recorded these species from Pakistan. *Choetospila* spp. are parasitoids of beetles associated with stored grain or anobiids.[21]

10. Eurytomidae

Systole albipennis Walker is phytophagous in the seeds of Daucaceae (Umbelliferae), including fennel and coriander.[21] The FDA has recorded *S. albipennis* from India and Syria.

11. Chalcididae

Dirhininae are pupal parasitoids of various Diptera, especially of synanthropic calliphorids, sarcophagids, muscids, certain tephritids and piophilids.[21]

12. Eucoilidae

Eucoilids are pupal parasitoids of cyclorrhaphous Diptera, including drosophilids, sarcophagids, sphaerocerids and muscids.[17]

13. Diapriidae

Diapriinae are pupal parasitoids mainly of Diptera, including stratiomyids, syrphids, tephritids, calliphorids, sarcophagids and muscids.[17]

14. Scelionidae

Telenominae are parasitoids primarily of the eggs of Lepidoptera, Heteroptera, Diptera and Neuroptera. Scelioninae are parasitoids of various families of Orthoptera, Hemiptera, Embioptera and spiders. *Baeus* spp. are strictly parasites of the eggs of various spiders, including Lycosiidae.[23] The FDA has recorded these species from dried mushroom from Korea.

15. Platygasteridae

Platygasteridae are parasitoids of the eggs or larvae of cecidomyiids, chrysomelids, Fulgoroidea, pseudococcids and aleyrodids.[17]

16. Bethylidae

Cephalonomia tarsalis (Ashmead) (Figure 9) is a parasitoid of *Oryzaephilus surinamensis* (L.) and *Sitophilus oryzae* (L.). *Cephalonomia gallicola* (Ashmead) is a parasitoid of *Stegobium paniceum* (L.), *Lasioderma serricorne* (F.), *Ptinus fur* L., *P. tectus* Bojeld, *Niptus hololeuca* (Fald.) (Ptinidae) and galls of *Andricus foliatus* (Cynipidae).[24] The FDA has recorded *C. tarsalis* from Mexico and Thailand and *C. gallicola* from Thailand, Korea, China, India and Hong Kong.

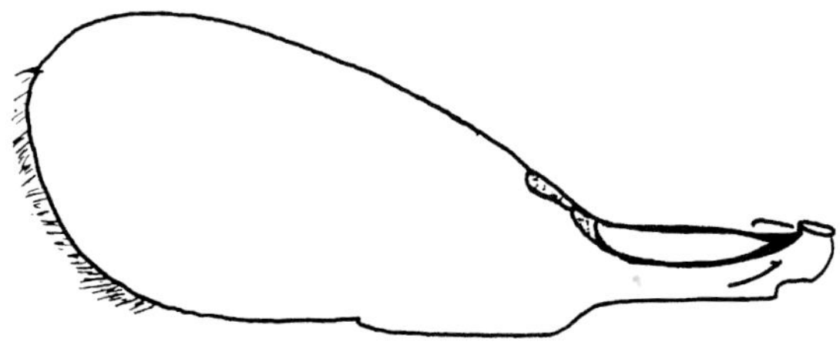

FIGURE 9. Parasitoid forewing, *Cephalonomia* sp.

V. OTHER HYMENOPTERANS

Other hymenopterans may contaminate food products. Honey, for example, is sometimes contaminated with honeybee fragments due to lapses in filtration, centrifugation or skimming during processing. A common fragment in honey is the bee hair, which is distinctively plumose.[25] Whole worker bees are occasionally found in honey, most likely due to poor windowscreening of honey houses to exclude bee workers that are attracted to the raw honey to reclaim it for the hive.

In the tropics, stingless bees, *Trigona* spp. (Apidae), produce a black, tarry honey that is prized as food. These small (3–4 mm) black bees resemble honeybees in general morphology, complete with pollen baskets on the hind legs, except they lack a sting. They often build their nests in the hollow spaces in walls of buildings and become a nuisance.[26] Whole adults, larvae and pupae, as well as their fragments, may be encountered in products that use honey from *Trigona* spp. as a sweetener.

A more serious nuisance are the yellowjackets, *Vespula* spp. (Vespidae), which are attracted to protein food sources including processed foods and garbage.[5,10] Among their foraging activities, these wasps prey on filth flies.[27] The cosmopolitan *V. germanica* (F.) visits garbage and excrement, where it captures adult and larval *Chrysomya megacephala* (F.) (Diptera: Calliphoridae) and other filth flies to provision its nest.[28] Because they possess a venomous sting, *Vespula* spp. are usually swatted when they are noticed, accounting for occasional fragment contamination from wasps that are in a food production area when they meet their untimely demise.

REFERENCES

1. Hölldobler, B. and Wilson, E.O., *The Ants*, Harvard Univ. Press, Cambridge, MA, 1-732, 1990.
2. Barber, M.A., Cockroaches and ants as carriers of the vibrios of Asiatic cholera, *Philos. J. Sci.*, B9(1), 1-4, 1914.
3. Beatson, S.H., Pharaoh's ants as pathogen vectors in hospitals, *Lancet*, 1(7747), 425-427, 1972.
4. Edwards, J.P., The biology, importance and control of pharaoh's ant (*Monomorium pharaonis* L.) infestations in hospitals, *Nursing Times*, Feb. 26, 2-4, 1981.
5. Harwood, R.F. and James, M.T., *Entomology in Human and Animal Health*, 7th ed., Macmillan, NY, 1-548, 1979.
6. Griffitts, S.D., Ants as probable agents in the spread of *Shigella* infections, *Science*, 96(2490), 271-272, 1942.
7. Singh, S., Sethi, M.S., and Sharma, V.D., The occurrence of salmonellae in rodent, shrew, cockroach, and ant, *Int. J. Zoonoses*, 7(1), 58-61, 1980.
8. Gorham, J.R., Ed., *Insect and Mite Pests in Food*, Vol. 2, U.S. Dept. Agric., Agric. Handbook No. 655, 311-767, 1991.
9. Metcalf, C.L., Flint, W.P., and Metcalf, R.L., *Destructive and Useful Insects*, McGraw-Hill, NY, 1-1087, 1962.
10. Hogue, C.L., *The Insects of the Los Angeles Basin*, Natur. Hist. Mus. Los Angeles Co. Sci. Ser. No. 27, 1-173, 1974.

11. Papp, C.S. and Swan, L.A., *Ouch! A Guide to Biting and Stinging Insects and Other Arthropods*, Okumura Biol. Inst., Sacramento, CA, 1-176, 1981.
12. Goetsch, W., *The Ants*, Univ. Mich. Press, Ann Arbor, MI, 1-173, 1957.
13. Essig, E.O., *College Entomology*, Macmillan, NY, 1-900, 1942.
14. Wheeler, W.M., *Ants, Their Structure, Development, and Behavior*, Columbia Univ. Press, NY, 1-663, 1910.
15. Beatson, S.H., Pharaoh's ants, *Health Soc. Serv. J.*, 83(4345), 14, 1973.
16. Eichler, W., Aspects of the injurious effects of insects on human health, with special regard to pharaoh's ant, *Wiss. Z. Humboldt Univ. Berlin (Math. Naturwiss. Reihe),* 13(1), 113-118, 1964 (in German, English summary).
17. Naumann, I.D. and Musner, L., Parasitic wasps of the proctotrupoid complex: a new family from Australia and a key to world families *(Hymenoptera: Proctotrupoidea Sensu lato)*, *Aust. J. Zool.*, 33, 761-783, 1985.
18. Marsh, P.M., Shaw, S.R., and Wharton, R.A., An identification manual for the North American genera of the family Braconidae (Hymenoptera), *Memoirs Entomol. Soc. Wash.* 13, 1-98, 1987.
19. Schauff, M.E., The Holarctic genera of Mymaridae (Hymenoptera: Chalcidoidea), *Memoirs Entomol. Soc. Wash.* 12, 1-67, 1984.
20. Grissell, E.E. and Schauff, M.E., *A Handbook of the Families of Nearctic Chalcidoidea*, Entomol. Soc. Wash., Washington, DC, 1-85, 1990.
21. Boucek, Z., *Australian Chalcidoidea (Hymenoptera): A Biosystematic Revision of Genera of Fourteen Families: With a Reclassification of Species*, C.A.B. International, Wallingford, UK, 1-832, 1988.
22. Clausen, C.P., Ed., *Introduced Parasites and Predators of Arthropod Pests and Weeds*, U.S. Dept. Agric., Agric. Handbook No. 480, 1-545, 1978.
23. Masner, L., Revisionary notes and keys to world genera of Scelionidae (Hymenoptera: Proctotrupoidea), *Memoirs Entomol. Soc. Canada*, 97, 1-87, 1976.
24. Evans, H.E., The Bethylidae of America north of Mexico, *Memoirs Am. Entomol. Inst.*, 27, 1-332, 1978.
25. Comstock, J.H., *An Introduction to Entomology*, 9th ed., Comstock Publ., Ithaca, NY, 1-990, 1940.
26. Curran, C.H., *Insects of the Pacific World*, Macmillan, NY, 1-317, 1946.
27. Akre, R.D., Greene, A., MacDonald, J.F., Landolt, P.J., and Davis, H.G., *The Yellowjackets of America North of Mexico*, U.S. Dept. Agric., Agric. Handbook No. 552, 1-102, 1981.
28. Bingham, Col. C.T., *Fauna of British India, Ceylon and Burma Hymenoptera—Vol. II. Ants and Cuckoo-Wasps*, Taylor & Francis, UK, 1903.
29. Olsen, A.R., Sidebottom, T.H., and Bennett, S.G., The Oriental latrine fly, *Chrysomya megacephala* (Fabricius, 1794) (Diptera: Calliphoridae), as an invading blow fly of public health importance, *Bull. Soc. Vector Ecol.*, 18, 133-146, 1993.

Chapter 9

MITES (ARACHNIDA: ACARINA)

Alan R. Olsen and Richard W. Potter

CONTENTS

I. Introduction 123

II. Acaroid Mites 124
- A. General 125
- B. Morphology 125
- C. Identification 126
- D. Discussion of Acaroid Families 128
  1. Acaridae (Grain and Mold Mites) 128
  2. Suidasiidae (Scaly Grain Mites) 129
  3. Lardoglyphidae (Driedfish Mites) 129
  4. Carpoglyphidae (Driedfruit Mites) 129

III. Other Common Families of Stored-Product Mites 129
- A. House Dust Mites, House Mites and Slime Mites 129
  1. Pyroglyphidae (House Dust Mites) 129
  2. Glycyphagidae (House Mites) 130
  3. Chortoglyphidae (Chortoglyphid Mites) 131
  4. Histiostomatidae (Slime Mites) 131
- B. Other Astigmatid Families 132

IV. Miscellaneous Food-Contaminating Mites 132

V. Additional Species 132

References 133

I. INTRODUCTION

Mites are small to minute arthropods of the class Arachnida, subclass Acari. They are characterized by their small size (1/2–3 mm), four pairs of walking legs and, in most cases, a complete lack of body segmentation. The mite body lacks distinct head, thorax and abdomen regions (insects) or cephalothorax and abdomen regions (spiders). The body also lacks segmented somites. Although diminutive in size, the Acari are not insignificant, comprising one of the larger groups of arachnids, with more than 20,000 species.[1] Approximately 100 of these species, distributed among 27 families, may occur in food products that have been exposed to insanitary conditions.[2] Table 1 lists the mite families that contain species associated with stored food products and provides examples of the commonest species of each family. Analytical keys are available for the identification of mites from these families.[1-3] Experience has shown that about 20 species of mites account for 70% of the incidents of adulteration of food products with mites.[4] This latter, compact group of pest species commands most of the attention of microanalytical entomologists and of this chapter.

TABLE 1
Examples of Food-Contaminating Mites

Family	Example	Common Name
Dermanyssidae	*Androlaelaps casalis* (Berlese)	
	Dermanyssus gallinae (De Geer)	Chicken mite
Ascidae	*Blattisocius tarsalis* (Berlese)	
Ameroseiidae	*Kleemania plumosus* (Oudemans)	
Phytoseiidae	*Amblyseius* spp.	
Digamasellidae	*Digamasellus* spp.	
Parasitidae	*Eugamasus butleri* Hughes	
Macrochelidae	*Macrocheles muscaedomesticae* (Scopoli)	
Uropodidae	*Leiodinychus krameri* (G. & R. Canestrini)	Tortoise mite
Cheyletidae	*Cheyletus eruditus* (Schrank)	
Bdellidae	*Spinibdella* spp.	
Cunaxidae	*Cunaxa setirostris* (Hermann)	
Tydeidae	*Tydeus interruptus* Thor	
Tetranychidae	*Bryobia praetiosa* Koch	Clover mite
Pygmephoridae	*Pygmephorus* spp.	
Pyemotidae	*Pyemotes tritici* (Lagreze-Fossat & Montane)	Straw itch mite
Tarsonemidae	*Tarsonemus granarius* Lindquist	Granary mite
Histiostomatidae	*Histiostoma* spp.	Slime mite
Carpoglyphidae	*Carpoglyphus lactis* (L.)	Driedfruit mite
Winterschmidtiidae		
Echimyopodidae	*Blomia freemani* Hughes	
Glycyphagidae	*Glycyphagus domesticus* (De Geer)	House mite
Aeroglyphidae	*Aeroglyphus robustus* (Banks)	Warty grain mite
Chortoglyphidae	*Chortoglyphus arcuatus* (Troupeau)	
Acaridae	*Acarus siro* L.	Grain mite
	Tyrophagus putrescentiae (Schrank)	Mold mite
Lardoglyphidae	*Lardoglyphus konoi* (Sasa and Asanuma)	Driedfish mite
Suidasiidae	*Suidasia pontifica* (Oudemans)	Scaly grain mite
Pyroglyphidae	*Dermatophagoides pteronyssinus* (Trouessart)	American house dust mite

There are three categories of insanitation that are the main sources of adulteration of food with stored-product mites. The primary source is infested ingredients that are not adequately inspected prior to admission into a food-handling establishment. Most infestations of stored-product mites are probably introduced into an establishment on infested incoming goods from which they spread to other products. Another source of adulteration is poor housekeeping that allows dust, dirt, static material and product residue to accumulate and become reservoirs of mite populations that can emerge unpredictably to infest products. Finally, gross neglect of sanitation can lead to the uncontrolled proliferation of mite pests and their spread to equipment and product throughout a facility. Some secondary sources of mite adulteration include phoresy by mites on stored-product insects, birds, bats and rodents; infested (often reused) product containers; shedding of ectoparasitic mites by vertebrate pests; and excrement.[2]

The systematics of the Acari above the taxonomic family level is not fully understood. There are various hypotheses for the higher classification of the Acari, each with its own group of supporters.[5] Discussion of the higher taxonomy of mites is not necessary for the purposes of this book because the mite families that contain stored-product species are each easily differentiated from other families by the unique combinations of morphological characteristics described in this chapter.[6]

II. ACAROID MITES

Acaroid mites belong to a larger group known as the astigmatid mites, recognized by their lack of stigmata or external breathing pores. Most other mites possess a pair of pores resembling spiracles that are located either at the base of the chelicerae (prostigmatid mites) or between the bases of one or another pairs of legs (mesostigmatid mites). In one group (cryptostigmatid mites), the stigmata are difficult to locate. The cryptostigmatids, however, are usually moderately to heavily sclerotized making them distinguishable from the soft-bodied astigmatids. The taxonomic status of this grouping of mites is often a subject of debate. Observing the configuration of stigmata (or lack of stigmata) retains a utilitarian value for recognizing broad groupings of mites regardless of the taxonomic status of the groups.

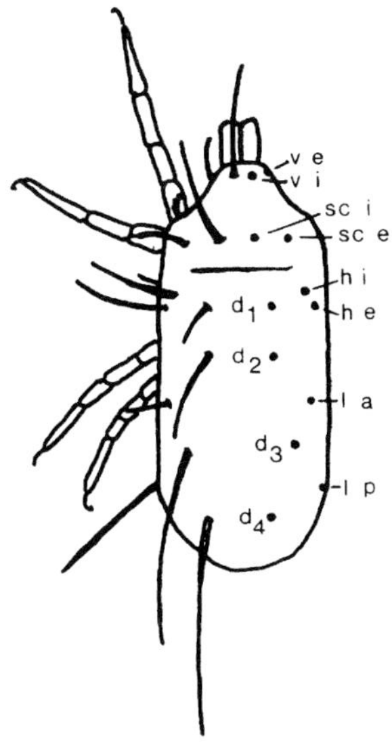

FIGURE 1. Mold mite, *Tyrophagus putrescentiae* (Acaridae), showing typical dorsal setation of acaroid mites. See Table 2 for explanation of seta designations.

A. GENERAL

Acaroid mites are mites of the related families Acaridae, Lardoglyphidae, Suidasiidae and Glycacaridae. The family Carpoglyphidae is included in this discussion of acaroid mites because it shares many morphological recognition characteristics with the other families even though it is not closely related to them. Glycacarids do not occur in food products. The other four families are those most likely to be encountered during the analysis of a food product. Many of the food-contaminating acaroid mites are also nidicolous, occurring naturally in the nests of mammals, birds and bees, but some species are known only from stored-products.(7)

B. MORPHOLOGY

At low magnifications (10–30×), as used for examining filth extraction filter papers, acaroid mites are recognized by a distinctive combination of morphological characteristics that differentiate them from other mites. The adult body is minute (<1 mm) and ovate, with the posterior margin entire and broadly rounded (Figure 1). The natural body color is pearly white. The integument is smooth to finely sculptured but lacking pronounced papillae, protuberances or other projections. Body setation (Figure 1) is sparse, with 16 or fewer pairs of dorsal and lateral setae combined. All setae, including leg setae, are smooth to slightly pectinate and generally short. A pair of opisthonotal glands are located dorsolaterally on the body, opening through an external pore behind the posterior pair of legs. The opisthonotal glands are filled with a colorless liquid that often becomes discolored shortly after death, darkening to appear as a pair of large brown spots located just behind the posterior legs.

At higher magnifications (100× or more), adult acaroid mites exhibit six pairs of distinctive appendages. There are four pairs of walking legs, each consisting of five articulated segments: the trochanter, femur, genu, tibia and tarsus (Figure 2). The tarsus lacks secondary segmentation. The coxae of acaroid mites are fused to the body ventrally and are not articulated segments. The distal tarsus terminates in a fleshy pretarsus, armed with a single claw. Legs I–IV each bear a separate, distinctive number and arrangement of setae. Of special taxonomic importance are the setae of the tarsus of leg I (Figure 3). By convention, the tarsal setae are designated with letter codes. In addition, the legs bear solenidia (sing. solenidion) that resemble setae except the internal core of a solenidion is occupied by protoplasm. The solenidia are designated with Greek alphabetic characters.

The remaining appendages are involved with feeding. They are the anterior chelicerae (Figure 4) followed by a pair of minute ventral two-segmented pedipalps. Together the chelicerae, pedipalps and internal mouthparts are called the gnathosoma. The chelicera of

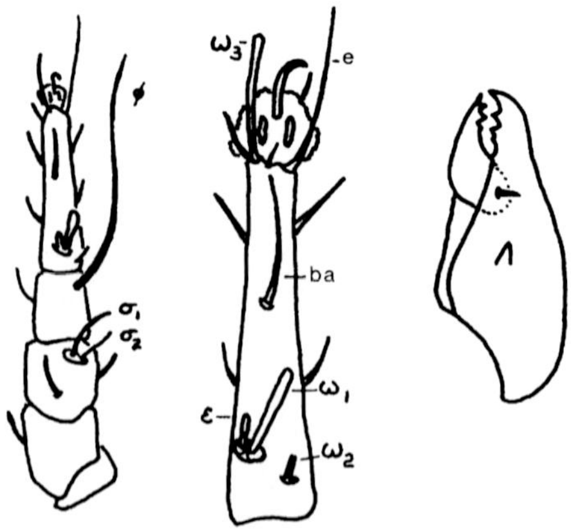

FIGURE 2. (Left) typical leg I of acaroid mites showing solenidia.
FIGURE 3. (Middle) typical tarsus I of acaroid mites showing solenidia and selected setae.
FIGURE 4. (Right) typical acaroid chelicera.

acaroid mites is a robust, pincerlike (chelate) appendage resembling a lobster pincer. The paired chelicerae project forward from the anterior margin of the body and, in true acaroid mites, each chelicera bears a distinctive mandibular (paraxial) spine on the interior surface near the articulation point of the moveable digit (Figure 4). The tiny ventral pedipalps are visible only at high magnifications. Their fused bases form the lower platform, or hypostome, of the gnathosoma. The pedipalps are cylindrical, segmented appendages that resemble legs except for their small size, segmentation reduced to two or three segments, and a lack of terminal claws.

Genitalia and secondary sexual characteristics of acaroid mites are distinctive. The female and male genitalia are located internally beneath a pair of genital flaps located ventrally between legs III and IV. The longitudinal slit between the flaps is flanked by two pairs of blunt, tubular sensory organs, sometimes called acetabulae. Females have a single central genital pore located on the posterior margin of the body that connects via a tubular duct to internal structures including a seminal receptacle and paired openings to the oviducts. Although largely internal, the genitalia are usually sclerotized and can be examined at high magnifications on specimens that have been cleared and mounted on microscope slides (see chapter 14). Genitalia are often visible on unmounted specimens at low magnifications, but their fine structure cannot be resolved without clearing and examination under a compound or phase-contrast microscope. Adult males of the families Acaridae and Lardoglyphidae have a pair of distinctive, round suckers flanking the anus. Some male Suidasiidae have rudimentary, padlike anal suckers.

C. IDENTIFICATION

The dorsal setation of acaroid mites provides a convenient means of identifying them, usually to the genus level. Figure 1 illustrates a typical adult configuration of the dorsal body setae. The designations of each setal pair are explained in Table 2. The configuration represented by the combined presence, absence and relative positions of these setae is diagnostic. The setae of leg I of acaroid mites are also important for identification. Figures 3 and 4 illustrate the setae and solenidia of the typical acaroid leg. The illustrations are of leg I. In acaroid mites, legs II–IV exhibit progressively reduced setation. The following analytical key differentiates the stored-product genera of acaroid mites, including the carpoglyphids, using the characteristics of the body and legs described above.

TABLE 2
Dorsal Body Setation of Adult Acaroid Mites

Setae	Designation	Remarks
External vertical	*v e*	*v e* rudimentary or absent
Internal vertical	*v i*	in some genera
External scapular	*sc e*	All four scapulars
Internal scapular	*sc i*	aligned in a transverse
		row except *Rhypoglyphus*
External humeral	*h e*	Third humeral present but
Internal humeral	*h i*	not visible dorsally
First dorsal	d_1	One or more pairs may be
Second dorsal	d_2	absent in some genera
Third dorsal	d_3	
Fourth dorsal	d_4	
Anterior lateral	*l a*	*l a* is absent in some
Posterior lateral	*l p*	genera
External sacral	*sa e*	
Internal sacral	*sa i*	

Rapid Key for the Identification of the Genera of Common Food-Contaminating Adult Acaroid Mites

1. Dorsal surface of idiosoma finely wrinkled or covered with a scalelike pattern (Figure 3) (Suidasiidae) .... *Suidasia*
   Dorsal surface of idiosoma smooth 2
2. Setae *sc e* located along the lateral margin and distinctly anterior to *sc i* .... *Rhypoglyphus*
   Setae *sc e* and *sc i* arranged in a transverse row 3
3. Length of idiosomal seta *v e* (external vertical) subequal to the length of genu I (*v e* usually curved) 4
   v e distinctly shorter than genu I, or *v e* absent 8
4. Idiosomal seta *sc e* longer than *sc i* 5
   sc e shorter than *sc i* 6
5. On genu I, the ratio of the lengths of σ_1:σ_2 is approximately 1:1 or 2:1 (Figure 4); tarsal claws simple; male homomorphic .... *Aleuroglyphus*
   On genu I, σ_1 (internal apical solenidion) is one-third or less the length of σ_2 (external apical solenidion); male always heteromorphic; female with bifid claws on all legs (Lardoglyphidae) .... *Lardoglyphus*
6. Hysterosomal setae d_l (first dorsal) and *l a* (anterior lateral) about equal in length 7
   l a four to six times longer than d_l .... *Tyrolichus*
7. Tarsus I with a dorsal apical spine (seta *e*) .... *Tyroborus*
   Tarsus I with seta *e* setiform, not spinelike .... *Tyrophagus*
8. Dorsal hysterosomal setation reduced, with at least setae d_l and *l a* absent 9
   Dorsal hysterosomal setation complete 10
9. On tarsus I, dorsal seta *ba* spinelike; male with a posterior hysterosomal projection forming four semicircular festoons, bearing four large, transparent leaflike flanges, arranged like a fan .... *Histiogaster*
   Seta *ba* absent on tarsus I; male with a simple shelflike posterior hysterosomal projection (posterior margin of hysterosoma apparently entire) .... *Thyreophagus*
10. Seta *v e* arising along the anterior one-fourth of the lateral margin of the dorsal propodosomal shield 11

v e arising midway along the lateral margin of the propodosomal shield, or *v e* occasionally absent 12

11. Tarsus I with no dorsal spinelike apical seta (seta *e* setiform); male with front femur enlarged and beating a ventral conical process .... *Acarus*
    Tarsus I with dorsal apical seta *e* spinelike; front femur of male *not* as above .... *Mycetoglyphus*
12. On tarsi I and II, seta *ba* is enlarged to form a conical spine close to solenidion ω_1 .... *Rhizoglyphus*
    On tarsi I and II, *ba* is setiform, not spinelike, and occurs distad to ω_1 .... *Caloglyphus*

D. DISCUSSION OF ACAROID FAMILIES

Table 3 lists the most common food-contaminating acaroid mites along with selected other mites that are of regulatory significance in food products.

1. Acaridae (Grain and Mold Mites)

This family (Figure 1) accounts for the majority of stored-product mite pests. The white body is divided dorsally by a transverse furrow located between legs II and III. The body region anterior to the furrow bears a dorsal, lightly sclerotized shield covering most of the body surface area between the furrow and the anterior margin. The integument is generally smooth. Body setae are smooth or slightly pectinate. The claw on legs I–IV is embedded in a fleshy pretarsus and pivoted between two stout rods or condylophores. Tarsi I and II bear a basal solenidion that is rodlike with a rounded tip. Males have two dorsal suckers on each tarsus IV and a pair of anal suckers.

Members of 12 genera are frequently found in food products. They fall within two groups (subfamilies) that are differentiated by their usual feeding habits. Mites of the acarine genera *Acarus* L., *Aleuroglyphus* Zakhvatkin, *Rhypoglyphus* Potter and Olsen,* *Tyrolichus* Oudemans and *Tyrophagus* Oudemans thrive in relatively dry habitats and feed directly on stored food products, whereas the rhizoglyphine genera *Caloglyphus* Berlese, *Histiogaster* Berlese, *Michaelopus* Fain, *Mycetoglyphus* Oudemans, *Rhizoglyphus* Claparede, *Thyreophagus* Rodani and *Tyroborus* Oudemans prefer damp habitats and are thought to feed mainly on fungi, molds and dead insects in a product.[7,8] The mold mite, *Tyrophagus putrescentiae* (Schrank) is found in foods more often than any other mite. It can be recognized by the male genitalia with a short recurved aedeagus. The grain mite, *Acarus siro* L., is another common pest. The males have enlarged legs I with a prominent ventral spine on the genu.

TABLE 3
Common Genera of Food-Contaminating Astigmatid Mites

Family	Common Genera
Histiostomatidae	*Histiostoma*
Carpoglyphidae	*Carpoglyphus*
Echimyopodidae	*Blomia*
Glycyphagidae	*Glycyphagus*, *Gohieria*, *Lepidoglyphus*
Aeroglyphidae	*Aeroglyphus*
Chortoglyphidae	*Chortoglyphus*
Acaridae	*Acarus*, *Aleuroglyphus*, *Caloglyphus* *Histiogaster*, *Michaelopus*, *Mycetoglyphus* *Rhizoglyphus*, *Rhypoglyphus*, *Thyreophagus* *Tyroborus*, *Tyrolichus*, *Tyrophagus*
Lardoglyphidae	*Lardoglyphus*
Suidasiidae	*Suidasia*
Pyroglyphidae	*Dermatophagoides*, *Euroglyphus*, *Pyroglyphus*

* We consider unwarranted the synonym *Rhypoglyphus* Potter and Olsen 1987 = *Cerophagus* Oudemans 1905 (*sic*), proposed by Fain in 1988 on the basis of a single deutonymph associated with a population of *Rhypoglyphus* adults. Affinities of the deutonymph's morphology, especially the tarsal claws, with the chaetodactylids indicate that it is unrelated to the acarine adults. Consequently, we treat *Rhypoglyphus* as a valid genus with two described species, *R. indicus* (Potter and Olsen) 1987 and *R. trigona* (Fain and Heard 1987).

2. Suidasiidae (Scaly Grain Mites)

The scaly grain mites (Figure 5) are placed in the family Acaridae by some authorities. The stored-product species, *Suidasia pontifica* (Oudemans) and *S. nesbitti* Hughes, are recognized by sculpturing over their entire body surface that resembles minute scales (Figures 5 and 6). The known taxa of Suidasiidae are nidicoles, some obligate, in the nests of birds, wasps or bees.[6]

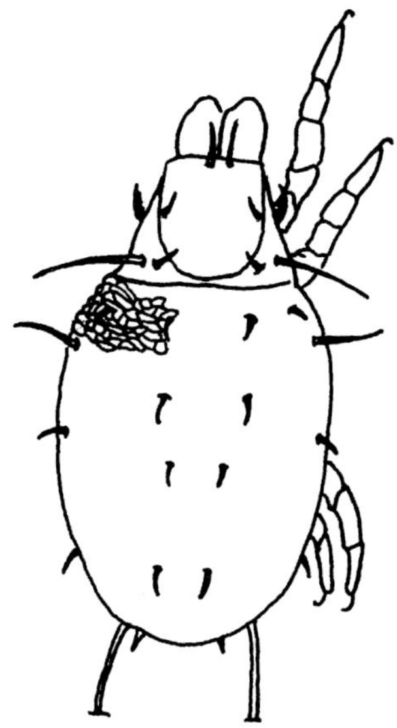

FIGURE 5. Adult scaly grain mite, *Suidasia* sp. (Suidasiidae).

3. Lardoglyphidae (Driedfish Mites)

This family has only one genus, *Lardoglyphus* (Figure 7). The driedfish mites are typical acaroid mites. They are distinguished by the obligate heteromorphic males the leg III of which is enlarged and modified to form an apical two-pronged fork that lacks a fleshy pretarsus and claw. Females have a bifid claw on legs I–IV. There are three stored-product species: *L. zacheri* Oudemans, *L. konoi* (Sasa and Asanuma) and *L. angelinae* Olsen. In *L. zacheri*, males also have bifid claws on legs I and II. Other species of *Lardoglyphus* that are not found in food products are nidicolous in bird and mammal nests. Immature driedfish mites (the deutonymph stage) are phoretic on stored-product beetles of the genera *Dermestes* (Dermestidae) and *Necrobia* (Cleridae).[9] Analytical keys are available for the differentiation of adult and immature *Lardoglyphus* spp.[10,11]

4. Carpoglyphidae (Driedfruit Mites)

The single genus of this family, *Carpoglyphus*, (Figure 8) is recognized by the distinctively interconnected ventral coxal fields. The apodemes of coxae I and II are joined to form a coxal–sternal skeleton that resembles the letter "v" on top of the letter "w," with the base of the "v" connected to the center peak of the "w" (Figure 8a). Unlike true acaroid mites, *Carpoglyphus* spp. lack a dorsal transverse furrow and sclerotized shield. The driedfruit mite, *C. lactis* (L.), has been found in numerous food products including rotten produce, fermented products and wine. The natural habitat of driedfruit mites is unknown.

III. OTHER COMMON FAMILIES OF STORED-PRODUCT MITES

A. HOUSE DUST MITES, HOUSE MITES AND SLIME MITES

In addition to the acaroid families discussed above, certain other astigmatid mite families contain species that may infest stored-products. Table 3 lists the astigmatid families and genera, including the acaroid mites discussed above, likely to be encountered by the microanalytical entomologist when examining food products for mite contamination.

1. Pyroglyphidae (House Dust Mites)

American house dust mites thrive in dust and other accumulations of static material. House dust mites are recognized by the unusual sculpturing of their body integument, the many fine wrinkles of which are arranged in patterns that resemble fingerprints. Other distinguishing characteristics include reduced dorsal body setation and minute tarsal claws embedded in a fleshy pretarsus lacking condylophores. A transverse furrow and dorsal shield are present. In some genera there is a second sclerotized shield posterior to the furrow. The female genital, opening between legs II and IV, is shaped like an inverted "u." Males have anal suckers, often on a sclerotized plate. House dust mites are a serious inhalant allergen hazard to sensitized individuals.[12]

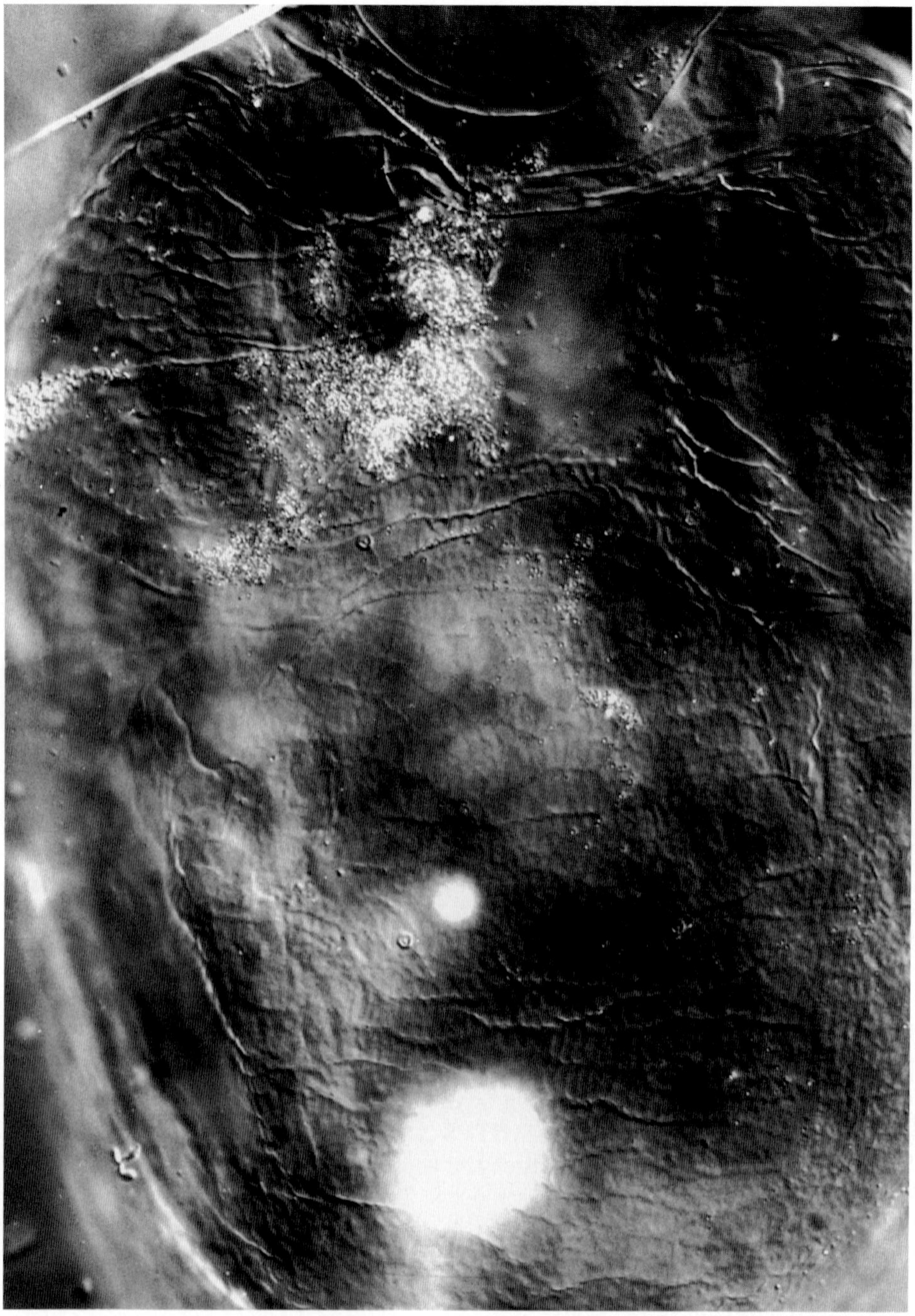

FIGURE 6. Photograph of dorsal integument of *Suidasia pontifica*, showing appearance of scale pattern at 200×.

2. Glycyphagidae (House Mites)

House mites lack a transverse furrow and dorsal shield. The tarsal claw rests at the end of an elongate-fleshy pretarsus and is joined to the tarsus by two thin rods. In many genera (except *Gohieria*) the body setae are very long and pectinate or else highly modified into leaflike or other distinctive shapes. Males lack anal suckers. The house mite, *Glycyphagus domesticus* (De Geer) is a common inhabitant of buildings and infests food products.

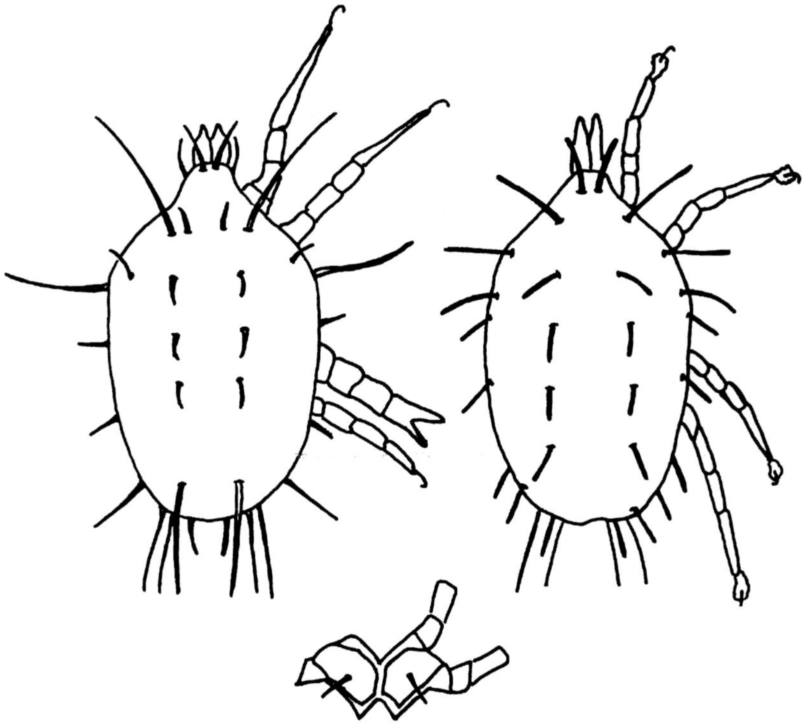

FIGURE 7. (Left) adult driedfish mite, *Lardoglyphus* sp. (Lardoglyphidae).
FIGURE 8. (Right) adult driedfruit mite, *Carpoglyphus lactis* (Carpoglyphidae).
FIGURE 8a. (Bottom) ventral coxal skeleton, *C. lactis*.

Lepidoglyphus spp. are distinguished by an elongate scale that originates at the base of the tarsus and envelopes the tarsus to the apical tip. Most glycyphagids, including the stored-product genera, are nidicolous.(2,8)

3. Chortoglyphidae (Chortoglyphid Mites)

Only one species of the family Chortoglyphidae, *Chortoglyphus arcuatus* (Troupeau), is known to infest stored-products. It is a free-living glycyphagoid mite characterized by distinctive male and female genitalia. The male aedeagus is long and located anteroventrally between coxae I and II. The female genital flaps, between coxae III and IV, are broad, sclerotized plates. The natural habitat of *C. arcuatus* is unknown but the other genera of this family are mostly nidicolous in rodent nests, where an immature stage (deutonymph) lives in hair follicles of its host.(6)

4. Histiostomatidae (Slime Mites)

Slime mites are easily distinguished from all others by their distinctive mouthparts. The chelicerae are reduced to a pair of slender, elongate, flexible digits with a serrated inner margin. The pedipalps are enlarged and flattened with two long, flexible setae at the apex of each pedipalp. Slime mites are filter feeders, using their modified mouthparts to extract microorganisms from their semiaqueous habitats in slime. Other distinctive characteristics include two pairs of chitinous rings located ventrally in coxal fields II and IV, and leg setae that are thickened into spines on all legs. The presence of slime mites in foods indicates insanitary conditions or the use of nonpotable water to process the food. *Histiostoma* spp. infest yeast cultures and semiaqueous habitats in and around buildings.

B. OTHER ASTIGMATID FAMILIES

The families Echimyopodidae, Aeroglyphidae and Winterschmidtiidae contain species that are occasionally encountered in stored-products. The winterschmidtiids resemble acaroid mites except the male lacks anal suckers. Echimyopodids resemble rotund glycyphagids. The stored-product species, *Blomia freemani* Hughes, lacks tarsal claws. The warty grain mite, *Aeroglyphus robustus* (Banks), is an aeroglyphid mite with densely pectinate setae, a dorsal shield and striated integument, it is a stored-product pest in Canada and other cool, temperate areas.

IV. MISCELLANEOUS FOOD-CONTAMINATING MITES

Another group of stored-product pests are the prostigmatid mites. These mites are primarily pests of grains and other raw agricultural products during postharvest storage prior to processing. The prostigmatids are characterized by a pair of breathing tubes, the peritreme, located at the anterior margin of the body. The stored-product species of the prostigmatid families Tydeidae, Tarsonemidae, Pyemotidae and Pygmephoridae are very tiny (0.2–0.5 mm) and frequently have one or another pair of walking legs highly modified. The tydeid mite, *Tydeus interruptus* Thor, has an ornately sculptured integument. An interesting species of the related family Tetranychidae, *Bryobia praetiosa* Koch, the clover mite, is a bright red mite that is known to invade buildings occasionally.

Predatory prostigmatids of the families Cheyletidae, Bdellidae and Cunaxidae are found in food products, where they prey on other mites and small insects. Each family has highly modified chelicerae and pedipalps that are diagnostic. The pedipalp of cheyletids is enlarged, robust and terminates in a large claw and several serrate setae. Bdellids have slender elongate pedipalps with long distal sensory setae at the end. The hypostome and chelicerae are also elongate. Cunaxids resemble bdellids except the pedipalps are more robust and the elongate chelicerae terminate in a short claw.

Gamasid mites are a diverse group which includes species that have been found in food products. They are large mites (>1 mm) with lateral stigmata (tracheal tubes) that open between legs III and IV. Gamasids have numerous dorsal, ventral and lateral sclerotized plates and shields and many stout setae on the body and legs. The families that are found in food products include Dermanyssidae, Ascidae, Ameroseiidae, Phytoseiidae, Digamasellidae, Parasitidae, Macrochelidae and Uropodidae. Most are also associated with the nests of mammals or birds and some are phoretic on insects. The macrochelid *Macrocheles muscaedomesticae* (Scopoli) is phoretic on the Oriental latrine fly, *Chrysomya megacephala* (F.) (Diptera: Calliphoridae), and other filth flies.[13]

V. ADDITIONAL SPECIES

There are many additional species of mites found in or around human habitations and occasionally in food products. A number of identification manuals are available that deal in detail with the mites that occur in buildings and therefore might come in contact with foods. These include the works of Smiley on mites of foods and of Mulla and Medina on domestic mites.[14,15] There are also a large number of agricultural pests that may account for incidental contamination of food products if these pests are not properly controlled in the field or not properly cleaned from raw agricultural products prior to processing for human consumption. For identifying these, the works of Krantz and of McDaniel are appropriate.[1,3]

REFERENCES

1. Krantz, G.W., *A Manual of Acarology*, 2nd ed., Oregon State Univ. Press, Corvallis, OR, 1- 509, 1972.
2. Hughes, A.M., *The Mites of Stored Food and Houses*, Ministr. Agric., Food and Fisheries Tech. Bull. No. 9, 2nd ed., HMSO, London, 1-400, 1976.
3. McDaniel, B., *How to Know the Mites and Ticks*, Wm. C. Brown, Dubuque, IA, 1-335, 1979.
4. Olsen, A.R., Food-contaminating mites from imported foods entering the United States through southern California, *Int. J. Acarol.*, 9, 189-193, 1983.
5. Smiley, R.L., The ordinal and subordinal names of mites with a list of the mite pests of stored products, *Proc. Internat. Working Conf. Stored-Prod. Entomol.*, 3, 37-43, 1983.
6. Woolley, T.A., *Acarology: Mites and Human Welfare*, Wiley Interscience, NY, 1-484, 1988.
7. Baker, E.W. and Wharton, G.W., *An Introduction to Acarology*, Macmillan, NY, 1-465, 1952.
8. O'Connor, B.M., Evolutionary ecology of astigmatid mites, *Annu. Rev. Entomol.*, 27, 385-409, 1982.
9. Hughes, A.M., The mites of the genus *Lardoglyphus* Oudemans, 1927 (= *Hoshikadania* Sasa & Asanuma, 1951), *Zool. Meded.*, 34, 271-285, 1956.
10. Olsen, A.R., A new pest of dried fish from the Orient, *Lardoglyphus angelinae* new sp. (Acarina: Acaridae), *J. Stored Prod. Res.*, 18, 181-188, 1982.
11. Baker, A.S., Two new species of *Lardoglyphus* Oudemans (Acari: Lardoglyphidae) found in the gut of human mummies, *J. Stored Prod. Res.*, 26, 139-147, 1990.
12. Wharton, G.W., House dust mites, *J. Med. Entomol.*, 12, 577-621, 1976.
13. Ho, T.M., Phoretic association between *Macrocheles muscaedomesticae* (Acari: Macrochelidae) and flies inhabiting poultry manure in peninsular Malaysia, *Exp. Appl. Acarol.*, 10, 61-68, 1990.
14. Smiley, R., Mites (Acari), in *Insect and Mite Pests in Food: An Illustrated Key*, Vol. 1, Gorham, J.R., Ed., U.S. Dept. Agric. Handbook 655, 4-43, 1987.
15. Mulla, M.S. and Medina, M.S., *Domestic Acari of Colombia*, COLCIENCIAS, Bogotá, 1-270, 1980.

Chapter 10

SPIDERS (ARACHNIDA: ARANEAE)

Alan R. Olsen, Warren E. Savary and Darrell Ubick

CONTENTS

I. Introduction 136

II. Synanthropic Spiders (Araneae) 136
- A. General 136
- B. Morphology 138
- C. Venomous Spiders 140
- D. Identification 141
- E. Discussion of Families 142
  1. Agelenidae (Grass and Funnel-Web Spiders) 142
  2. Amaurobiidae (White-Eyed Spiders) 142
  3. Araneidae (Orb Weavers) 142
  4. Clubionidae (Sac Spiders) 143
  5. Ctenidae (Wandering Spiders) 143
  6. Desidae 144
  7. Dictynidae 144
  8. Dysderidae 144
  9. Filistatidae 144
  10. Gnaphosidae (Hunting Spiders) 144
  11. Heteropodidae (Giant Crab Spiders) 145
  12. Linyphiidae (Sheet-Web Weavers) 145
  13. Mimetidae (Pirate Spiders) 145
  14. Nemesiidae 145
  15. Nesticidae (Cave Spiders) 145
  16. Oecobiidae 146
  17. Oonipidae (Minute Jumping Spiders) 146
  18. Philodromidae (Crab Spiders) 146
  19. Pholcidae (Cellar Spiders) 146
  20. Salticidae (Jumping Spiders) 146
  21. Scytodidae (Spitting Spiders) 147
  22. Selenopidae 147
  23. Sicariidae (Recluse Spiders and Brown Spiders) 147
  24. Tetragnathidae (Silk Spiders) 147
  25. Theraphosidae (Tarantulas) 148
  26. Theridiidae (Cobweb Spiders) 148
  27. Thomisidae (Crab Spiders) 148
  28. Uloboridae 148

References 149

I. INTRODUCTION

This book is primarily concerned with the insects and mites that infest food products and the facilities where foods are manufactured, processed or stored. The careful observer soon learns that other arthropods besides stored-product insects and mites may occur under insanitary conditions in a food-handling facility. These include a diversity of synanthropic species representing many of the major orders, and a few minor ones, of the phylum Arthropoda. Among the most frequently encountered are spiders (Araneae), book scorpions (Pseudoscorpiones) and house centipedes (Scutigeromorpha). Fragments of these arthropods are also found in processed, comminuted food products.[(1)] Less frequently, the microanalytical entomologist can expect to encounter certain species of scorpions (Scorpiones), sunspiders or windscorpions (Solifugae), tailless whipscorpions (Amblypygi) and harvestmen (Opiliones) in food products or buildings where food is handled or stored. Table 1 lists examples of synanthropic spiders. (The other arthropods are discussed in chapter 11.)

Spiders are carnivores that are attracted to the insects and mites that infest a building or food product rather than to the food products themselves. The observation that food-infesting predators do not cause economic damage by directly feeding on food products, but prey instead on the arthropods that consume food products, leads to the common misperception that the infestation of a food-handling area or storage facility by synanthropic predatory arthropods is somehow beneficial in controlling pests and, therefore, tolerable.[(2)] Predatory arthropods, spiders included, never occur indoors in sufficient numbers to be economically beneficial.[(3)] On closer examination of the issue, it is evident that infestations by synanthropic predatory arthropods occur only under the same grossly insanitary conditions that also allow infestations of buildings and products by the pests that are their intended prey. The concept of characterizing filth contaminants as avoidable versus unavoidable (see chapter 2) applies equally to predators as to the insects and mites that are primary consumers of a product.[(4)] Arthropod pests that thrive under insanitary conditions are avoidable, whether the pest causes direct feeding damage to a food product or occupies a different trophic level in the ecology of an infestation. The etiology of the contaminant determines avoidability, not the trophic level of the pest species. The ability of the analyst to differentiate avoidable synanthropic predators from incidental predators is as important as the ability to make the same differentiation for the insects and mites that are primary consumers of foods.

Within each taxonomic group discussed here, there is a limited number of species that live or are attracted indoors to prey on stored-product insects and synanthropic arthropods. The first step in identifying these synanthropic predators is to use the analytical keys for the identification of arthropod orders presented in standard textbooks. Then, use the analytical keys and technical descriptions provided in this book to determine the identity, and possible etiology, of a particular contaminant. For asynanthropic species, which are not covered here, refer to a general text on entomology or arachnology.[(5,6)] Van der Hammen provides full discussions of the morphologies of each of the groups of arachnids discussed here.[(7)]

II. SYNANTHROPIC SPIDERS (ARANEAE)

A. GENERAL

There are about 30,000 known species of spiders.[(8,9)] Of these, about 50 are synanthropic. The filth that synanthropic spiders produce includes webbing, exuviae and discarded remnants of their insect prey. Spider webbing often adheres tenaciously to products and product containers, where it is difficult to remove completely. Spiders molt from three to ten times during the first few months of life, and their cast skins may drop into anything that happens to be below their lair or web.[(10)] Even the spider's method of eating has the potential to produce

TABLE 1
Examples of Synanthropic Spiders

Family	Genus/Species	Common Name
Theraphosidae	*Aphonopelma* spp.	Tarantula
Nemesiidae	*Calisoga longitarsis* (Simon)	
Filistatidae	*Kukulcania hibernalis* (Hentz)	
Oecobiidae	*Oecobius annulipes* Lucas	
	O. parietalis (Hentz)	
Uloboridae	*Uloborus glomosus* (Walckenaer)	Featherlegged spider
Dictynidae	*Dictyna sublata* (Hentz)	
Amaurobiidae	*Amaurobius ferox* (Walckenaer)	
Sicariidae	*Loxosceles laeta* (Nicolet)	South American brown spider
	L. reclusa Gertsch & Mailuk	Brown recluse spider
Scytodidae	*Scytodes thoracica* (Latreille)	Tropical dust spider
Oonipidae	*Oonops domesticus* deDalmas	
	Orchestina saltitans Banks	
Dysderidae	*Dysdera crocata* C.L. Koch	
Pholcidae	*Holocnemis pluchei* (Scopoli)	A cellar spider
	Pholcus phalangioides (Fuesslin)	Longbodied cellar spider
	Physocyclus globosus (Taczanowski)	Tropical cellar spider
	Psilochorus simoni (Berland)	A cellar spider
	Spermophora senoculata (Duges)	Shortbodied cellar spider
Theridiidae	*Achaearanea tepidariorum* (C.L. Koch)	American house spider
	Latrodectus geometricus C.L. Koch	Brown widow
	L. hesperus Chamberlin & Ivie	Western black widow
	L. mactans (F.)	Black widow
	Steatoda bipunctata (L.)	
	S. grossa (C.L. Koch)	Boreal combfoot
	Nesticodes rufipes (Lucas)	A cobweb weaver
Nesticidae	*Eidmannella pallida* (Emerton)	A cave spider
	Nesticus cellulanus (Clerck)	A cave spider
Linyphiidae	*Lepthyphantes* spp.	A money spider
	Pityohyphantes costatus (Hentz)	Hammock spider
Araneidae	*Argiope* spp.	A garden spider
	Nuctenea umbratica (Clerck)	
Tetragnathidae	*Nephila clavipes* (L.)	Golden silk spider
	N. maculata (F.)	Giant wood spider
	Nephilengys cruentata (F.)	
Agelenidae	*Coras medicinalis* (Hentz)	A funnel weaver
	Tegenaria domestica (Clerck)	European house spider
	T. gigantea Chamberlin & Ivie	Hobo spider
	T. parietina (Fourcroy)	Cardinal spider
Desidae	*Badumna longiquus* L. Koch	
Mimetidae	*Ero furcata* (Villers)	A pirate spider
Gnaphosidae	*Herpyllus ecclesiasticus* Hentz	Parson spider
	Scotophaeus blackwalli (Thorell)	Mouse spider
Clubionidae	*Cheiracanthium inclusum* Hentz	A sac spider
	C. mildei C.L. Koch	A sac spider
Ctenidae	*Cupiennius salei* (Keyserling)	Banana spider
	Phoneutria fera (Keyserling)	A wandering spider
Heteropodidae	*Heteropoda venatoria* (L.)	Huntsman spider
Selenopidae	*Selenops* spp.	
Philodromidae	*Philodromus rufus* Walcknenaer	Ceiling spider
Thomisidae	*Xysticus* spp.	A crab spider
Salticidae	*Phidippus formosus* (Peckham)	Red jumping spider
	Salticus scenicus (Clerck)	Zebra spider

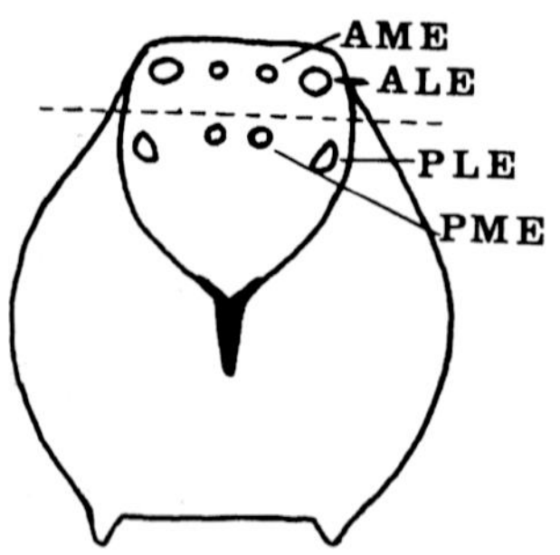

FIGURE 1. (Left) diagram of a spider carapace, showing anterior median eyes (AME), anterior lateral eyes (ALE), posterior median eyes (PME) and posterior lateral eyes (PLE).
FIGURE 2. (Middle) front view of face of a house spider (Pholcidae), showing chelate chelicerae.
FIGURE 3. (Right) example of a fanglike chelicera from a huntsman spider (Heteropodidae).

filth contamination because spiders do not consume the exoskeleton of their insect prey. The prey's exoskeleton is sometimes macerated during feeding, but ultimately, the spider discards the entire exoskeleton from the web either intact or as a wad of fragments. If spiders are allowed to occupy positions above or near food products or food preparation areas, they may discard insect carcasses into the product. Basic good sanitation measures such as sweeping and cleaning premises, screening doors and windows and inspecting incoming shipments of raw materials and supplies are effective in excluding spiders from a facility.[2]

B. MORPHOLOGY

Most spiders can be recognized without magnification by a combination of characteristics. The body is unsegmented and divided into two distinct regions, the cephalothorax (Figure 1) and abdomen, narrowly connected by a slender pedicel. The cephalothorax bears six pairs of appendages along with the eyes (Figure 2) and the mouth. The first pair of appendages are the chelicerae, which function as mouthparts, and the next pair are the pedipalps, which serve a variety of functions. Chelicerae are short appendages that terminate in a sharp fang or claw used to attack and envenom prey (Figure 3). Pedipalps resemble legs but are usually shorter and have six (not seven) segments. Next are four pairs of walking legs, each leg with seven segments (Figure 4). The leg segments from the body outward are coxa, trochanter, femur, patella, tibia, metatarsus and tarsus. The tarsus lacks secondary segmentation. The eyes are simple and located on the top cover, or carapace, of the cephalothorax. The abdomen may bear ventral external genitalia and tracheal spiracles. At least two pairs of posteroventral spinnerets are also found on the abdomen.[6,11]

Spiders and spider fragments can be precisely identified by comparing them with authentic specimens for characteristics that are visible at low magnifications (10–30×). Many of these characteristics are unique to spiders and render spider fragments easily recognizable when the fragments occur in food products. The morphology of minute structures on the legs, such as the various leg setae, claws and specialized hairs, is a weighty identification characteristic. Most spiders exhibit at least one microscopic leg characteristic, or a combination of characteristics, that is diagnostic at the family level (e.g., Figures 5 and 6). Leg autospasy (the breaking of an appendage by extrinsic force) is often a good clue to identity. The specific joint where autospasy occurs is characteristic for certain groups of spiders.[12] When autospasy occurs, it is often at the coxa-trochanter (C-T) joint but occasionally at the patella-tibia (P-T) joint, or, rarely, the patella will divide. Autospasy can be a useful observation because specimens are often abused either by an inexperienced collector or by manufacturing machinery that was operating when the spider fell in. Other taxonomically important leg structures

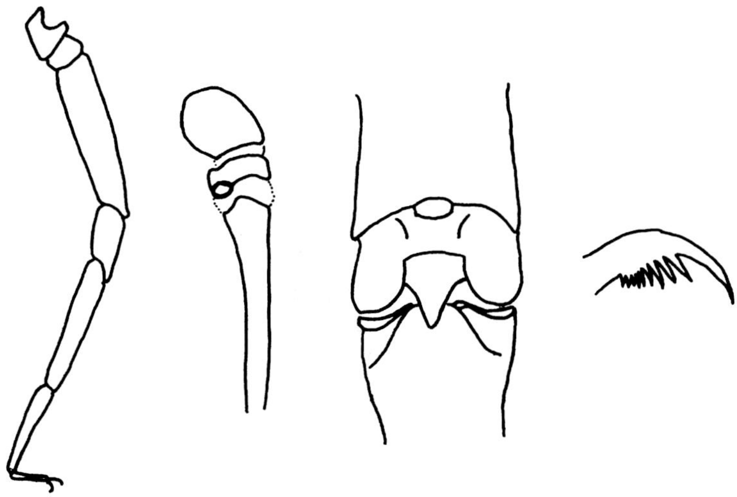

FIGURE 4. (Far left) diagram of a spider leg.
FIGURE 5. (Middle left) trochanter showing cuneal disk (Pholcidae).
FIGURE 6. (Middle right) tibial protusion (Heteropodidae).
FIGURE 7. (Far right) typical serrated spider tarsal claw.

are the number of tarsal claws and the presence or absence of hair tufts between the claws. Spider claws are typically serrated (Figure 7). The metatarsal segment, especially on the fourth pair of legs (metatarsus IV), often bears structures that are found only on spiders. Ventrally on metatarsus IV (as well as on other metatarsi and tarsi), a dense brush of short stiff hairs, called the scopula, may occur. Dorsally on metatarsus IV, certain spiders may have one or two rows of characteristic short, curved bristles. This series of bristles is the calamistrum and its presence indicates that the fragment originated from a cribellate spider. On the distal leg segments there may be a series of dorsal hairs (trichobothria) that are arranged in distinctive sequences and are recognizable by their slenderness and the relatively small sockets where they originate. Specialized types of leg setae are also diagnostic for specific groups of spiders.

The carapace (Figure 1) is a durable fragment that frequently survives food processing to be recovered in recognizable condition from a product. The configuration of the eyes on the carapace is a diagnostic characteristic in synanthropic spiders. The eyes of a spider are typically arranged in two rows of four eyes each across the front margin of the carapace. Depending on the family, one or another row of eyes may arch forward (procurve), backward (recurve) or be straight. The type of curvature and degree of curvature of each row of eyes combine in different ways for different families. Usually, each eye is separated from other eyes in the row, but in some families the eyes are grouped in adjoining pairs called dyads, or adjoining triplets called triads. In these latter cases, the total number of eyes is often reduced to six instead of eight. Another diagnostic characteristic is enlargement or reduction in size of the eyes in one or more pair. The combination of number, position, arrangement and relative size of the eyes is sufficient to identify a carapace fragment to the family, and, for many synanthropic spiders, the eye configuration supports a genus determination. The shape of the carapace is a good confirmatory characteristic when used in combination with eye configuration.

The chelicerae of spiders are an important source of taxonomic information.[13] The way in which the paired chelicerae are oriented in relation to each other delineates two major groups of spiders. In one group, the orthognathous spiders, the chelicerae move vertically, or up and down, in parallel motion. In the other group, labidognathous spiders, the chelicerae move horizontally and in opposition to each other (Figure 2). The synanthropic species of spiders are largely labidognathous. A chelicera has two segments: a stout basal segment and an articulated, hollow fang or claw that rests in a mesal groove in the basal segment. When biting, the spider extends its fang outward from the basal segment. The margins of the mesal groove are designated as the promargin (upper side) and the retromargin (lower side). The margins are often armed with chitinous teeth, the number, sizes and arrangement of which are diagnostic. The mesal margin may bear a straight, transparent ridge (lamina) that is sometimes modified distally to form a single toothlike protrusion that meets with the tip of the fang when the latter is at rest. This pincerlike configuration of the chelicera is known as a chelate chelicera (Figure 2). On the basal segment, some spiders have a localized series of ridges on the outer side that is used for stridulation. In these groups, notably the Linyphiidae and Sicariidae, the pedipalp bears a stridulatory peg on the femur that is raked across the ridges on the chelicera to produce sound.

The spinnerets and associated structures on the abdomen are useful for identification. The size, shape, number and arrangement of the spinnerets can be characteristic and distinctive for many spider families. Some spiders have a primitive silk-spinning structure in front of the spinnerets called the cribellum that is flat and platelike. Most synanthropic spiders are ecribellate; that is, they lack a cribellum. Some ecribellate spiders have a conical protuberance, the colulus, in front of the spinnerets. Synanthropic spiders are known among the cribellate families Amaurobiidae, Desidae, Dictynidae, Filistatidae and Oecobiidae. The size and shape of the cribellum is diagnostic for these spiders.

In mature male spiders, the distal segment of the pedipalp is enlarged to form an intromittent organ, or copulatory bulb. This organ is borne on the spoon-shaped distal segment, the cymbidium, of the pedipalp. In some groups, the cymbidium also bears a lateral projection known as the paracymbidium. The male copulatory organ is complex in structure and, like the genitalia of most arthropods, its micromorphology is diagnostic at the species level. In female spiders, the genitalia occur anteriorly on the venter of the abdomen. Two types of female external genitalia are found among the ecribellate synanthropic spiders. The haplogyne type consists of an epigastric furrow that stretches across the abdomen approximately between or behind the anterior breathing apertures. The genital pore occurs medially in the epigastric furrow. The entelegyne type of genitalia has an additional sclerotized plate, the epigynum, anterior to the epigastric furrow. The epigynum is flat and bears a pair of aperatures to receive the emboli, or tips of the male copulatory bulbs, during copulation. A system of internal ducts transports sperm from the emboli to storage sacs and the uterus. The micromorphology of the epigynum and the epigastric furrow is used to identify female spiders at the species level.

Other micromorphological characteristics that are taxonomically useful are the endites and the labium. Endites are enlargements of the basal segment of the pedipalps that extend ventrally under the chelicerae. They are used during feeding to crush the body of the captured prey. In some spider groups the labium is rebordered, or thickened along the anterior margin. This characteristic is helpful in identifying aerial web-spinners.

C. VENOMOUS SPIDERS

Most spiders are venomous to some degree. The most dangerously venomous spiders found in association with human habitations are the widow spiders (*Latrodectus* spp.) (Theridiidae), the recluse spiders (*Loxosceles* spp.) (Sicariidae), some sac spiders (*Cheiracanthium* spp.) (Clubionidae), hobo spiders (*Tegenaria* spp.) (Agelenidae) and the wandering spiders (Ctenidae).[14] If encountered alive in food products or during the inspection of a food-handling establishment, extreme caution is necessary. There are two principles of action to avoid envenomation by spiders or other venomous arthropods such as scorpions: (1) never thrust your hand or kick your foot where the eye cannot see, and (2) if you see or feel something crawling on your body, do not swat, *brush*. Swatting will guarantee envenomation, but a quick brushing action will catch an arthropod by surprise and remove it before envenomation can occur.[15]

D. IDENTIFICATION

The following analytical key is for the identification of synanthropic spiders. A companion key for the identification of venomous spiders of medical importance is provided by Milliken.[16]

Rapid Key to Families of Adult Food-Contaminating Spiders

1\. Chelicerae orthognathous 2
  Chelicerae labidognathous 3
2\. Large, up to 8 cm; abdomen with dorsal patch of urticating hairs; tarsal claw tufts present .... Theraphosidae
  Abdomen lacking urticating hairs; tarsal claw tufts absent; carapace with dorsal median transverse groove .... Nemesiidae
3\. Legs slender and long; tarsi flexible; chelicerae chelate with toothlike laminae; six or eight eyes .... Pholcidae
  Legs shorter; tarsi not flexible 4
4\. Six eyes 5
  Eight eyes 8
5\. Eyes in a compact cluster, body small (<5 mm) .... Oonipidae
  Eyes in three dyads; body medium to large (>5 mm) 6
6\. Two-clawed tarsi 7
  Three-clawed tarsi .... Scytodidae
7\. Tarsi with claw tufts; labium notched distally; abdomen with four distinct lung slits .... Dysderidae
  Tarsi without claw tufts .... Sicariidae
8\. Anterior row with six eyes, posterior row with two .... Selenopidae
  Anterior and posterior rows other than above 9
9\. Cribellum and calamistrum present; three-clawed tarsi 10
  Cribellum absent (ecribellate); two- or three-clawed tarsi 13
10\. Large anal tubercle bearing a dense fringe of long, stiff hairs .... Oecobiidae
  Anal tubercle small or absent 11
11\. Eyes compactly arranged on a central hump; chelicera chelate .... Filistatidae
  Eyes arranged in two rows, not in a central compact group; posterior median eyes not enlarged 12
12\. Cribellum divided medially .... Amaurobiidae, Desidae
  Cribellum entire .... Dictynidae, Uloboridae
13\. Two claws; tarsal claw tufts usually present 14
  Three claws; tarsal claw tuft absent; body more than 5 mm long 19
14\. Eyes in three rows with anterior median eyes greatly enlarged facing forward .... Salticidae
  Eyes in two rows; anterior median eyes not more than two times anterior lateral eyes 15
15\. Legs laterigrade, at least legs I 16
  Legs prograde 17
16\. Large (10–25 mm) spiders; metatarsus with dorsal, trilobed extension .... Heteropodidae
  Medium-sized (<10 mm) spiders .... Philodromidae, Thomisidae
17\. Eye rows strongly recurved, appearing as three rows consisting of two, four and two eyes, respectively .... Ctenidae
  Eyes arranged in two rows 18
18\. Anterior spinnerets conical and contiguous; claw tufts with lamelliform setae .... (part) Clubionidae
  Anterior spinnerets cylindrical and widely separated at base .... Gnaphosidae

19. Front margin of tibia and metatarsus of legs I and II with series of long, curved spines interspersed with three to four short spines .... Mimetidae
Tibia and metatarsus of legs I and II lacking distinctive series of long spines interspersed with short spines 20
20. Trichobothria present in a single dorsal row; spinnerets paired, not elongate or slender, lacking long spines on tibiae I and II .... Agelenidae
Trichobothria absent; other characteristics variable .... see Table 2

E. DISCUSSION OF FAMILIES

1. Agelenidae (Grass and Funnel-Web Spiders)

This family of three-clawed ecribellate entelegyne spiders has a single row of two or more dorsal tarsal trichobothria that progressively increase in length toward the tip of the tarsus. The epigynum and male palpus are diagnostic for species in this group.[12] The synanthropic species of the genera *Coras* and *Tegenaria* have eight eyes arranged in two rows, with the anterior row slightly procurved. In *Coras* spp., the anterior median eyes are somewhat enlarged. In both genera, the carapace is pear-shaped with the eyes located near the narrower, anterior margin. The synanthropic species are relatively large, 12–18 mm, with spiny legs. *Tegenaria* spp. have distinctive plumose setae that are present on the legs only but are easily broken off. These setae distinguish *Tegenaria* spp. from all other agelenids. Chelicerae are fanglike with four or more teeth on the promargin. The cosmopolitan European house spider, *T. domestica* (Clerck) overwinters in unheated buildings by storing body fat in the autumn to be metabolized in winter when prey is scarce.[13] This species is also a potential passive vector of food-borne microbial pathogens.[17] The hobo spider, *T. gigantea* Chamberlin and Ivie (= *T. saeva* Blackwall), is a dangerously venomous synanthropic species.[18] In eastern North America, *Coras medicinalis* (Hentz) is a synanthropic species with robust, geniculate chelicerae with three promarginal teeth. Webs of *C. medicinalis* were once used as a folk remedy to dress wounds and were thought to possess narcotic properties useful for treating fevers.[11] The webs of both genera are dense and sheetlike with a tubular retreat.

2. Amaurobiidae (White-Eyed Spiders)

The amaurobiids resemble the synanthropic funnel-web spiders (Agelenidae) in morphology and size. They have three-clawed tarsi and a row of dorsal tarsal trichobothria that progressively increase in size toward the tip of the tarsus. The amaurobiids have an obvious cribellum and calamistrum that are absent in the Agelenidae. Amaurobiids also have less conspicuous spines on their legs and the posterior row of eyes is procurved with the anterior row straight, on a pear-shaped carapace. The cribellum is divided medially, which, in addition to the larger size (10–12 mm), distinguishes synanthropic amaurobiids from members of the Dictynidae and Uloboridae. It is best to refer to detailed technical descriptions to differentiate between these families. *Amaurobius ferox* (Walckenaer) is a 12–16-mm synanthropic species that builds a loose, irregular web around a retreat in a crack or crevice. It has geniculate chelicerae but the anterior median eyes are not enlarged as is the case with a similar funnel-web spider, *C. medicinalis* (Agelenidae).

3. Araneidae (Orb Weavers)

The orb weavers are a diverse group of spiders noted for their characteristic aerial webs. The chelicera is usually distinguishable by a smooth, shiny prominence, or boss, located at the outside angle of the base.[6] On the tip of tarsus IV, araneids have ventral auxiliary "claws" that are actually modified setae. Use Table 2 to distinguish araneids from spiders of the ecribellate, entelegyne families Linyphiidae, Nesticidae and Theridiidae. In the tropics, large (up to 25 mm) *Argiope* spp. spiders often construct webs near light fixtures where they remain active during evening hours capturing flying insects that are attracted to light. Although they

TABLE 2
Family Identification of Aerial Web Spinners

Characteristics	Araneidae	Linyphiidae	Nesticidae	Theridiidae
Rebordered labium	Present	Present	Present	Usually not present nor swollen distally
Endites	Parallel	Parallel	Parallel	Strongly convergent in many spp.
Cheliceral margins	Both margins toothed	Both margins toothed	Promargin with 2–3 teeth; retromargin with 6–12 small denticles	Usually no cheliceral teeth; few with 1–3 on promargins; rarely 1–3 teeth on retromargins
Stridulating structures	Absent	Usually present on side of chelicerae	Absent	If present, on carapace and abdomen around pedicel
Tarsus IV				
Serrated bristles	Absent	Absent	Present	Rarely absent
Auxiliary claws	Present	Absent	Absent	Absent
Colulus	Present	Present	Present	Absent or consisting of only 2 setae; present in a few general
Abdominal scuta	Absent	Present or absent	Absent	Present or absent
Cymbial process of male (Paracymbium)	Usually present or absent	Present or absent	Absent	Present or absent
Palpal bulb attachment in alveolus of cymbium	Centrally	Centrally	Centrally	Proximally
Web type	Usually an orb	Irregular	Irregular	Irregular
Autospasy	Occurs at C-T joint	Occurs at P-T joint	Occurs at C-T joint	Occurs at C-T joint

From Roth, V.D., *Spider Genera of North America*, Portal, AZ, 1-172, 1985.

rarely become contaminants in food products, they can cause contamination by their habit of routinely and frequently discarding the exoskeletons of their prey onto the floor and equipment located beneath their webs. These orb weavers are known to inflict a venomous bite.[19,20] Other orb weavers that are occasionally found in or near buildings include *Araneus* spp. and *Nuctenea umbratica* (Clerck).

4. Clubionidae (Sac Spiders)

These ecribellate entelegyne spiders resemble the gnaphosids except, in general, clubionid legs are longer and their eyes are evenly spaced. The anterior spinnerets of clubionids are conical and contiguous, not cylindrical. Most species are asynanthropic, but one group, *Cheiracanthium* spp., occurs in buildings. They are differentiated from other sac spiders by their light color (abdomen yellow to pale green), smooth carapace, eyes of uniform size and legs I noticeably longer than legs IV. *C. inclusum* (Hentz) occurs in the United States and *C. mildei* C.L. Koch is found in both Europe and North America. Other species are found in the tropics and subtropics. *Cheiracanthium* spp. are 5–6 mm long and are dangerously venomous.[20]

5. Ctenidae (Wandering Spiders)

The eye arrangement is diagnostic for this group. Both rows are strongly recurved with the posterior eyes enlarged. The smaller anterior lateral eyes are aligned with the posterior median eyes, giving the appearance of three rows of two, four and two eyes, respectively. The body has distinctive dorsal tufts of hair. This group is mentioned because the dangerously venomous

Phoneutria fera (Keyserling) and *Cupiennius salei* (Keyserling) are occasionally encountered in produce from South America.[21]

6. Desidae

The desids resemble the agelenids and amaurobiids in general appearance. Their enlarged chelicerae are a useful recognition characteristic. It is best to refer to technical descriptions to identify these spiders. Most species live in intertidal zones, but one species, *Badumna longiquus* L. Koch, has invaded the San Francisco Bay area where it lives in buildings, and another *Badumna* sp. has been collected in buildings in Australia.

7. Dictynidae

This is the largest family of cribellate spiders, with most of the species belonging to the genus *Dictyna*. A synanthropic species, *D. sublata* (Hentz) builds webs in window frames. It is small (2–4 mm), with three-clawed tarsi, and lacks dorsal tarsal trichobothria. This lack of dorsal trichobothria distinguishes dictynids from the Uloboridae, which have dorsal trichobothria on femurs II–IV. The carapace is pear shaped with eight eyes arranged in two rows, the anterior row recurved so that the anterior lateral and posterior lateral eyes appear paired. The cribellum is not divided, unlike that of the Amaurobiidae. The cheliceral retromargin has a single tooth.

8. Dysderidae

The six-eyed dysderid spiders are distinguished from the recluse spiders (Loxoscelidae) by the position of the median pair of eyes, which, in dysderids, is located posterior to the lateral dyads. *Dysdera crocata* C.L. Koch, usually feeds on woodlice (Isopoda) in fields and suburban yards or gardens but may also occur inside buildings. It is 12–13 mm long, with a dark carapace and lighter, slate-colored abdomen. The legs are dark reddish brown to black. Legs III are directed backward, unlike spiders of the related family Segestriidae, which have legs III directed forward.[8] The chelicerae are quite large and ventral endites are pronounced. Dysderid filth has been found in food products.

9. Filistatidae

These subtropical and tropical cribellate spiders are distinguished by the compact grouping of eight eyes on a rounded mound. The carapace is oval, with the anterior margin slightly projecting over the chelicerae and a longitudinal, median thoracic groove following the anterior ocular mound. The eyes on the mound are closely grouped in two rows, with the anterior row procurved. Tarsi are three clawed. The chelicerae are fused at their base and chelate with a small claw opposed by an extension of the basal lamina. Ventrally, the labium is fused to the sternum. The cribellum is small and inconspicuous. Females lack an epigynum. The legs of the male are longer than the female's legs. Autospasy occurs at the patella-tibia joint, an occurrence unique to filistatids among the cribellates. The synanthropic species *Kukulcania hibernalis* (Hentz), 10–19 mm long, constructs a tubular retreat with snare threads radiating from the entrance.

10. Gnaphosidae (Hunting Spiders)

Gnaphosids can be recognized by their large, cylindrical, separated anterior spinnerets. Most are large in body size, 10–14 mm, with oval, posterior median eyes set at an angle with the front margins more widely separated than the back margins. The carapace is oval with a truncated, straight anterior margin bearing the eight eyes in two rows. Gnaphosids are entelegyne and ecribellate. There are two tarsal claws, sometimes with claw tufts. Chelicerae are not chelate. Two genera, *Scotophaeus* and *Herpyllus*, contain synanthropic species. These genera are recognized by the combination of equidistant posterior eyes, toothed promargin and retromargin of the chelicera and a trochanter that is slightly notched distally. The mouse spider, *Scotophaeus blackwalli* (Thorell), is a widespread synanthrope, 8–10 mm long, with greasy-appearing gray and brown pubescence.[22] Another synanthrope, the parson spider, *Herpyllus ecclesiasticus* Hentz, is distinguished by a light, median longitudinal dorsal band on the anterior two-thirds of the abdomen.[23]

11. Heteropodidae (Giant Crab Spiders)

This group is the family Sparassidae of some authors. These spiders are differentiated from other two-clawed ecribellate entelegyne spiders by their large size (up to 25 mm), laterigrade anterior legs and a unique dorsal trilobed process at the tip of the tibia (Figure 6). Tarsal claw tufts are present and both margins of the chelicera are toothed. The eight eyes are arranged in two rows at the truncate anterior margin of the circular carapace. The huntsman spider, *Heteropoda venatoria* (L.), is a cosmopolitan synanthropic species that preys on roaches in buildings. The anterior median eyes are moderately reduced. Females carry a large, lenticular egg sac beneath them until hatching. Huntsmen spiders do not construct webs for snaring insect prey.

12. Linyphiidae (Sheet-Web Weavers)

This is a large family of aerial web spinners. Table 2 is useful in differentiating this group from other ecribellate, entelegyne web spinners. The dwarf spiders, subfamily Erigoninae, are often encountered as field contaminants simply due to the phenomenally large numbers that can occur in field and agricultural areas. It is estimated that nearly two million dwarf spiders per acre can occur.[8] These diminutive spiders (1–2 mm) are difficult to identify. In the subfamily Linyphiinae, *Lepthyphantes* spp. and *Pityohyphantes costatus* (Hentz) enter buildings to construct indoor webs.[6] They are larger spiders, 3–7 mm in length, with a stridulatory file on the outside of the chelicera.

13. Mimetidae (Pirate Spiders)

The ecribellate entelegyne pirate spiders prey only on other spiders. They have a distinctive series of long, curved setae interspersed with three to four shorter setae on the tibia and metatarsus of legs I and II. The eyes are uniform in size, with the anterior and posterior lateral eyes on each side fused to form a dyad on either side of the median eyes. *Ero furcata* (Villers), 4 mm, preys on *Achaearanea* spp. and other house spiders.[8]

14. Nemesiidae

Aside from the tarantulas (Theraphosidae), the nemesiid, *Calisoga longitarsis* (Simon), is the only mygalomorph included in this chapter. *C. longitarsis* is a sizable spider, 2–5 cm, that is occasionally encountered around buildings in California. It is distinguished from the tarantulas by its lack of tarsal claw tufts. The chelicerae lack the series of short, robust distal spines that is present in many other mygalomorphs. Nemesiids are also distinguishable by a medial transverse furrow on the carapace and by posterior spinnerets with the distal segment longer than any other segment.

15. Nesticidae (Cave Spiders)

These spiders are similar to comb-footed spiders (Theridiidae) except lighter in color. Two species, *Nesticus cellulanus* (Clerck) and *Eidmannella pallida* (Emerton), are found in cellars where they construct cobwebs.[8,22,23] The two genera are differentiated by their genitalia. The eye configuration of nesticids has lateral dyads similar to the pirate spiders (Mimetidae) but with the lateral eyes enlarged and the anterior median eyes reduced in size. Use Table 2 to separate nesticids from similar aerial web weavers.

16. Oecobiidae

Members of this cribellate family have an enlarged anal tubercle fringed by long, stiff hairs. The carapace is flat and circular with eight eyes grouped nearly in the center. The posterior row of eyes is procurved, with the posterior median eyes elongate. Chelicerae are fanglike with a relatively long claw. Tarsi are three clawed. The small (2 mm) *Oecobius parietalis* (Hentz) is synanthropic as are most species in the genus. They construct delicate sheetlike webs over crevices or slightly protruding objects and retreat underneath. When disturbed, they can run very rapidly.

17. Oonipidae (Minute Jumping Spiders)

The minute jumping spiders have only six eyes, unlike the true jumping spiders (Salticidae), which have eight. The posterior median eyes are enlarged. The carapace is nearly circular and slightly convex, with the eyes occurring anteriorly, usually in a darkly pigmented ocular area. The chelicerae lack teeth and tarsi are three clawed. *Orchestina saltitans* Banks, is a small (3 mm), pinkish synanthrope with a black line along the lateral margin of the carapace. The enlarged posterior median eyes are aligned with the anterior eyes, forming a straight or slightly procurved line. This species often builds webs in light fixtures and will readily jump backwards when disturbed.

18. Philodromidae (Crab Spiders)

The philodromid crab spiders are ecribellate entelegyne spiders with eight eyes and two tarsal claws. The anterior legs are turned so that the ventral surface faces forward. They can walk sideways, a behavior known as laterigrade movement. Philodromids have a flattened body covered with soft recumbent setae. The carapace is circular with a short, truncate anterior projection bearing eight eyes in two recurved rows. Most philodromid crab spiders are differentiated from thomisid crab spiders (Thomisidae) by the prone body hairs, all four pairs of legs roughly equal in length, a claw tuft with distinctive spatulate setae, a ventral scopula on tarsi I and II and promarginal teeth on the chelicera. *Philodromus* spp. often enter buildings, where they crawl on walls and ceilings searching for prey. They are 4–8 mm long, with the posterior median eyes widely separated and the posterior row of eyes only slightly recurved.

19. Pholcidae (Cellar Spiders)

Also called daddy-long-legs spiders and longlegged spiders, the pholcids include several genera that are synanthropic. Cellar spiders are ecribellate haplogyne spiders with long, flexible legs and three-clawed tarsi with numerous pseudosegments for flexibility. Pholcids have a unique cuneal disk on the trochanter (Figure 5).[12] The body is 2–9 mm long with a circular, flat carapace. Chelicerae are chelate with a complete mesal lamina. Most pholcids have eight eyes arranged on a central prominence in two lateral triads and one median dyad (Figure 2). In *Spermophora* spp., the median dyad is absent and the carapace is flat, lacking an ocular prominence. Synanthropic pholcids construct their webs in buildings and under protected eaves where they typically hang upside down. Females hold the egg sac in their chelicerae until hatching. The longbodied cellar spider, *Pholcus phalangioides* (Fuesslin), is a cosmopolitan species with an elongate abdomen. *P. phalangioides* is an opportunistic web invader that will prey on other spiders. *Holocnemis pluchei* (Scopoli) is occasionally found in buildings in California's central valley.

20. Salticidae (Jumping Spiders)

These small to minute spiders are easily recognized by the greatly enlarged, forward facing anterior median eyes. The eight eyes are arranged in three (rarely four) rows. They are ecribellate entelegyne two-clawed spiders with a claw tuft. The chelicerae have robust claws and promarginal teeth. The zebra spider, *Salticus scenicus* (Clerck), occurs indoors and has been found in food products. It is 4–6 mm long with conspicuous black and white bands. Males have elongate chelicerae, extending as far forward as the extended pedipalps. *S. scenicus* is found throughout the northern hemisphere. A favorite prey of *S. scenicus* is spiderlings of other species.[13] In the western United States, the red jumping spider, *Phidippus formosus* (Peckham), occurs in buildings. Both species are mildly venomous.[19]

21. Scytodidae (Spitting Spiders)

Morphologically, spitting spiders resemble recluse spiders (Sicariidae), except for their smaller size (6–8 mm) and three-clawed tarsi. The six eyes are arranged in three dyads on a distinctively dome-shaped carapace. The carapace is contoured to house large glands that secrete sticky threads that the spider spits at its prey to immobilize it. Spitting spiders are ecribellate, with chelate chelicerae that are used to carry the egg sac. *Scytodes thoracica* (Latreille) is a cosmopolitan synanthropic species, 3–4 mm long, that is yellow with dark markings.

22. Selenopidae

These tropical crab spiders are easily recognized by the configuration of the eyes. The anterior row consists of six eyes aligned in a single row. The remaining two posterior eyes are positioned posterolaterally and usually are enlarged. Legs are laterigrade. Although relatively large (17 mm), *Selenops* spp. can flatten the body to hide in very narrow crevices in buildings.[23]

23. Sicariidae (Recluse Spiders and Brown Spiders)

The recluse spiders are large, 11–25 mm, synanthropic ecribellate spiders with long, slender legs and two-clawed tarsi that lack claw tufts. The six eyes are arranged in a characteristic pattern of three dyads with the median dyad advanced anteriorly from the lateral dyads. Each lateral dyad has the two eyes aligned nearly longitudinally. The carapace is round and flat with a projecting anterior margin bearing the eyes. Chelicerae are chelate and have a stridulatory file of coarse grooves on the outside. The palpal femur has a short, basal stridulation peg. Recluse spiders are similar to the scytodid spiders except for being larger and having two, rather than three, tarsal claws. The brown recluse spider, *Loxosceles reclusa* Gertsch and Mailuk, constructs irregular webs in buildings. The largest of the recluse spiders, *Loxosceles laeta* (Nicolet), is found in the tropical and subtropical Western hemisphere.[24] Populations of *L. laeta* (Nicolet) reached levels in downtown Los Angeles in 1990 that required a general extermination effort. All *Loxosceles* spp. are dangerously venomous.[25–27]

24. Tetragnathidae (Silk Spiders)

These orb weavers resemble garden spiders (Araneidae) except that the basal prominence (boss) on the chelicera of tetragnathids is always rudimentary and their femora may bear trichobothria, unlike araneids, where femoral trichobothria are usually absent. The microanalytical entomologist may encounter *Nephila* spp. because of their potential commercial and scientific uses as silk producers and experimental animals.[11] The golden silk spider, *Nephila clavipes* (L.), is a New World species that has spread to other regions through commerce. It is frequently found in buildings in the gulf states. The female is 2–3 cm long and has tufts of long setae on the femur and tibia of legs I–IV. Most immature *Nephila* spp. females have these leg "gaiters," but *N. clavipes* retains this characteristic as an adult whereas other species do not. One of the largest web-inhabiting spiders in the world is *N. maculata* (F.), with a body length up to 5 cm. In the tropics, this spider sometimes lives very near buildings and is a food item in some cultures.[28,29] The males of *Nephila* spp. are small, ranging from one-third to one-tenth the size of mature females. In tropical Africa, *Nephilengys cruentata* (F.) lives in buildings.[30]

FIGURE 8. Examples of urticating hairs (Theraphosidae).

25. Theraphosidae (Tarantulas)

Tarantulas are unique among mygalomorphs in having tarsal claw tufts. This, combined with their large size (3–7 cm), makes tarantulas easy to recognize. Tarantulas are occasionally found in produce shipments and are also popular as pets. They are mildly to dangerously venomous depending on the species. In addition, tarantulas have urticating hairs (Figure 8) on their abdomen that they shed readily. These hairs are a potentially hazardous adulterant, so pet tarantulas should not be housed near food-manufacturing areas. *Aphonopelma* spp. are widespread in the western United States.

26. Theridiidae (Cobweb Spiders)

Use Table 2 to differentiate the cobweb spiders from other aerial web spinners. They are ecribellate entelegyne spiders with three serrated tarsal claws. Cobweb spiders are also known as combfooted spiders because the tarsus on legs IV bears a distinctive, ventral row of serrated setae that is not found in any other spider family except the pale-colored Nesticidae. The chelicerae are fanglike, without marginal teeth. The carapace is oval with eight anterior eyes arranged in two rows, the anterior row sometimes recurved. The females of synanthropic species have spherical abdomens, as do many other cobweb spiders. The American house spider, *Achaearanea tepidariorum* (C.L. Koch), is a cosmopolitan pest, 6–8 mm long, with mottled brown and white coloration. It constructs cobwebs in buildings. The larger (10–14 mm) widow spiders, *Latrodectus* spp., are also found indoors. The black widow, *L. mactans* (F.), is shiny black, with a characteristic red hourglass mark on the venter of the abdomen. Most *Latrodectus* spp. can inflict a dangerously venomous bite.[31] They occur in foods, where people have inadvertently eaten them.[24] *Steatoda grossa* (C.L. Koch), another venomous species, occurs in buildings, where it preys on roaches and sometimes on black widows.[21] The abdomen is dark with a dorsal yellow median line anteriorly on the dorsum. The dorsal anterior margin of the abdomen is also yellow. *S. bipunctata* (L.) is a potential passive vector of pathogens.[17] Males of *Steatoda* spp. have a stridulating organ located at the dorsal base of the abdomen that is rubbed against the margin of the carapace to produce sound.[8, 13] *Nesticodes rufipes* (Lucas) and *Theridion* spp. are also found in buildings.[32,33]

27. Thomisidae (Crab Spiders)

The thomisids are typical crab spiders with laterigrade legs. Tarsal claw tufts are either sparse or absent, usually the latter. Thomisids are distinguished from philodromid crab spiders by the presence of a colulus and absence of a scopula on tarsi I and II. In philodromids, the colulus is absent and scopulae are present. One group of crab spiders, *Xysticus* spp., occurs on low vegetation including many agricultural crops. These spiders may be encountered in food products as an incidental field contaminant.

28. Uloboridae

The synanthropic *Uloborus* spp. are readily recognized by a distal tuft of long setae on tibia I (only) of the female. The tibial tuft is similar to the "gaiters" found on immature *Nephila* spp. (Tetragnathidae) of the same approximate size (3–5 mm) as adult *Uloborus* spp., but in *Nephila* spp. tibial (and femoral) tufts occur on all legs, not exclusively on leg I. Uloborids are distinguished from other cribellates by a dorsolateral concavity on metatarsus IV adjacent to the calamistrum. Trichobothria are present on femora II–IV. The cribellum is entire, not divided. The featherlegged spider, *Uloborus glomosus* (Walckenaer) is a widespread synanthrope. Uloborids are unique among spiders in lacking venom.

REFERENCES

1. Kurtz, O.L. and Harris, K.H., *Microanalytical Entomology for Food Sanitation Control*, Assoc. Off. Agric. Chem., Washington, DC, 1-576, 1961.
2. Metcalf, C.L., Flint, W.P., and Metcalf, R.L., *Destructive and Useful Insects*, 4th ed., McGraw-Hill, NY, 1-1087, 1962.
3. Dobie, P., Haines, C.P., Hodges, R.J., and Prevett, P.F., *Insects and Arachnids of Tropical Stored Products*, TDRI, Slough, Berks, U.K., 1-273, 1984.
4. FDA, *The Food Defect Action Levels*, Natl. Tech. Inform. Serv., Springfield, VA, 1-18, 1989.
5. Borror, D.J., Triplehorn, C.A., and Johnson, N.F., *An Introduction to the Study of Insects*, 6th ed., Saunders Coll. Publ., Philadelphia, PA, 1-875, 1989.
6. Kaston, B.J., *How to Know the Spiders*, 3rd ed., Wm. C. Brown, Dubuque, IA, 1-272, 1978.
7. Van der Hammen, L., *An Introduction to Comparative Arachnology*, SPB Academic Publ., The Hague, 1-576, 1989.
8. Levi, H.W. and Levi, L.R., *Spiders and Their Kin*, Golden Press, NY, 1-160, 1987.
9. Platnick, N.I., *Advances in Spider Taxonomy 1981-1987. A Supplement to Brignoli's A Catalogue of the Araneae Described between 1940 and 1981*, Manchester Univ., NY, 1-673, 1989.
10. Snow, K.R., *The Arachnids: An Introduction*, Columbia Univ., NY, 1-84, 1970.
11. Comstock, J.H., *The Spider Book* (revised), Comstock Publ., Ithaca, NY, 1-729, 1948.
12. Roth, V.D., *Spider Genera of North America*, Portal, AZ, 1-172, 1985.
13. Foelix, R.F., *Biology of Spiders*, Harvard Univ., Cambridge, MA, 1-306, 1982.
14. Southcott, R.V., Disease and Arachnids in the Tropics, in *Mammalian Diseases and Arachnids,* Vol. 2, Nutting, W.B., Ed., CRC Press, Boca Raton, FL, 15-58, 1984.
15. Stahnke, H.L., *Scorpions*, rev. ed., Ariz. State Univ., Tempe, AZ, 1-36, 1956.
16. Milliken, L.E., Biology of Spiders (Araneida), in *Mammalian Diseases and Arachnids*, Vol. 1, Nutting, W.B., Ed., CRC Press, Boca Raton, FL, 59-81, 1984.
17. Sramova, H., Daniel, M., Absolonova, V., Dedicova, D., Jedlickova, Z., Lhotova, H., Petras, P., and Subertova, V., Bacterial contamination by arthropods in public hospitals, *Czech. Epidem. Microbiol. Immunol.*, 41, 223-232, 1992.
18. Vest, D.K., Emergent patterns in the occurrence and severity of probable hobo spider (*Tegeneria agrestis*) envenomation in humans, *Toxicon*, 25, 175-184, 1987.
19. Brodie, E.D., *Venomous Animals*, Golden Press, NY, 1-160, 1989.
20. Gorham, J.R. and Rheney, T.B., *J. Amer. Med. Assoc.*, 206, 1958-1962, 1968.
21. Ebeling, W., *Urban Entomology*, Univ. Calif. Press, Berkeley, 1-695, 1975.
22. Jones, D., *The Larousse Guide to Spiders*, Larousse, NY, 1-320, 1983.
23. Gertsch, W.J., *American Spiders*, 2nd ed., Van Nostrand, NY, 1-285, 1978.
24. Papp, C.S. and Swan, L.A., *Ouch! A Guide to Biting and Stinging Insects and Other Arthropods*, Okumura Biol. Inst., Sacramento, CA, 1-176, 1981.
25. Baerg, W.J., The black widow and five other venomous spiders in the United States, *Univ. Ark. Agric. Exp. Sta. Bull.*, 608, 1-43, 1959.
26. Hogue, C.L., *The Insects of the Los Angeles Basin*, Natur. Hist. Mus. Los Angeles Co., Los Angeles, 1-173, 1974.
27. Gorham, J.R., *The Brown Recluse*, U.S. Public Health Service, Washington, DC, 1-5, 1970.
28. Olsen, A.R., Observation of unusual courtship behavior of the spider *Nephila maculata* (Fabricius) (Araneae: Tetragnathidae) in Palau, *Micronesica*, 26, 221-225, 1993.
29. Robinson, M.H. and Robinson, B., The ecology and behavior of *Nephila maculata*: A supplement, *Smithson. Contrib. Zool.*, 218, 1-22, 1976.
30. Robinson, M.H. and Robinson, B., Comparative studies of the courtship and mating behavior of tropical araneid spiders, *Pac. Insects Monogr.*, 36, 1-217, 1980.
31. Thorpe, R.W. and Woodson, W.D., *Black Widow*, Univ. N. Carolina, Chapel Hill, 1-222, 1945.
32. Emerton, J.H., *The Common Spiders of the United States (1902)* (rev. S.W. Frost), Dover, NY, 1-227, 1962.
33. Preston-Mafham, R. and Preston-Mafham, K., *Spiders of the World*, Blandford, Oxford, 1-191, 1984.

Chapter 11

OTHER ARTHROPODS

Alan R. Olsen

CONTENTS

I. Introduction 151

II. Pseudoscorpions (Pseudoscorpiones) 151
A. Morphology 151
B. Significance 152

III. Scorpions (Scorpiones) 153

IV. Sunspiders (Solifugae) 153

V. Tailless Whipscorpions (Amblypygi) 154

VI. Harvestmen (Opiliones) 154

VII. Centipedes (Chilopoda) 154

References 155

I. INTRODUCTION

One consequence of insanitary conditions is that, under conditions of neglect, infestations of stored-product pests can increase until their populations become large enough to attract and support arthropod predators. The most frequently encountered commensal predators are the spiders (chapter 10). Other arachnid predators, however, are also found in buildings where suitable prey are present. Included in this group of miscellaneous commensal arachnids are certain pseudoscorpions, scorpions, sunspiders, tailless whipscorpions and harvestmen. These intruders into the facilities where food products are handled or stored comprise a varied group, but each type of predator is easily recognized by a distinctive (and often familiar) body form. Figure 1 provides a diagrammatic comparison of the recognizable body forms of these predators and compares them with the more familiar arachnids, the spiders and mites. Included in the discussion is the house centipede, which, although not an arachnid, exhibits indoor predatory habits similar to the commensal arachnids.

II. PSEUDOSCORPIONS (PSEUDOSCORPIONES)

A. MORPHOLOGY

Pseudoscorpions are diminutive arachnids, rarely longer than 5 mm.[1] The body (Figure 1b) is flat with an ovoid to elongate outline. The general appearance is scorpionlike except that pseudoscorpions lack a slender tail and sting. The body is divided into two main regions: the cephalothorax (prosoma) and abdomen (opisthosoma). Dorsally, the cephalothorax consists of a shieldlike carapace, usually with smooth margins. The carapace shape is rectangular to broadly triangular and may bear one or two transverse furrows. The cephalothorax also bears

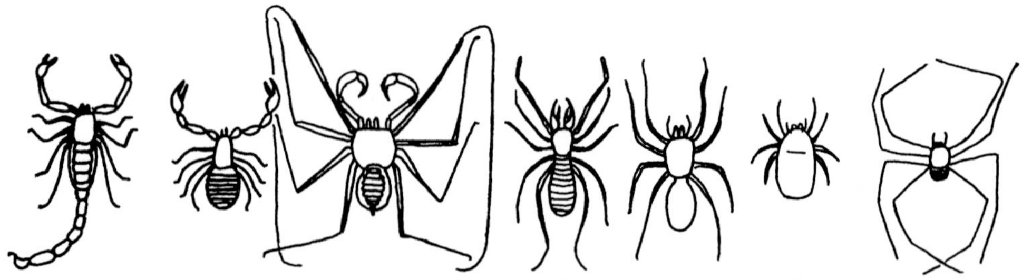

FIGURE 1. Diagrammatic comparison of body forms of various commensal arachnids: (a) scorpion (Scorpiones); (b) pseudoscorpion (Pseudoscorpiones); (c) tailless whipscorpion (Amblypygi); (d) sunspider (Solifugae); (e) spider (Araneae); (f) mite (Acarina); (g) harvestman (Opiliones).

simple eyes, located anterolaterally near the front margin of the carapace. In most pseudoscorpions, there are one or two pairs of eyes, although some groups lack eyes. The abdomen is broadly rounded posteriorly and has twelve distinct segments.

The appendages of pseudoscorpions consist of a pair of chelate chelicerae, a pair of pedipalps that are modified into scorpionlike pincers, and four pairs of walking legs. The chelicera projects anteriorly from under the front margin of the carapace and terminates in a fixed digit and a movable digit that function as a grasping organ. The chelicera's movable digit has a membranous comblike serrula attached to its inner margin. The movable digit also contains the organ for producing silk, a single spinneret that opens through a pore on the outer side near the tip of the digit. The silk is used for constructing chambers for brooding, molting and overwintering in cooler climates.(2) The fixed digit is approximately the same size as the movable one and there is a tuft of pectinate seta, the flagellum, at its base. The pedipalps are composed of six segments and are developed into a pair of large terminal pincers, or chelae. The pincer bears the venom apparatus when present. The terminal pincer of the pedipalp is the pseudoscorpion fragment most frequently found in processed food products. Each walking leg has six segments. The distal segment, the tarsus, may be secondarily divided into two segments. Tarsal segmentation varies within groups of pseudoscorpions. Some taxa have all tarsi segmented, whereas other groups are characterized by segmentation on the posterior two pairs of legs only or, in some cases, no segmentation.

B. SIGNIFICANCE

Several species of pseudoscorpions are likely to be encountered in food products and around food-handling facilities.(3) The commonest of these synanthropic species is the cosmopolitan book pseudoscorpion, *Chelifer cancroides* (L.) (family Cheliferidae), also known as the house pseudoscorpion.(4) It is uniformly brown in color and may attain a length of 4 mm. All tarsi are unsegmented. The book pseudoscorpion feeds on the larvae of clothes moths (Lepidoptera: Tineidae) and carpet beetles (Coleoptera: Dermestidae), as well as booklice (Psocoptera: Liposcelidae) and other stored-product insects and mites.(2) The granary pseudoscorpion, *Larca granulata* (Banks), is reported from grain elevators in Ontario.(5) It is a smaller teardrop-shaped pseudoscorpion with a light body and dark, reddish brown pedipalps. Even smaller, *Cheiridium museorum* (Leach) (Cheiridiidae), which barely exceeds 1 mm in length, is sometimes found in and around food products. In the tropics, *Withius niger* (Simon) (Withiidae) is the most commonly encountered pseudoscorpion in stored-products.(6) Other pseudoscorpions that are distributed by commerce and may be found in stored-products include *Chthonius terrachelatus* (Chthoniidae); *Paisochelifer callus* Beier (Cheliferidae); *Allowithius* spp. (Withiidae); and *Pselaphochernes* sp. *Allochernes powelli* (Kew), *Dinocheirus panzeri* (C.L. Koch) and *Lamprochernes minor* Hoff (Chernetidae).(7,8)

III. SCORPIONS (SCORPIONES)

Scorpions are relatively well known and easily recognized (Figure 1a) by their large, pincerlike pedipalps and their slender, five-segmented tail terminating in a bulbous segment that bears a venomous sting. The body has two main divisions: the cephalothorax and abdomen. The cephalothorax is covered dorsally by an unsegmented carapace that bears a pair of centrally located simple eyes and two to five pairs of lateral eyes. The abdomen is broad basally, abruptly narrowing posteriorly into a slender tail. The basal portion of the abdomen is divided into seven distinct segments and the tail has five segments. The total body length can range from very small (10 mm) up to 4 or 5 in. (125 mm). The anterior chelicerae are chelate and the pedipalps are elongate, terminating in a large pincer. There are four pairs of walking legs, each with segments. The tarsi are secondarily segmented.

Scorpions enter buildings to seek shelter and prey. Cockroaches and silverfish serve as food for scorpions in buildings where these pests are allowed to thrive.[9] Members of the cosmopolitan family Buthidae are the scorpions most often associated with human habitations and storage facilities. Buthid scorpions are distinguished from other scorpions by a lack of slender, elongate, sensory setae (trichobothria) on the ventral surface of the pedipalp patella (fourth segment) and by the presence of two enlarged teeth on the ventral margin of the movable digit of the chelicera. Some buthids are potently venomous. The buthid genus *Centruroides* is attributed with hundreds of envenomation deaths per year.[10] *Centruroides* spp. have been found as stowaways in shipments of dried chili peppers and bananas from Mexico and Central America and are frequently encountered in buildings in the southwestern United States. The cosmotropical spotted isometrus, *Isometrus maculatus* (De Geer), enters buildings and has been introduced in the Neotropical and Palearctic regions through human activity. This 5–7 cm long buthid is yellow with black marbling.[2,4] Although not as dangerously venomous as *Centruroides* spp., care should be taken in collecting and handling the spotted isometrus and all scorpions.

IV. SUNSPIDERS (SOLIFUGAE)

Sunspiders or windscorpions (Figure 1d) are generally straw colored and range from 10 mm to more than 30 mm in size. They are unusual among arachnids in having rudimentary dorsal segmentation of the cephalothorax, which is divided dorsally into an anterior carapace (propeltidium) and two posterior "segments" (mesopeltidium and metapeltidium) that are visible from a dorsal view. The anterior carapace bears one anterior medial pair of simple eyes on a raised tubercle. The abdomen is elongate ovoid and distinctly segmented. The abdomen narrows slightly near the anterior attachment to the cephalothorax.

The pincerlike chelicerae are large and project forward from the anterior margin of the cephalothorax. The chelicerae articulate vertically. The pedipalps are leglike, clawless and terminate in adhesive organs. There are four pairs of walking legs. The first pair of walking legs are slender and are not used to support the weight of the animal's body. The tarsi may be secondarily segmented on some or all of the walking legs, depending on the taxon.

Sunspiders are attracted to lights (or to the movements of insects attracted to the lights) and are thus frequently found in and around buildings. In Egypt, *Galeodes* sp. (family Galeodidae) has been seen climbing onto tables to hunt flies.[2] *Gluvia* sp. is reportedly common in the streets of Madrid, Spain.[8] *Eremobates* sp. and *Eremorhax* sp. (family Eremobatidae) are commonly found inside buildings in the arid regions of North America. In California, *Ammotrechula* sp. prob. *peninsulana* (Banks) (family Ammotrechidae) was found in a nut-packing facility, where it was trapped on a glue board while presumably foraging for prey. Although sunspiders lack venom glands, they are nonetheless regarded as venomous in many parts of the world, and can inflict a painful bite.[11]

V. TAILLESS WHIPSCORPIONS (AMBLYPYGI)

Amblypygids, or tailless whipscorpions, (Figure 1c) are spiderlike arachnids ranging in size from 1 to 10 cm. The carapace bears eight simple eyes on the anterior margin. The compact abdomen is segmented, narrowed anteriorly, and bears no caudal appendage. The chelicerae are small and chelate. The pedipalps are raptorial and bear a series of spines on their inner margins. The first pair of legs are slender, elongate, highly flexible, sensory, and are often described as "antenniform." Legs II–IV are more robust and are used for locomotion, which is frequently accomplished through sideways (laterigrade) movement. Amblypygids are nonvenomous. Several genera occur in buildings in the tropics and subtropics. Members of the New World genus *Phrynus* readily enter houses.[8] In South Africa, *Damon* sp. is responsible for chronic infestations of subfloor ventilation areas in buildings.[12] An examination of museum specimen records reveals that other amblypygids have been found in buildings, including the genera *Paraphrynus*, *Phrynichus*, *Paracharon* and *Masicodamon*.

VI. HARVESTMEN (OPILIONES)

Harvestmen, or daddy-long-legs, are easily distinguished from other arachnids by their long, slender legs (Figure 1g). When walking or standing, their eight legs are arched, suspending the compact, oval body above the ground. The body consists of an unsegmented cephalothorax that is broadly joined to a short, segmented abdomen. The carapace of the cephalothorax bears a single pair of simple eyes (ocelli) located on a central tubercle. Laterally, the cephalothorax bears a pair of excretory coxal glands that open between legs III and IV. The adult's abdomen has ten tergites and nine sternites. The second abdominal sternite bears a pair of tracheal openings. The walking legs are similar in size and have the complete arachnid segmentation of coxa, trochanter, femur, patella, tibia, metatarsus and tarsus. The coxae and trochanters on all legs are swollen, much larger in circumference than the spindly distal leg segments. The tarsus has many segments, is flexible and may terminate in one to three claws. The pedipalp is leglike, often terminating in a single claw. The chelicera is slender, chelate and composed of three segments.[12]

Harvestmen are not venomous. They are known to invade buildings.[13] A common indoor species, *Mitopus morio* (F.) (Phalangiidae) is found throughout temperate Europe and North America.[14,15] *M. morio* is capable of passively transporting pathogens, making its presence in a food-handling facility a public health concern.[16] Harvestmen are also known to phoretically transport mites and pseudoscorpions.[8]

VII. CENTIPEDES (CHILOPODA)

The house centipede, *Scutigera coleoptrata* (L.) (Scutigeromorpha: Scutigeridae) is not an arachnid, but it is included in this chapter for convenience. It is a distinctive centipede, having a gray body with darker longitudinal stripes, legs with white bands, and compound eyes. The head is distinctive and, in addition to the compound eyes, bears a pair of mandibles (each of which terminates in a venomous fang), two pairs of maxillae, and a pair of long, slender antennae. The adult body is up to 4 cm long and bears 15 pairs of long, slender legs. In the female, the hindmost pair of legs are at least twice as long as the other legs.

S. coleoptrata is able to reproduce indoors and in colder regions is found only inside buildings.[13] It occurs throughout the United States and Canada. Indoors, it feeds on flies and spiders.[1] It is mildly venomous.[17] Immature scutigerids have a reduced number of legs (as few as four pairs in newly hatched larvae). Scutigerids are unique among centipedes in possessing true compound eyes.

REFERENCES

1. Borror, D.J., Triplehorn, C.A., and Johnson, N.F., *An Introduction to the Study of Insects*, 6th ed., Saunders Coll. Publ., Philadelphia, 1-875, 1989.
2. Comstock, J.H., *The Spider Book*, rev. ed., Comstock Publ., Ithaca, NY, 1-729, 1948.
3. Kurtz, O.L. and Harris, K.L., *Microanalytical Entomology for Food Sanitation Control*, Assoc. Off. Agric. Chem., Washington, DC, 1-576, 1961.
4. Levi, H.W. and Levi, L.R., *Spiders and Their Kin*, Golden Press, NY, 1-160, 1987.
5. Sinha, R.N. and Watters, F.L., *Insect Pests of Flour Mills, Grain Elevators, and Feed Mills and Their Control*, Agric. Can. Publ. No. 1776, 1-290, 1985.
6. Dobie, P., Haines, C.P., Hodges, R.J., and Prevett, P.F., *Insects and Arachnids of Tropical Stored Products.* TDRI, Slough, Berkshire, U.K., 1-273, 1984.
7. Gorham, J.R., Ed., *Insect and Mite Pests in Food*, Vol. 2, U.S. Dept. Agric., Agric. Handbook No. 655, 311-767, 1991.
8. Hickin, N.E., *Pest Animals in Buildings*, Longman, NY, 1-385, 1985.
9. Stahnke, H.L., *Scorpions*, Poisonous Animals Res. Lab., Tempe, AZ, 1-36, 1956.
10. Simmard, J.M. and Watt, D.D., Venoms and Toxins, in *Biology of Scorpions*, Polis, G.A., Ed., Stanford Univ. Press, Palo Alto, CA, 64-160, 1990.
11. Muma, M.H., Solpugida, in *Synopsis and Classification of Living Organisms*, Vol. 2, Parker, S.B., Ed., McGraw-Hill, NY, 1982.
12. Savary, T.H., *The Arachnida*, Edward Arnold, London, 1-218, 1935.
13. Ebeling, W., *Urban Entomology*, Univ. Calif. Press, Berkeley, 1-695, 1975.
14. Cokendolpher, J.C. and Lee, V.F., *Catalogue of the Cyphopalpatores and Bibliography of the Harvestmen (Arachnida, Opiliones) of Greenland, Canada, U.S.A., and Mexico*, Vintage Press, Lubbock, TX, 1-82, 1993.
15. Crawford, R.L., *Catalogue of the Genera and Type Species of the Harvestmen Superfamily Phalangioidea (Arachnida)*, Burke Museum, Seattle, 1-60, 1992.
16. Sramova, H., Daniel, M., Absolonova, V., Dedicova, D., Jedlickova, Z., Lhotova, H., Petras, P., and Subertova, V., Bacterial contamination by arthropods in public hospitals, *Czech. Epidemiol. Microbiol. Immunol.*, 41, 223-232, 1992.
17. Busvine, J.R., *Insects and Hygiene*, 3rd ed., Chapman and Hall, NY, 1-568, 1980.

Chapter 12

HAIRS AND FEATHERS

Dana Hayakawa Ludwig and John R. Bryce

CONTENTS

I. Hair Identification 157
A. Hairs as Adulterants 157
B. Sources of Hair Contaminants 158
C. Modes of Contamination 158
D. Regulatory Significance by Taxon 159
1. General Significance 159
2. Commensal Rodents 160
3. Asian Musk Shrew 161
4. Bats 162
5. Cats and Dogs 162
6. Rabbits 162
7. Noncommensal Rodents 162
8. Humans 163
9. Domestic Livestock 163
E. Hair Morphology 163
1. Hair Growth and Development 163
2. Hair Structure 163
3. Types of Hairs 165
F. General Diagnostic Hair Characteristics 168
1. Hair Characteristics Tables 168
2. Hair Identification Key 169
3. Identification Problems with Hairs 173
4. The Reference Collection 176
G. Microscopic Techniques for Mounting and Clearing Hairs 176

II. Feather Identification 177
A. Feathers as Adulterants 177
B. Sources of Feather Contaminants 177
C. Feather Morphology 177

References 215

I. HAIR IDENTIFICATION

A. HAIRS AS ADULTERANTS

Hairs are adulterants of major regulatory significance in food, drug and cosmetic products. The discussion in this chapter emphasizes food, but applies equally to drug and cosmetic products. Hairs may indicate contamination of the material by mammals and their metabolic byproducts[1] through direct contact, being in close proximity or, occasionally, the animals accidentally ending up as an ingredient in the product. The significance of these hairs can range from incidental contaminants to indicators of insanitation. These hairs may be shed by the source animal or may be a fecal indicator, as is the case with rodent hairs, the presence of which often indicates adulteration by feces.[2] Consequently, analysts in the microanalytical laboratory must be familiar with mammalian hair morphology in order to properly identify the source animal and evaluate the legal and sanitary significance of these contaminants in food.

B. SOURCES OF HAIR CONTAMINANTS

A variety of animal hairs can contaminate foods, and these are informally divided into two main groups, striated and nonstriated hairs, corresponding to groups of mammals typically having one type or the other. Striated hairs are those that have a discontinuous medulla; that is, the central core of the hair shaft contains an alternating series of cells and air spaces similar in appearance to a ladder. Some animals in this group include rodents such as commensal rats and mice, as well as cats, dogs, shrews, rabbits, bats, weasels, squirrels, field mice and voles. The geographical distribution of selected mammals whose hairs are commonly found in food products are discussed later in the chapter. Although nonstriated hairs may also occur, the predominant body hairs on these mammals are striated. For identification purposes, commensal rat and mouse hairs may be grouped together under the category rat/mouse because of their similar appearance and the fact that they have the same regulatory significance.[3] In the same manner, dog and cat hairs are included under the category cat/dog because the large number of breeds results in tremendous variation in hairs and often makes separation of the two taxa difficult.

Nonstriated hairs are those that do not have a distinct series of alternating cells and air spaces in the medulla, but rather exhibit a continuous medullary channel with an amorphous or matrix type of structure. Mammals with nonstriated hairs that may contaminate foods include humans, sheep, cows, swine, goats and horses.

C. MODES OF CONTAMINATION

Animal hairs contaminate foods via a number of routes. The most common way is by shedding. Hairs may be shed by direct contact with foods or surfaces on which foods are prepared or processed. For example, rats are constantly shedding hairs during their normal activities, and these hairs can then become attached to or mixed with the food on which the animal is feeding.[4] The workers themselves may serve as a source of contamination if they fail to use good manufacturing practices such as wearing hairnets and gloves while handling foods. According to Vazquez, hairs from a variety of animals including pets and livestock become airborne and adulterate food products or food contact surfaces in their vicinity.[2] Domestic cats in a food preparation or storage facility shed their hairs, which may subsequently end up in foods or on food contact surfaces.[4] In the same manner, dogs kept in a food plant will contaminate foods with their hairs from shedding as well as from other insanitary habits.[4] Pet hairs can also become attached to workers' clothing at home and then get transferred to the food in the processing facility.

Another mode of contamination is via feces. Many mammals such as commensal rats and mice groom themselves by licking their fur and, in the process, swallow hairs, which survive digestion and are passed out in the feces. Rats and mice are known to deposit fecal pellets wherever they go and may regularly visit food storage facilities where they can adulterate unprotected stored foods.[1] Typically, the concentration of fecal pellets is greater where the rats and mice stop to feed. Embedded hairs removed from these intact fecal pellets can be used to identify the source animal.[4] Rodents can contaminate everything from raw grain to ingredients and finished products.[5] In foods that are extensively processed or comminuted, the significance of hairs is that they may be the one constituent of rodent fecal pellets readily identifiable in the food after processing. A direct relationship exists between the number of hair fragments in a milled product and the number of rodent pellets ground up with the grain.

Small hair fragments in foods containing milled ingredients probably indicate premilling contamination originating from pulverized rodent pellets, whereas long, whole hairs suggest postmilling contamination where hairs defiled the product after grinding.[5] According to Frantz and Davis,[6] one rat pellet may contain up to 200 hair fragments. Cats may defecate directly on food substances in storage facilities[7] or bury their feces in open food containers, so the presence of their hairs may indicate fecal contamination of the food.[4] Another example of fecal contamination occurs with bats, which can inhabit buildings and contaminate foods with their metabolic wastes.[8] Bat hairs in foods indicate that the product is adulterated and has been stored or processed in a building where bats roost.[9] As bats groom themselves with their tongues, they ingest their hairs.[10] As with rodent fecal pellets, bat hairs can be recovered from intact fecal material found in unprocessed foods. Often, the only evidence of bat contamination is the presence of their hairs.[9] Consequently, the presence of hairs in food may be an indication that the product was originally adulterated with feces.[2,4]

There are several other less common routes for hair to contaminate foods. Clothing made from animals can serve as a source of hairs and includes items such as fur coats, wool products, angora rabbit items, rabbit fur felt and fur-lined gloves.[2] An animal may fall directly into the food manufacturing machinery and become part of the finished product. For example, Eisenberg[4] states that the presence of several whole rodent hairs with roots intact, or clumps of hair, attached to skin fragments in a dairy product strongly suggests that a rodent fell into the product, where it eventually decomposed, releasing skin and hair.[4]

D. REGULATORY SIGNIFICANCE BY TAXON

1. General Significance

Hair contaminants generally fit the definition of filth as explained in chapters 1 and 2. People do not like the idea of hairs in their food, or, even less, the association of these hairs with insanitary conditions during preparation, packing or holding of the product. The regulatory significance of hair filth or contaminants is based on these concerns as well as any possible health risks involved.

In general, hair contamination is divided into categories of unavoidable and avoidable. Unavoidable contamination refers to conditions in the field where rodents and other animals cannot be expected to be totally excluded, thus resulting in a limited number of hairs or fecal pellets in the commodity. Most of the defect action levels (DALs) set by the Food and Drug Administration (FDA) for allowable levels of hairs in various products are based on unavoidable contamination (see chapter 2). Extra precaution should be followed for those products undergoing no further bactericidal steps in processing to control possible harmful microorganisms transmitted by the hairs and animals they come from.

Avoidable contamination is that which is preventable under good manufacturing practices and indicates a breakdown of sanitary operations. There are generally no guidelines for this kind of contamination and cases are usually evaluated individually (see chapter 2). Hair contamination that is avoidable, such as from active infestations of rodents in a facility or lack of necessary gloves and hairnets on workers, is evaluated much more strictly and with much less, if any, allowable amounts than with unavoidable contamination. Not only is the hair considered filth, and conditions in the facility filthy, but the possibility exists that such insanitary conditions may lead to health-threatening situations of microbial contamination, etc. Hair filth found in products may indicate more serious breakdowns of proper food handling.

Animals may acquire human pathogens from various sources and mechanically transmit these either internally or externally to foods or food contact surfaces.[7] Of particular interest is hair from mammalian commensal pests, which live in close association with man and contaminate food stored or manufactured in structures that they frequent.[1] Feces may harbor microorganisms, which are then transferred to food and can cause diseases in humans. Another health concern involves urine, which is considered soluble filth, that is absorbed into the food and cannot be removed. This metabolic byproduct is insanitary and can harbor potential disease causing organisms. Adulteration by feces or urine is of special concern in food items such as spices, nuts and dried foods, which go directly to the consumer's table without further processing or cooking, thereby eliminating the opportunity to kill any microorganisms on the food.

Foods contaminated by direct contact with a source animal rather than from metabolic wastes may suggest another potential health concern. By contacting the food itself, the animal may mechanically transmit microorganisms from contaminated to uncontaminated products and surfaces with its paws or other body parts, or it may shed external parasites, which are possible disease vectors of humans.

2. Commensal Rodents

When commensal rodents actively search for food and water, they may come into contact with possible sources of pathogens such as floors, refuse areas, sewers and animal wastes, as well as foods and food-contact surfaces.[(7)] Rodents are known to adulterate unprotected food products through chewing, shedding, defecating and urinating,[(4,5)] and these habits can lead to acquiring and transmitting disease. Rats can acquire and carry disease-producing organisms from these sources on their feet and/or in their intestinal tracts.[(11)] Pathogens on the external body parts are mechanically transmitted and internal pathogens are deposited via metabolic byproducts onto uncontaminated food sources. Because hair contaminants may be accompanied by feces or urine,[(4)] their presence suggests the potential for disease transmission.[(5)] The more common rodent-borne diseases are rat-bite fever, murine typhus fever, salmonellosis, trichinosis and the Black Death plague.[(12)] Rats are known to harbor a number of *Salmonella* serotypes associated with food-borne diseases in humans, such as *S. typhimurium*, *enteritidus* and *newport*.[(11)] According to Gorham, *Salmonella* bacteria were found to remain viable in rodent pellets for at least 148 days at room temperature.[(7)]

Rodents leave small amounts of urine wherever they go. Rodent urine is highly concentrated but has a surface tension half that of distilled water, which enables it to spread readily and soak into surfaces. This increases the likelihood of urine penetrating the packaging material to the food rather than remaining on the packaging surface.[(13)] Thrasher cites a study by Laakso et al. in which a rat was restricted to a container of wheat with a surface area of 0.23 m^2 for 48 hours.[(13)] During this time the rat was able to adulterate up to 10,000 kernels per day to a depth of 7.8 cm with its urine. The urine of rats and mice may contain leptospirosis pathogens, which may infect humans who come into contact with infected urine. This microbial disease is common in the Norway rat and less common in the house mouse and roof rat.[(7)] Gorham states that up to 52% of rats are naturally infected with this microorganism.[(7)] The shed leptospires can survive anywhere from 30 min to 13 days depending on environmental conditions.

The three most common commensal rodents are the house mouse (*Mus musculus*), roof or black rat (*Rattus rattus*) and Norway or brown rat (*Rattus norvegicus*).[(6,10)] *M. musculus* is believed to occur naturally from Sweden and the Mediterranean region to Japan and Nepal. It is now found partly as a human commensal throughout the world. *R. rattus* is believed to originate in the Malaysian area and is now distributed worldwide as a human commensal. *R. norvegicus* may have originated in northern China and now has worldwide distribution as a human commensal. [(10)] These three species are closely associated with storage ecosystems, where they are very successful because of their tremendous adaptability, high reproductive rate and omnivorous eating habits.[(14)] The Polynesian Rat, *Rattus exulans*, is another human commensal that is distributed (mostly by introduction) in an area extending from Bangladesh to Vietnam, and including New Zealand, the East Indies, Hawaii and many other Pacific islands. It occurs in buildings and refuse heaps, feeds on a wide variety of animal and plant material and is associated with health problems in humans.[(10)] Norway and black (roof) rats and the house mouse are responsible for spreading diseases such as bubonic plague, murine typhus, *Salmonella* (food poisoning), leptospirosis and tularemia. Additionally, Norway and roof rats transmit trichinosis and rat-bite fever and the house mouse spreads rickettsial pox.[10]

A fifth commensal rodent that is distributed from Pakistan to Burma, Sri Lanka, and via accidental introduction by humans in Penang Island off the west coast of the Malay Peninsula, Sumantra and Java is the lesser bandicoot rat, *Bandicota bengalensis*.[10] This rat has become the dominant commensal rat in India in recent decades, having spread from the fields into villages and cities. Its reproductive rate exceeds that of any other rat—one litter per month per female, seven pups per litter.[15] Lesser bandicoot rats infest warehouses, and their destructive behavior includes consumption of food, damage to bags and contamination of food with feces, urine, hair and ectoparasites.[6]

3. Asian Musk Shrew

The commensal or Asian musk shrew, *Suncus murinus*, is an insectivore that is distributed primarily in eastern Asia[16,17] but has been introduced to various other regions of Asia and East Africa.[16] Specifically, it occurs from Egypt to Tanzania, Zanzibar, the Comoro Islands, Madagascar, Arabia to Japan, Sri Lanka, Taiwan, the Philippines and islands of the East Indies and was probably introduced into many of these areas.[10] *S. murinus* ranges in size from 30 to 100 g, is nocturnal and can occur in or around human dwellings.[10,16] Although the diet of *S. murinus* consists primarily of insects, it also eats meat and meat products, processed seafoods, other human foods and refuse.[10, 17] As a result of its spread via commerce, the commensal shrew is becoming a significant pest of stored-products.[17]

In addition to eating stored-products, these shrews can serve as hosts for internal and external parasites. Viruses and bacteria responsible for diseases in man and his domestic animals have been isolated from shrews. For example, a number of cases have been reported in Southeast Asia and China where *S. murinus* has been found harboring the rat flea, *Xenopsylla cheopis*, a primary vector of the plague bacillus, *Yersinia pestis*.[16,18,19] Plague isolates have been found in the commensal shrew in Burma and plague was found extensively throughout urban and suburban habitats there.[20]

In general, shrews may deposit feces to delineate borders of their territory, and *S. murinus* exhibits territorial marking behavior in the form of urinating.[16] *S. murinus* fecal material has been observed to lack embedded hairs, suggesting that this shrew may not always exhibit the same type of grooming behavior found in rodents.[17] According to Churchfield,[16] shrews groom their fur primarily by scratching it with their hind feet. Sometimes they may lick areas of the body that they can reach, especially the feet and anal region.[16] Thus, it would appear that *S. murinus* hairs contaminate food via either direct contact by the animal or the shedding of hairs in the vicinity of food. Direct contact increases the likelihood of defecating and/or urinating directly on the food and provides routes of contamination by microorganisms and internal parasites through these waste materials. It also increases the opportunity for mechanical transmission of microorganisms from contaminated to uncontaminated foods and the possible transfer of external parasites to foods. A number of *Salmonella* serotypes were found in shrews, indicating their importance as a reservoir of *Salmonella*, and shrews can vector salmonellosis by contaminating stored grains, food, feeds and water when they enter human and animal dwellings and access food and filth.[21] Internal parasites include many species of helminths (both tapeworms and flukes), for which the shrew serves as an intermediate host.[16]

The correct identification of commensal shrew hairs is important for several reasons. There is a considerable lack of literature pertaining to hair morphology of commensal pests from Asia. *S. murinus* hairs in particular are insufficiently described in the current literature. This scarcity of information may cause the unfamiliar analyst to mistakenly identify commensal shrew hairs as rodent hairs. Determination of *S. murinus* hairs is also essential because of the shrew's ability to physically defile food as well as its role as a potential vector of human diseases, which makes it of considerable public health importance. Also, because *S. murinus* is spread via commerce, its distribution and importance as a stored-product pest will probably increase, thereby further emphasizing the need for its correct identification in foods.[17]

4. Bats

The bat (Chiroptera) hairs most commonly found in foods belong to species of the cosmopolitan family Vespertiliontidae.[9] The 42 genera and 355 species in this family are distributed in temperate and tropical areas worldwide,[10] and members of this family include little brown bats (*Myotis*), pipistrelles, as well as others.[9] They will infest buildings seeking roosting sites and contaminate food stores with their waste droppings,[8,9] which may contain pathogens that can also become air-, or water-borne[4] and then get into foods or on food-contact surfaces.[7] A number of diseases have been associated with bats, but rabies and histoplasmosis are the ones of most concern to humans. Rabies are transmitted via bites, so one should avoid bat bites.[8] Histoplasmosis, a fungal disease, is caused by *Histoplasma capsulatum* and exists naturally in the soil in many warm humid areas in the world. In some regions of the world, certain bats release the microorganism from small lesions in their intestines. *H. capsulatum* multiplies to very high levels in fecal material and infects an individual who inhales or ingests the spores.[8]

Other microbes that occur in association with bats are *Leptospira* in urine and *Salmonella* in feces. Bats also serve as host to a number of parasites including protozoans, helminths, mites, ticks, bat "bed bugs," streblid and nycteribiid flies (see chapter 6) and fleas. The arthropods can increase to large numbers in highly populated bat roosts, and they attach themselves to bats during foraging flights. Diseases are rarely associated with some of these arthropods, however, mites can cause dermatitis.[8]

5. Cats and Dogs

Pets, particularly domestic cats (*Felis catus*) and dogs (*Canis familiaris*), have been found to carry staphylococci in the skin and nose.[11] They also harbor serotypes of *Salmonella*. Dog-borne *Salmonella* occurs by dogs eating *Salmonella*-contaminated, underprocessed dog food.[11] Domestic cats may defecate directly on food substances in food storage facilities[7] and can serve as a reservoir for *Toxoplasma gondii,* which causes toxoplasmosis.[7,22] The *Toxoplasma* oocytes are found in their feces; thus, disease can be spread by direct contact of the feces in food or by the oocytes being carried to food by both filth flies and cockroaches.[22]

6. Rabbits

Rabbits (Leporidae) shed hairs easily, but unavoidable contamination from wild rabbits is less frequent than contamination from clothing, where hairs are used for fur trim on coats, felt fabrics, fur-lined gloves and other clothing.[2] Rabbit hairs come in various natural colors and lengths and, when found in food products, are usually dyed, indicating that they were from clothing rather than rabbits in the field. Rabbits in Africa, North and South America and parts of Central Asia can serve as reservoirs of plague.[23]

7. Noncommensal Rodents

Squirrels (Sciuridae) are semicommensal rodents but are not primary food pests.[2] They are usually opportunistic feeders and are a source of field contamination. Although they do not occur in large numbers, they may occasionally infest products. They may contaminate foods left in the open to dry, such as fruits or spices.[2] Because their fur is rarely used in clothing, the presence of squirrel hairs suggests contamination by live animals, which, in a food facility, indicates very insanitary conditions.[2] Squirrels have been associated with the plague in Asia, Africa and North and South America.[23]

Noncommensal rodents that contaminate foods differ in various parts of the world. For this discussion, we will use as examples *Peromyscus* and *Microtus*, two genera in the family

Muridae. *Peromyscus* spp. include the white-footed mice and deer mice, which may invade structures from nearby woodlots or fields. The species *Peromyscus leucopus* (white-footed mouse) occurs from southeastern Alberta and Nova Scotia to Arizona and the Yucatan Peninsula.[10] Occasionally meadow mice and voles (*Microtus* spp.) enter structures from outside sources;[6] this genus occurs in North America, Libya, Europe and Asia [10] and has been associated with the plague.[23]

8. Humans

Human hair in foods is a problem because its presence is objectionable and it carries a heavy dose of bacteria.[12] Human hairs found in food products usually consist of smaller hairs from the face and extremities and larger hairs from the scalp. Their presence indicates lack of good manufacturing practices such as wearing head coverings and gloves in the food processing facility.

9. Domestic Livestock

Ungulates or hoofed animals include sheep (Bovidae), cows (Bovidae), goats (Bovidae), swine (Suidae) and horses (Equidae). *Salmonella* has frequently been found in farm animals such as horses, cattle and hogs and these animals may serve as carriers.[11]

E. HAIR MORPHOLOGY

1. Hair Growth and Development

A hair follicle is created from an invagination of the epidermis, which encircles an evagination of the dermis known as the dermal papilla, from which the cells forming the hair fiber are made.[24] Cells forming the different layers of the hair (cuticle, cortex and medulla) are formed along the papilla, joined together, pushed up from the follicle and through the outer layer of epidermis, creating the shaft of the hair. As it extends past the skin the cells in the shaft have died and keratinized. Nuclear material becomes less evident as the hair keratinizes, but may be visible in the medullary cells of the apical portions of the hairs of some mammals under transmitted light microscopy. The nuclear remnant, often referred to as the nuclear ghost, can be seen near the tip of the hair, where medullary cells are still visible but where medullary pigment has stopped, and appears as a faint, slightly darker circle or oval against the lighter unpigmented area. Nuclear ghosts may be observed in many rodents, cats and rabbits, for example.

Varying factors in the follicle affect the type of hair to be created and characteristics such as thickness, length, constrictions, color, etc. In addition to genetic coding, factors such as health, diet, age and environmental conditions can affect hair characteristics such as thickness, length and color. As the hair and follicle grow, they will curl in two planes at right angles to each other, forming a spiral. This condition may explain the protruding scales on one side of shrew hairs alternating from one side to another at internodes. There is no mammalian hair that is absolutely straight.[25] Another factor influencing hairs is the molting of the coat that occurs for many species of mammals, being seasonal for some and independent of seasons for others. In *R. norvegicus* and *M. musculus*, for example, there is no distinct molt cycle and hairs are shed irregularly throughout the year.[26] Animals having a definite molt cycle will shed hairs regardless of their stage of development. Thus, for example, a half grown hair may be shed with a pigmented medulla extending to the root, which, if allowed to mature, would have developed a nonmedullated portion at the root. Knowing this may aid in the identification of some atypical hairs.

2. Hair Structure

a. Cuticle

The three cellular components of the hair are the cuticle, cortex and medulla. The cuticle is the outermost layer and is composed of cells forming overlapping scales, with the exposed portion pointing toward the apex of the hair. The cuticle is unpigmented and transparent in the mammals covered here, except for bats, which may have well-pigmented cuticular scales. Scales may be coronal, in which one scale fully encircles the hair, with the next scale fitting within as in stacked paper cups. Coronal scales commonly occur in bats but may occasionally be found in very thin hairs or tips of hairs of other animals associated with foods. The majority of mammals generally have imbricate scales, overlapping each other as in shingles of a roof. Scale margins may range from smooth to rippled in different species. The cuticle helps protect the inner structures of the hair against the invasion of liquids and external abrasion.

In general, structural characteristics of cuticular scales are more related to hair-shaft diameter than to the mammal species from which the hair came.(2,27,29) Hausman observed that the length (along the hair axis) of the free or visible portion of the cuticular scale varied inversely with hair diameter.(27) That is, the coarser the hair, the finer the scales and vice versa.(27) The amount of scale overlap or imbrication often varies between different kinds of mammals and may be diagnostic. As an example, overlap range may be one-sixth or less for wool to five-sixth or more for human hair.(2) Thus, human hairs are often described as heavily imbricated. Measurement of scale overlap is generally not used for identification purposes in the food laboratory because statistically significant numbers of hairs are required and seldom found in food samples. In contrast to scale length, scale width changes in proportion to shaft diameter so that those at the tip of the hair are small and get progressively wider as the hair diameter increases.(30)

Cuticular scales may vary in shape between species, different parts of the body, and along different parts of the hair itself. The shape of the scales will usually change along the length of a hair, becoming wider at the shield portion. With respect to increasing hair diameter, scale shape generally changes from elongate petal-like to broad with wavy margins.(2) Day uses the terms *lanceolate* and *chevron* to describe two extremes in the range of scale types.(31) Lanceolate refers to individual scales that, side by side, make up a pattern (Figures 1 and 2), whereas chevron describes scales whose edges of irregular waves, rather than separate scales, form the pattern (Figure 3). Actually, scale patterns are not always that distinctive, and most hairs have both types or variations somewhere in between.(31) Scale pattern also varies along the length of the same hair.(2) Those at the base are simpler and, along with those in the tip and shield, vary little among species.(30) In contrast, scale patterns in the shaft are much more distinctive and diagnostic for determining species and other taxonomic groups.(30,31) Scale patterns may change over time as a result of wear.(2) Hairs undergoing much abrasion may show an erosion of the cuticle. Human hairs will often show evidence of cuticle erosion.

b. Cortex

The cortex lies inside next to the cuticle and is composed of longitudinal spindle-shaped cells providing most of the strength of the hair. In some animals, such as the bovines, these cells are still visible as longitudinal striations under magnification in transmitted light. For most of the mammals presented here, this characteristic is not visible under normal conditions, and the appearance of the cortex may range from clear colorless to stained to containing varying amounts of pigment granules. The distribution, size, shape and color of the pigment granules are often diagnostic.

Cortical fusi (spindle-shaped air pockets) may occur in bovine hairs. Other characteristics that may be diagnostic include the thickness of the cortex in relation to the medulla; the nature of the edge between the cortex and the medulla (for example, smooth vs. serrate); annular intrusions of the cortex into the medulla, giving medullae a pinched or hourglass appearance; or cortical material separating multiple medullary rows. Previous literature incorrectly described this cortical intrusion as pegs because of its two-dimensional appearance under microscopic examination. The cortex may be dyed, indicating an origin in clothing fabric or fur, felt, or from coloring in the product it was found in. Alkali treatment, such as from sodium hydroxide, will swell and eventually destroy the cortex leaving it a gel and freeing up the inner medullary cells. This treatment has been used as a means of studying medullary cells. Excessive heat will scorch the hair and, in a medium such as glycerin jelly (see chapter 14), will cause the cortex to expand and distort the appearance of the hair, making identification difficult. Hairs that have been broken or cut or that have rounded ends from abrasion may give clues to identity or to past treatment. Hair from a decomposing animal may show a swollen area at the level of the skin surface.

c. Medulla

The medulla is the innermost layer, with cells forming a channel in the center of the hair and having varied characteristics in the different animal groups. In very thin hairs the medulla may be lacking or appear to be lacking. In the fur hairs, the fine hairs composing the undercoat of animals such as most rodents, shrews, domestic cats and dogs, rabbits, and mustelids such as weasels, the medulla is visible as a discrete row of cells, with pigmentation usually concentrated at one end of each cell and a clearer area at the other. The cells will be separated from each other by an air space. The combination of pigmented and clear areas and air spaces gives the typical fur hairs of the animals mentioned above a striated appearance, and hence the informal identification category of striated hairs. This striated appearance is readily observed at the 30× magnification of a stereoscopic microscope (see chapter 3), normally used for examining light filth extractions from food and other products. Striated hairs are distinguished from the hairs of other mammals, which have typical medullae described as absent or continuous. The wool of sheep is an example of absent medullae. The fur hairs of many ungulates also lack medullae and are often nondescript, making identification difficult. Continuous medullae describes the condition where the medullary units are tightly compressed against each other, often forming what appears to be stacked plates. The guard hairs of cattle, goats, swine and horse body hairs are common examples. Hairs with continuous medullae are often called nonstriated hairs to distinguish them from striated hairs. When no other identification features are present, this rough classification may assist in describing a broad group of animals and gives more information than just the identification of "hair." Fragmentary or interrupted medullae are terms used to describe irregularly present medullae along the length of the hair and may occur in both striated and nonstriated hairs. Fragmentary medullae are often encountered in human hairs and are visible even at the 30× magnification of a stereoscopic microscope.

3. Types of Hairs

The coat of most species of mammals can be divided into the two main groups of guard hairs and fur hairs. This is in large part a distinction of convenience, with hairs actually falling within a range of the thicker, longer, guard hairs to the thinner, shorter, fur hairs.

a. Guard Hairs

Guard hairs, also known as overhairs, are longer, stiffer hairs and are classified by some researchers into different types according to thickness and length or placement of a broadened shield region along the shaft. A major diagnostic characteristic for striated hairs is the internode. Moore et al. define an internode as a narrow portion of the hair next to the shield or in the shaft that exhibits any two of the following qualities: a noticeable decrease in hair shaft diameter, an alteration of medulla formations, a change in scale pattern, and a definite bending of specific areas along the hair giving it obvious angles.[32] Guard hairs may or may not have internodes resulting from constrictions along their length. The medullae may be absent, striated, with multiple cells across in many animals, nonstriated, or fragmentary. Cross-sections include circular, oval, bean shaped, dumbbell shaped, "H" shaped and others.

Guard hairs are typically more differentiated between different taxa of animals than are fur hairs and are more often used as a means of identification. They are not as numerous on the coat as fur hairs, however, and not as often encountered in consumer products.

b. Fur Hairs

Fur hairs, or underhairs, are the thinner, shorter hairs on the coat of most mammals, usually functional as insulation. The medullary channel, when present, is usually a single cell wide. Conspicuous internodes, where pronounced constrictions separate wider node sections, are common in some groups, giving the hairs a zigzag appearance, such as in rodent and insectivore hairs. In some rodents, internodes may appear either as pronounced reductions in hair shaft diameter or more gradual reductions in diameter with an abrupt narrowing of medullary channel. Fur hairs are not often well studied because they tend to be nondescript in many groups of animals, such as domestic livestock, with guard hairs offering more diagnostic features for identification.

Coloration per se may not be a diagnostic characteristic, but pigmentation patterns may be useful in narrowing down the origin of a hair to certain taxonomic groups. For example, many animals exhibit a banding or agouti pattern in which the hair exhibits alternating light and dark areas along its length. This occurs in many striated animal hairs, including rat and mouse.[(2)]

The most accurate level of identification is desired because the significance of the contamination is evaluated based on a number of factors, including identity of the animal. Most hair contaminants are found in limited numbers, making statistical measures of length, type or thickness impractical. For these reasons, it is necessary to mount hairs on microscope slides for viewing under transmitted light microscopy up to approximately 400×. Any entrapped air in the medulla, which occludes internal structures, is first cleared by replacement with the mounting or other medium to allow possible identification.

c. Human Hair

Human hair is covered in the identification table later in the chapter but requires some additional discussion. It occurs fairly often in microanalytical food samples, and situations arise when it may be confused with either nonmedullated striated or nonstriated animal hairs (which include hoofed animals). Because the literature on microanalytical analysis does not really address potential identification problems between human and animal hairs, we have included a detailed discussion of human hair and some of the features that distinguish it from other taxa.

The length of time that a human hair remains rooted in its follicle is proportional to its diameter. That is, coarser hairs grow more slowly and fall out less frequently than finer hairs.[(33)] This is substantiated by the fact that the human hairs frequently found in regulatory samples are small, fine hairs such as those from the face and extremities rather than the long, large hairs from the scalp.

Human hairs can be distinguished from animal hairs by a number of characteristics. Human hairs usually have a gradually tapered point, unless they have been either slightly abraded to form a rounded tip or heavily abraded or damaged causing a split or frayed tip. A freshly cut hair will exhibit a partially compressed tip if cut by scissors or will have a long tail on one side if razor cut. After two to three weeks a cut hair will become rounded at the tip due to wear.[(33)] Animal hairs usually possess a pointed tip.

The scales of human hairs are flattened, heavily imbricated, overlap tightly and are nonpigmented.[(33,34)] The overall shape of the scales may be similar to a wave with crests and valleys, and the edges may have finely and irregularly jagged margins. In contrast, animal scales are quite variable with a wide variety of cuticular patterns ranging from finely to heavily serrated margins.[(33)] However, situations can occur where human hair scales range from invisible to very prominent, and size, pattern and contour of the edges may vary depending on location on body, wear and chemical damage. For instance, cuticular scales on the shaft near the scalp normally have smooth and unbroken edges. Typically, hairs show increasing abrasive damage with increasing distance from the base in the form of broken scale edges, partially removed scales and hair splits indicating cortical damage.[35] This damage may be due to abrasion from hair grooming, backcombing, treatment or weathering, which forces the distal edges of the scales up and allows the hair to be puffed and tangled.[33,35] In fact, some scales may be elevated from the shaft similar to wool, and overlapping scales may be near or widely spaced.[36] Human hair cuticle is normally four to ten layers thick, whereas most wool fibers are one to two scales thick and some mammal fur hairs are up to 20 scales thick.[35]

An example of human hairs that can pose identification problems are the very fine hairs such as forehead and knuckle hairs. As mentioned previously, the smaller hair diameter corresponds to longer and narrower cuticular scales. The scales do not overlap tightly over the length of the hair as in the coarser hairs, but rather exhibit both tightly overlapping and irregularly spaced scales near the distal end due to wear. In some areas, it may appear that some scale layers are completely missing and the diameter over that portion of the hair decreases. Scales near the base are spaced at more regular intervals.

Human hair can be distinguished from wool by several features. Uncut human hair diameter tapers from its widest point at the middle to the two ends. Under magnification of 100–400×, the hair does not exhibit any noticeable changes in diameter along the shaft for the naturally straight hairs of Mongoloid or Caucasian individuals. On the other hand, wool appears to have a uniform diameter under lower magnification but at magnification of 100–400× contains small variations in diameter over its length. It is analogous to making a rope of modeling clay and then gently pinching it at irregular intervals over its length. This gives the hair an appearance of slightly narrowed and bulged areas along its length. Another difference is that wool scales are spaced far apart (about as far as the scale width) and the scales are much more prominent in profile (i.e., they appear to protrude from the hair surface). The edges of the wool scales are fairly smooth and lack the small indentations found in human hairs. Also, as mentioned previously, human hair scales are four to ten layers thick, and wool is only one or two layers thick.

Human scalp hair contains tiny air vesicles in the cortex called cortical fusi. These structures occur in large numbers near the base of the hair, although they can be scattered throughout the cortex or clustered near the medulla.[37] In uncleared hairs, the fusi may vary from small and round to long and oval, or irregularly shaped. The shape, size and distribution are important for comparison.[33]

Although cortical fusi and the medulla affect hair coloring to some extent, the main determining factor is the type and amount of cortical pigment.[33] The cortex contains pigment granules that are arranged in specific patterns to give hairs their color. In general, the color and pigmentations of human hairs are consistent throughout their length, whereas animal hairs may have large color changes or banding over their length. Human hair pigment is evenly distributed or may be concentrated toward the periphery of the cortex. In contrast, animal hair pigmentation is often concentrated around the medulla.[33,38] Human hairs are distinguishable from nonmedullated striated hairs by the fact that the cortical pigment granules are very fine and either finely dispersed or in aggregates in the cortex. In contrast, striated animal hairs have coarser cortical pigment granules arranged in parallel rows or streaks. In human hairs, the size and shape of these granules, along with their density and distribution along the shaft differ for each individual. Pigment granules in Caucasian hair are normally fairly evenly distributed, whereas those in Negroid hair are often clustered in large clumps. This may assist in determining racial origin. Because melanin and medulla formation simultaneously stop as the follicle nears the end of its growth cycle, the last portion of the hair that grows is unpigmented and unmedullated.[33]

Medullae do not occur in fine animal hairs such as wool fibers.[35] When present, animal and human hair medullae may appear dark due to the presence of air. Once the air is cleared, the medullae may appear translucent in human hair and translucent to dark in animal hairs depending on the amount of pigment present. The medulla is usually broad in many animals and can occupy more than half to almost the entire hair shaft diameter, and it often has a distinctive pattern.[33,38] In contrast, the human medulla is rarely greater than one-third the width of the hair shaft and is amorphous in appearance.[33,38] The medulla of human hairs varies based on diameter of the hair shaft.[37] Finer hairs usually lack a medulla whereas coarser hairs usually possess one; intermediate sized hairs have variably fragmented medullae.[28] A number of different classification schemes have been used for characterizing human hair medulla. Hausman, for example, uses the terms absent, fractional, fragmentary, broken or continuous.[27]

There is tremendous variation among hairs from different parts of the body. Head hairs are usually longer than those from other body regions, with greater uniformity in diameter (30–80 μm), texture and appearance. The medullation of head hairs, if present, is generally narrower and less apparent than that of other body hairs. Beard hairs range from the finer ones on the cheeks (rarely thicker than 100 μm) to the coarsest ones on the chin (up to 200 μm). The small hairs of the limbs (20–40 μm in diameter) are arcuate and sinuous in appearance and frequently contain a wide medulla. The tips may be rounded as a result of friction. Eyebrow and eyelash hairs are coarse (60–100 μm), with eyebrow hairs sinuous and eyelash hairs strongly arched. Both have fine tapering tips and prominent medullae.[2] Forensic scientists categorize human hair on the basis of texture, shape of hair in cross-section and pigmentation into three major racial groups: Caucasoid, Mongoloid and Negroid.[33] In general, Negroid head hairs have prevalent undulations (kinky) and tend to be flat in cross-section with the pigment dense and clumped; Caucasoid head hair ranges from straight to kinky, although usually wavy, is oval in cross-section with pigmentation evenly distributed and typically has fragmentary medullae that may be off center; Mongoloid head hair is coarse, straight, round in cross-section and has dense auburn pigmentation.[33,35,36] The rule of thumb is, the curlier the hair, the more elliptical its cross-section; the straighter the hair, the more circular its cross-section.[27]

Major differences exist among the different racial groups, but diagnostic characteristics used are based primarily on head hairs and are not absolutes.[33] The ability to distinguish hairs of one group from another is somewhat limited because of racial mixtures, sample size and poorly defined characteristics. In addition, the high mobility of human populations as well as increased interracial mixtures increases the range of hair variation and makes individualization more difficult.[33,39]

F. GENERAL DIAGNOSTIC HAIR CHARACTERISTICS

1. Hair Characteristics Tables

The hair characteristics tables located at the end of the chapter provide diagnostic characteristics for hairs most commonly found in food products. Because of the high volume and nature of regulatory sample work at the FDA, expediency of analyses while still obtaining precise identifications are of primary concern to the analyst. Consequently, the authors felt that hair characteristics in table form would be much more useful than in text form. The hair characteristics tables include striated and nonstriated hairs and both fur and guard hairs when applicable. The tables are restricted to generalized body hairs because they are the most commonly encountered and, according to Day,[31] hairs from main body parts show a consistent scale pattern and medulla type. In contrast, hairs from the head, snout and extremities such as lower legs, paws and tail have characteristics that may be similar to, very different from or simplified versions of those found on the body.[31,40] It is recommended that examples of the head and extremity hairs of the most common animals contaminating products be kept in the collection of authentic hairs. Tactile hairs (whiskers) also exhibit very uniform structure over different body regions and across species, so they have little taxonomic use.[34] The information provided in the tables is based on observations by the authors as well as data in the literature.[1-3,9,30-33,35,37,38,40-45]

Table 1 lists the combinations of characteristics of fur hairs that can be used to differentiate hair contaminants from commensal murine rodents, *Suncus murinus* and bats. Table 2 lists the combinations of differentiating hair characteristics for *Rattus* spp., *Mus musculus* and *Bandicota bengalensis*. Table 3 differentiates fur hairs from common noncommensal rodents and Table 4 treats other commonly encountered fur hairs. Tables 5 and 6 differentiate types of guard hairs. Table 7 lists characteristics for distinguishing the nonstriated hairs of humans and livestock.

The tables refer to figure numbers that correspond to photographs of hairs mounted in glycerin jelly (see chapter 14) and taken at microscope magnifications ranging from 43—630× (630× with dry lens). The orientation of these photographs is with the root end to the left and the tip end to the right. Much of the available hair literature contains photographs of hairs mounted in balsam, salicylic acid or some other medium. Often, the air is not completely cleared from the medullary channel in the hairs pictured in these other references, obscuring some key diagnostic characteristics.

A few points should be made regarding the tables. When discussing size and shape of the medullary cell, width refers to the dimension perpendicular to the hair shaft, whereas length refers to the dimension parallel to the hair shaft. Distal refers to the apical or tip end of the hair and proximal refers to the basal or root end of the hair. One type of air space that is quite diagnostic is an I-beam air space (typically found in some rodents and the Asian musk shrew), which resembles a capital letter "I" and is oriented perpendicular to the direction of the hair shaft. When describing the shape and relative size of these air spaces, the shape of the main shaft is referred to as thick or thin to describe its width and tall or short to describe its height. These terms are used instead of wide and long so as to avoid confusion with these terms, because they are used with medullary cells.

2. Hair Identification Key

Fur hairs make up the majority of the animal's pelage and are the most commonly encountered type in food products, whereas sensory and other types of specialized hairs (such as those from appendages, muzzle and tail) are much less frequently encountered in foods. Identification of fur hairs is more difficult than guard hairs due to structural similarities among the hairs of widely separate taxonomic groups. In fact, scientists who study mammalian hairs usually use guard hairs to identify the source animal because they exhibit consistent structure, and they are larger and have more easily observable characteristics than fur hairs.[34] Much of the identification literature for guard hairs is used primarily for identifying hairs from animal pelts or as a diagnostic tool in animal ecology and wildlife biology and management to determine predator diet from hairs found in their feces. In these cases, the availability of numerous hairs from the source animal aids in its identification. In contrast, microanalytical samples often contain a single or a very limited number of hairs and hair fragments.

There are a number of identification keys for guard hairs, but few keys have been published for fur hairs.[30-32,43,44,46] Two exceptions are Mathiak and Brown, whose works include both fur and guard hair keys.[40,41] However, the only mammal covered in Brown's key that is usually found in foods is the weasel, and Mathiak's key briefly describes fur hairs for only a limited number of the taxa that are of interest to microanalytical analysts. Even the other previously mentioned guard hair keys provide only limited coverage of the taxa whose hairs occur in microanalytical samples; some include commensal rodents, very few include domestic cats and dogs and none include the Asian musk shrew. Consequently, and due to the scarcity of identification keys on fur hairs, the key in this chapter focuses only on striated fur hairs that are found as contaminants in foods.

The key is based only on generalized body hairs rather than specialized body hairs for the reasons discussed in the introduction to the hair identification tables. In the key, some species are grouped together because of the close resemblance among their hairs and similarities in their regulatory significance (e.g., cat/dog, rat/mouse). The key is based on the observations of the authors and others.[1–3,9,30–33,38,40–45,47,48]

When using the key, hairs should be compared with all of the characteristics in the couplet to ensure correct identification. Also, the whole hair or a large sized fragment containing the necessary diagnostic characteristics should be used. Damaged or treated hairs may not key out properly (see Section F.3). Supplemental comments and characteristics are included in certain couplets in the key to assist in hair determinations. Photographs corresponding to figure numbers in the key are located at the end of the chapter. It would be useful to familiarize oneself with the key by first using it with known specimens before trying to identify unknowns found in food samples.

The key, photographs and tables can be used to aid in the identification of hairs, but ultimately positive identifications should be verified with authentic material. Analysts who perform hair identifications should establish their own reference collection so that they can confirm their identifications by comparison with authentic specimens. Due to the high degree of variability of hair characteristics, one should make up several microscope slide mounts of authentic material from each of the different body regions (i.e., dorsal, ventral and lateral, extremities, head) of each specimen. The age of the mammal, its sex, regional and seasonal differences can also affect microscopic structures. It is very important to compare the same portion of the unidentified hair and the authentic specimen (i.e., base with base, middle with middle or tip with tip[34]) because many characteristics vary with hair shaft diameter. In addition, for many mammals, the dorsal hairs have more cortical pigment and are darker than the ventral hairs that make up the lighter belly area of these animals.

Key to the Identification of Striated Fur Hairs Commonly Found in Foods

(*Note:* When discussing medullary cells in a hair, width refers to the dimension perpendicular to the hair shaft and length to the dimension parallel with the hair shaft. Characteristics are listed in descending order of importance.)

1. Cuticular scales coronal; may be simple or ornate. Pigment granules fine to large, in parallel strings in scales (Figures 20, 21); pigment evenly distributed throughout hair, but decreases at tip and base of hair. Medulla usually absent (Vespertiliontidae) .... Bat

    Hairs of food-contaminating bats have hastate coronal scales that encircle entire hair shaft; may be unequal and cause the hair to look asymmetrical (characteristic of bats, especially most Vespertiliontidae. Rarely do they occur in other orders, and then only occasionally in very thin hairs). Some bats have medullated hairs including a few of the fruit bats that get into buildings (Molossids, etc.), but their hairs still have coronal scales and are not found in food very often.

    Cuticular scales not coronal; variable [petal lanceolate (Figure 1) to chevron (Figure 3)]. Pigment granules not in scales but may be in medulla and/or cortex (Figures 6, 9, 25, 26, 30). Medulla present 2

    Medulla may be absent, interrupted or fragmentary in some cases such as in the finer hairs of animals such as rabbit, dog and some rodents, and in specialized hairs; however, if so, scales not coronal.

2. Cortex clear to hyaline (Figures 6, 9, 17, 35); may have large, dark, round to elliptical/rod-shaped pigment granules scattered or in streaks in distal portion. Distal end may have stainlike pigmentation or small, dark pigment granules often in longitudinal streaks and may be heavily pigmented. Medullary pigment granules large, brown to black, and round or oval (Figures 6, 9, 10, 17, 35) to elliptical or rod-shaped (Figure 23); in

compact mass (some hairs may be unpigmented, but other characteristics still apply). Air space I-beam shaped (Figures 6, 9, 17), biconcave (Figure 35), square or rectangular with distal end concave (Figure 36) or bowl shaped (Figure 22). Rodents, rabbit, Asian musk shrew (*Suncus murinus*) 3

> Scales may be symmetrical (rodent) (Figure 6), have parallel oblique arrangement (rabbit) (Figure 34) or be asymmetrical with scalloped appearance that changes sides at internode (shrew) (Figure 16); never fingerlike. In the distal portion of rabbit hairs, the air spaces may appear linear with a series of semicircular indentations on the distal edge (Figure 37). In the distal portion of rodent and shrew hairs, the I-beam type air spaces may be distorted and curved with cross-beams of I extending apically.

Cortex somewhat clear to cloudy (Figures 27, 28, 31) to containing fine cortical pigment scattered (Figure 32) or in parallel rows throughout length of hair (Figures 26, 30). Medullary pigment granules fine, elliptical, scattered (not in tight packet); may appear in an X- or chromosome-shaped arrangement (Figures 25, 26, 29). If nonpigmented, may appear granular (Figures 27, 31). Air spaces linear (Figures 26, 27), bowl (Figures 30, 31) or crescent shaped (Figure 33); never I-beam shaped. Cat/ dog, weasel 8

> Scale pattern variable, may be prominent, fingerlike (Figures 32, 33).

3. Scales asymmetrical—scales on one side protrude more prominently on all nodes except apical node, giving a "scalloped" pattern that switches sides at internode (this is more visible at lower magnification) (Figures 16, 18). Apical node without evident scallop pattern; the side that would be scalloped has what appears to be a double edge, as seen by focusing up and down at higher magnification (Figure 15). Sometimes nonmedullated area near base of hair may express this asymmetrical pattern as an enlargement on one side that narrows down to an apparent internode and then continues as the asymmetrical scale pattern on the opposite side (Figure 19). Tip of hair more abruptly tapered than in rats and mice. Internodes abrupt, gives hair zigzag appearance at lower magnification; medullary units become narrow but increase two to three times in length, and pigment packets appear to run together (Figures 16, 18). Medullary pigment cells near base may run together similar to that at internode (Figure 19). At node, medullary pigment densely packed in oval or bean shape. No cortical intrusion apparent, no hourglass shape (Figures 15, 17, 18). Unlike in rats and mice, the medulla in the distal node/shield is usually pigmented (Figure 15) .... Asian musk shrew (*Suncus murinus*)

   Scales not distinctly asymmetrical; no scalloped pattern (Figures 5, 11, 12, 13, 14, 34). Nonmedullated area at base of hair may be slightly enlarged (Figure 8), but not asymmetrical as in *Suncus*. Internodes variable; absent, gradual or abrupt (Figures 4, 5, 11). If abrupt, hair has zigzag appearance (apparent at lower magnification) (Figure 4), and medullary units at internodes remain relatively the same length or may increase about two times, but pigment packets remain discrete and never run together (Figures 5, 11). Pigment in medullary cells near base remain in discrete units. At node, medullary pigment in compact packet; shape variable, may be irregular in outline (often with proximal portion of packet narrower than distal portion) (Figures 6, 9, 10, 12, 14). Cortical intrusion, if present, pinches packet; if prominent, gives medullary cell an hourglass shape (Figures 6, 9, 10) or evenly pinches packet at midpoint (Figures 14, 35). Rodents and rabbit 4

   > Scale pattern usually petal lanceolate; chevron in rabbit, *Microtus* and squirrel.

4. Scales appear as series of parallel lines at an oblique angle, especially visible along periphery of hair shaft over cortex (focus up and down) (Figure 34). Air spaces thick I-beam or biconcave (Figure 35), square or rectangular with concave distal end (Figure 36); in distal portion of hair linear with series of semicircular indentations on distal edge (Figure 37). Internode absent or gradual. Medullary pigment packet very dense and compact; cannot usually distinguish individual granules. Pigment granules dark and round .... Rabbit

   Domestic rabbit hair may be white or dyed; if so, parallel oblique scale pattern may not be apparent, perhaps due to thinness of hairs commonly encountered. Air spaces throughout are usually square or rectangular with concave distal ends.

   Scales do not appear in parallel oblique orientation. Air spaces I-beam shaped (can be short with thick main shaft) or bowl shaped. Internode present or absent; if present, usually abrupt and with zigzag appearance. In some cases, the overall hair diameter at the internode may not decrease abruptly, but the medullary channel may narrow to about one-half of its width at a node (Figures 12, 13). Medullary pigment packet compact, individual granules usually distinguishable, at least along perimeter of packet. Pigment granules dark, round to oval to rod-shaped. Rodents 5

5. Medullary pigment large, rod or elliptical shaped (Figure 23). Rod- or elliptical shaped pigment may occur in cortex as longitudinal striations (characteristic of squirrel). Pigment may appear as a fine staining at distal end. Internode absent. No cortical intrusion. Air spaces bowl shaped or linear with concave distal end (Figure 22) .... Squirrel

   Medullary pigment large and round to oval. If cortical pigment present, round to oval, scattered and not abundant except near the tip; may have streaks in distal one or two nodes. Internodes present, usually abrupt and with zigzag appearance (apparent at lower magnification) (Figure 4). Cortical intrusion present, usually pinches pigment packet (except *Microtus*). Air spaces usually I-beam shaped but may have proximal portion of cross-bars less prominent (*Bandicota*). Other rodents 6

6. Cortical intrusion does not pinch pigment packet. Medullary pigment packet width at node one to three times length; square to rectangular with rounded corners (shape may be slightly irregular). Pigment granules very large, dark and round, densely packed so that only granules along perimeter of packet are discernable. Cortical pigment in distal portion may be heavy .... Vole (*Microtus*)

   I-beam air space with main shaft of I-beam fairly thick (up to diameter of three pigment granules) except in apical and subapical nodes, where the main shaft is very thin (thickness of one pigment granule).

   Cortical intrusion prominent, pinches medullary cell (Figures 6, 9, 10). Medullary pigment in compact packet, shape usually irregular (Figures 6, 9, 10). Pigment granules large, dark and round; individual granules throughout packet discernable (except *Peromyscus*, whose granules are very dark and densely packed and only discernable around perimeter of packet). Cortical pigmentation in distal portion variable. Diamond petal or lanceolate scale pattern (Figure 1). Rat/mouse 7

7. Medullary pigment packet in nodes irregular in shape (often with proximal portion of packet narrower than distal portion due to prominent cortical intrusion) (Figures 6, 9, 10). Gives cell an hourglass shape, but because clear area at proximal end of cell occupies up to half of cell length, hourglass consists of pigmented distal end and proximal clear end. Individual granules may be discernable throughout packet. Air spaces usually I-beam shaped but may have proximal portion of cross-bars less prominent (*Bandicota*). Main

shaft of I-beam may vary from thin (diameter of one pigment granule) to thick (up to three pigment granules) .... Rat/mouse (*Rattus*, *Mus*, *Bandicota*)
Medullary pigment packets in nodes divided equally by cortical intrusion so that portions distal and proximal to intrusion are about equal in width and length (Figure 14). Clear area up to typically one-fourth or more of medullary cell length (can be up to one-third), so that most of cell is occupied by pigment packet. Pigment granules so densely packed that only granules around perimeter are discernable. Main shaft of I-beam air space very thin (never more than 1-1/2 times the diameter of a pigment granule); can appear dumbbell shaped .... *Peromyscus*

> *Peromyscus* is not commonly found in food establishments and is normally only found in rural areas. Its geographical distribution is North America, Mexico, Central America south to Panama and adjacent islands, but not the West Indies.[10]

8. Scales in proximal portion of shaft fingerlike with scales along edge of hair appearing sharp, prominent and angular in profile (Figure 32). Medullary pigment fine. If cortex pigmented, in fine scattered rows or a stain. Air spaces narrow and linear, concave distally (Figure 33) or rectangular with up to three-fourths of distal end of air space concave. May appear as a very distorted I-beam with cross-bars short on proximal side .... Weasel
Scales in proximal portion of shaft not fingerlike; edges may be prominent but not sharp and angular in profile (Figure 25). Medullary pigment fine, elliptical, in loose to dense mass. May be X-shaped (Figures 25, 26, 29). If cortex is pigmented, appears in scattered parallel rows (Figure 26). Cortex wide relative to medulla (Figures 30, 31). Air spaces are bowl, lenticular, crescent (Figures 30, 31) or linear shaped with concave distal end (Figures 26, 27, 28). Cat and dog 9

> It may be difficult to separate these into different taxa. Refer to table and photos.

9. Medullary cell usually longer than wide; however, if medulla width greater than one-half the overall hair width, cell is wider than long. Adjacent X-shaped medullary cells may appear to form a ball and socket joint (see also Table 4, footnote 1) (Figure 26). Combined width of cortex is one-half or less than medulla width (Figures 26, 28). Scales near base may be prominent (Figure 25). Air spaces bowl, crescent, or linear with concave distal end (Figures 25, 26, 27). Internodes may be gradual (Figure 29). Cortical pigment fine stain to small, elliptical granules in faint to dense parallel rows; corresponds to banding in medulla. Cortical fusi absent. Nuclear ghosts may be present in medulla. Root shaft usually straight (Figure 24) .... Cat
Medullary cell usually wider than long. Cells have no distinct shape. Combined width of cortex one-half or more of medulla width; sometimes asymmetrical with one side of cortex thicker than other (Figures 30, 31). Scales near base not prominent as in cat. Air spaces bowl, crescent or lenticular shaped (Figures 30, 31). Internodes absent. Cortical pigment fine (coarser than cat); can occur in dark patches (Figure 30). Cortex lightly striate (Figure 30), stained or cloudy (Figure 31); may be banded. Cortical fusi may be present. Nuclear ghosts absent. Root usually spade shaped, however may be straight as in cat .... Dog

3. Identification Problems with Hairs

Factors causing difficulties in identification of hairs may originate from the analysis to recover and identify the hair, or from treatment the hair may have undergone during processing of the product, or from atypical hairs, or from a combination of the above.

a. Handling and Technique Problems

Analytical problems in general are covered in chapter 15. Specific to hairs, analytical problems may occur with the microscope slide mount. If the hair is not cleared of air, internal medullary structures may be occluded, making identification difficult for most of the hairs normally found contaminating products. Overheating of the hair in an attempt to deaerate it may cause a swelling of the cortex and a distortion of internal and external features. Overheating may occur during deaeration on the slide when the glycerin jelly mount is allowed to boil too vigorously or for too long at high temperatures. Other treatments during analysis that may affect the hair include too strong of an acid solution during digestion of the product, which may breakdown medullary cells, and use of bases such as sodium hydroxide in too high a concentration, which swell and eventually break down the cortex. Enzymatic digestion treatments may also damage hair structure. Many hairs are relatively flat and tend to curl on edge when mounted. Identifying features and typical illustrations are based on the hair lying flat. When viewed on edge, a hair may appear quite different from when viewed flat and this may cause misidentification. A special problem may occur if the hair is very long and is difficult to lay flat. Pressure on the cover slip while the mounting medium is still liquid may help. If not, cutting the hair into manageable lengths may offer a solution, but care should be taken to preserve the order of the sections from root to tip if this is needed for identification or future study. Mounting a thick guard hair with a fur hair may prevent the fur hair from lying flat.

b. Damaged and Treated Hairs

Damaged or treated hairs may cause identification problems. Treatment received during processing may alter characteristics of the hair. Milling of a product such as wheat will cut the hairs into short fragments or occasionally create a series of zigzags or sharp corners that may confuse the analyst. Milling may also cause some damage to medullary structures, especially near the cut ends. Care should also be taken during the analyses not to cut or crush hairs. Dyed or discolored hairs may mislead the analyst. If dyed very dark, the hair may not resemble a hair at all. Even at the 30× magnification of a stereoscopic microscope, however, the hint of a scale pattern may be discerned when light reflects off the wet surface of the hair. For this purpose, it is necessary that the hair is not completely submerged in liquid. Wool and rabbit hairs are examples of dyed hairs originating from clothing or felting that are frequently found in food products. Commercially dyed fur may also be encountered. The food, an ingredient or the processing may also discolor a hair to make it look atypical. The analyst should be aware of this possibility to first, recognize a contaminant as a hair and then, to make a correct identification. Abrasion, both before and after being shed, may affect the ease of identification. Humans and some draft animals may show heavy wear of the cuticle and quite rounded, blunt ends. Milling and various other processing may also cause abrasion. Highly acidic solutions used during processing may produce the same effect on hair medulla as mentioned in analytical problems above, but this is rare. More common are hairs swallowed during grooming behaviors by animals such as rats and mice and passed through the digestive tract with the feces. These embedded hairs are a convenient way to identify the origin of the fecal pellet. Medullary cells may be disrupted but a long enough section may be present for identification. Loose hairs originating from fecal pellets may be found in products when processing has broken up or dissolved the pellet. The fecal origin of the hair may be inferred when processing or analysis would not account for the "digested appearance."

c. Hair Fragments

Hair fragments may pose a number of problems for identification. It is at the shield or widest point of the hair that identification characteristics are usually most typical. A fragment of the apical tip of the hair will often exhibit a rudimentary scale pattern and medullary structure not typical of the shield or widest point. Pigment granules may be finer or more stainlike both in the cortex and medulla. Cortical pigment may be more profuse than in other parts of the hair, as occurs in *Rattus*. Again, it should be stressed to compare unknown hair specimens with authentic material of similar hair section, thickness, coloration, etc. The tips of the hairs of *Bandicota bengalensis* may be unmedullated for a relatively long distance, and a fragment of this section may not provide enough clues and be difficult to identify. The same situation may occur with fragments from other parts of the hair as well. Dog hairs, for instance, may have unmedullated sections, which, if found as fragments, not only would be difficult to identify as dog but may not even be recognizable as a "striated hair." Of course, if identifiable dog hairs were also found along with such a fragment, then a search of authentic dog hairs with unmedullated sections may prove helpful. As a reminder, a classification of "striated hair" assists in narrowing down the source to a group of mammals including rodents, shrews, rabbits, cats, dogs and others and excluding humans and livestock such as sheep, cattle, swine and horses. It must be understood, as well, that the shorter the fragment the less information may be contained, making identification correspondingly more difficult.

d. Variants

Variants from typically described hairs may occasionally be found that may not lend themselves to easy identification. Most typically described in the literature are dorsal hairs. This is justifiable because the dorsum probably has the most representative pelage and, taking an animal such as a rat, the back and sides, which are similar in coloration, make up the bulk of the coat. Taking the same rat, however, the belly may be lighter and hairs of the head and legs may be shorter and more spindle shaped. These hairs from other body regions may appear differently microscopically and the analyst should be familiar with such hairs from study of an authentic collection. The whitish unpigmented or lightly pigmented hairs may be easily missed against the white background of the filter paper on which extraction results are filtered. If perceived at all the hair may somewhat resemble a nonhair fiber. Some clues that may suggest a hair even at the 30× of a stereomicroscope are a gradual tapering instead of parallel sides, the presence of a root or tip, the recognition of a scaly versus a smooth surface with reflected light as long as the hair is not totally submerged in liquid, and the hint of internal "stuff." Sensory and other specialized hairs are rarely found in food samples, are very difficult to identify and are not covered in this study.

Other variants in terms of thickness deserve mention. Occasionally, very thin hairs will be found that may look very different from the norm. Such a rat or mouse hair, for instance, because of its thinness, may have ill-defined medullary cells, air spaces that if present at all may not appear I-beam shaped, may have an interrupted medulla with unmedullated portions, may not have an internode, and if thin enough may even appear to have coronal scales in some areas. Corresponding effects may be noted in other animals. Thicker hairs, excluding specialized hairs, tend to be less troublesome to identify but may be confusing when they are transitional between fur and guard hairs. A hair with a single row of medullary units, for example, that changes to two rows and back to one may have medullary units in the single sections much wider than typically encountered in normal fur hairs and should be taken into account in identification.

The presence or absence of internodes when not typical may also cause confusion. In animals typically having internodes, a lack of internodes may occur in a very thin hair, as mentioned above, but this condition may also occur with some hairs that are transitional between fur and guard, and in hairs of the head and legs or paws. The presence of internodes may occur in animals that do not typically exhibit them. This generally occurs as a gradual thinning and return to normal thickness. The thinner section may or may not have medullary units affected, with the thinning often affecting only the cortex. Common animals exhibiting this condition are cats, dogs and rabbits.

4. The Reference Collection

As explained in chapters 3 and 14, accurate identifications of specimens require comparison with authentic material. Keys, descriptive text and illustrations aid in making tentative identifications, but a reference collection of authentic material is essential for scientifically valid identifications. Personally examining known pelts or numerous hairs and making numerous slide mounts will familiarize the analyst with the variety of hairs that may be encountered as well as amass a collection of authentic material for future reference. The goal is to have representation of the different varieties and colorations of hairs likely to be found as contaminants. A number of hairs from any particular taxon is necessary including examples of different hair types such as guard or fur, different body regions, different strains or breeds, geographic or seasonal variations, and age and sex differences. Authentic material should be mounted in the same medium or in media of similar refractive index as that used in mounting unknown hairs. Hairs specimens may appear different in mounting media of different refractive indices. All reference specimens must be authenticated and properly labeled to retain their authenticity.

G. MICROSCOPIC TECHNIQUES FOR MOUNTING AND CLEARING HAIRS

Glycerin jelly may be used to mount individual hairs on microscope slides. Glycerin jelly provides advantages over other media, being a semipermanent medium that can last a number of years if the coverslip is ringed with a sealant such as fingernail polish to prevent drying. Hairs mounted in glycerin jelly need no elaborate pretreatment as with some permanent media and many finer hairs can be deaerated quite quickly in the medium itself on a microscope slide. Filth analysts thus have a rapid means of mounting and identifying a contaminant hair.

The medullary region in hairs from rodents, shrews and many other animals contains entrapped air. Mounted on a microscope slide and viewed under magnification, this air appears as dark bubbles in the hair, occluding medullary structures and making identification difficult. Direct heating of the microscope slide with a hair mounted in glycerin jelly is often sufficient to deaerate fur hairs but not very thick guard hairs. A quick method for deaerating thick hairs was devised by Don Vail, formerly of the FDA District Laboratory in Atlanta. The Vail method is recommended as follows: Pour 2–3 ml of an isopropyl alcohol–propylene glycol solution (1 + 3) in a 4–5-ml porcelain crucible. Place the hair in the solution and heat the contents on a small hot plate until the solution fumes. Remove from heat for fine fur hairs. Coarser guard hairs may require a longer period of heating or repeated periods of heating. Make a permanent slide by mounting the hair directly into glycerin jelly. Avoid prolonged overheating, which will cause hairs to curl and swell. This method consistently clears even the thickest rat guard hairs. A slight variation of this method is to use a temporary mount with propylene glycol to determine if all the air was removed. If not completely deaerated, the hair is easily removed and returned to the heated solution. The hair is mounted directly in glycerin jelly when deaeration is complete.

Care must be taken not to lose the limited number of hairs found in food samples in the transfer from one source to another. The following slide mounting technique is one method that may prove helpful. A small, cleanly cut piece of glycerin jelly is placed in the middle of a microscope slide and gently heated to melting on a hot plate, avoiding formation of bubbles. A hair is taken from a source (e.g., filter paper disk or crucible) using fine-tipped forceps under a stereomicroscope. Holding the specimen stationary above the microscope stage, replace the original source with the microscope slide, place the hair and press it to the bottom of the drop of glycerin jelly. There is less chance of the hair drifting away when placing the cover slip if it is at the bottom of the drop. For optimum optical definition, number 1-1/2 thickness cover slips should be used, as most microscope manufacturers correct objective lenses for spherical aberration at this thickness. The coverslip should also be glass, not plastic. While holding one edge of the coverslip with forceps, place the opposite edge on the glass slide (use the forefinger from your free hand as a stop for the edge of the coverslip to rest against so that it will not slide) and gently lower the coverslip onto the glycerin jelly. This will avoid trapping air bubbles under the coverslip. After placement, the cover slip may be gently pressed to make the hair lie flat. The slide is then heated on a hot plate to remove entrapped air, if the Vail method is not used, and to spread the medium to fill the area under the coverslip. It may be allowed to reach boiling, but care should be taken that the medium does not boil too vigorously and throw off the cover slip, move the hair to the edge, or leave bubbles after cooling the slide. Excessive heating may also distort the cortex of the hair, making identification difficult. Only gentle heating is needed for cleared hairs. After removal of interior air, which can be checked under a compound microscope, the slide can be set aside overnight and then ringed with nail polish or equivalent to prevent eventual drying of the medium. Slides prepared with glycerin jelly are best stored flat, not on edge, to avoid migration of specimen or medium. Slide cabinets that provide for flat storage of slide mounts are necessary to contain these reference collections. Useful slide labels to identify the specimen are 15/16″ square and "pathology thickness," providing appropriate writing space and protection to the cover slip.

II. FEATHER IDENTIFICATION

A. FEATHERS AS ADULTERANTS

Feather fragments are the most common bird contaminants found in food products, and their presence may be indicative of insanitary practice or noncompliance with good manufacturing practices. Because birds have been proven to be carriers of such human pathogenic organisms as *Salmonella*, *Listeria monocytogenes*, *Vibrio cholerae* and numerous other bacteria, viruses, protozoa and parasites,[7,49-52] it is important from a health standpoint, as well, to recognize their defilement of food products.

B. SOURCES OF FEATHER CONTAMINANTS

The three most common pest birds associated with food establishments in the United States are feral pigeons, starlings and English sparrows.[49] These pests or close relatives are found worldwide. Determination of origin of feather fragments to a specific taxon is generally not done. For example, starlings and English sparrows belong to the order Passeriformes whose members exhibit a general similarity in the micromorphology of feather fragments, making differentiation between the two groups difficult. In addition, identification depends on knowing the origin of the fragment in terms of location on the body and on the feather itself. Similar parts of a feather from different body regions will differ. For instance, downy barbules from a breast feather will differ from the same type of barbules from the back or wings. Thus, identification of the source bird is generally not done except in cases where it is necessary to tie a particular type of bird found in an establishment with feather contaminants in a product. The task is made more difficult because feather contaminants are not usually found in large enough numbers to reliably identify a passerine bird beyond order. Other orders, including Columbiformes, to which pigeons belong, have greater micromorphological diversity of feather structures between subgroups and are easier to identify. Some references exist to assist in the identification of the fragments of feathers that may commonly be found in food products.[9,31,50,53-57]

The same light filth methods (see chapter 15) used to isolate hair and arthropod contaminants from food products will also recover feather contaminants, because these are oleophilic as well.

C. FEATHER MORPHOLOGY

Although whole feathers are occasionally found in products, much more common are fragments consisting of substructures of the feather. To recognize the fragments of feathers usually encountered, it is necessary to have some knowledge of the types and morphology of feathers. The three main types of permanent feathers are the contour feathers, down feathers and hairlike filoplumes.[54] Filoplumes are not normally found and will not be covered in this discussion. The contour feathers form the contour of the bird's body and include the flight and tail feathers. Down feathers are the soft underplumage used for insulation. A downy structure may also exist at the basal portion of many contour feathers.

The general structure of a feather is described by Chandler.[53] The main stem of a feather is called the quill and consists of the shaft or rachis, to which the vanes are attached, and the calamus, which is the hollow base. An accessory or secondary feather called the aftershaft or afterfeather may also be present, originating near an indentation above the calamus. The two vanes are attached on opposite sides of the shaft and consist of barbs, usually with barbules. The barb is the primary branch of the shaft and consists of the rachilla or ramus to which are attached the barbules.[57] The barbules, whole or fragmentary, are the most commonly found feather contaminants in the light filth analyses of foods, with barbs, whole or fragmentary, having at least one barbule, being second. Barbules are mostly microscopic in size and thus not easily discovered in macroscopic, or unaided visual, examinations. The descriptions that follow will cover the commonly encountered contaminants.

Barbules are basically of two types: downy and pennaceous. Downy barbules are the more commonly encountered of the two, being more easily detached and airborne. They are relatively long, filamentous and gradually tapering (Figures 60, 61 and 62). They consist of a series of interlocking nonliving cells with a widened, flat basal cell that attaches to the rachilla of the barb and a filamentous section called the pennulum divided into units of internodes and nodes. The internode is a section longer than wide, roughly parallel sided, often unpigmented and at the end of which is a node, which is the attachment point of the succeeding internode. When pigmentation is present in the internode it is usually light brown in color and diffuse. The shape of the node may range from a gently expanded end of the internode to an abrupt structure at the end of the internode appearing as oval, knobby, triangular, or modified as prongs. The nodes may be pigmented or not. In the case of passerine birds, pigmentation is present and consists of a compact black mass becoming thinner in distal units (Figures 60 and 61). At the 30× magnification of a stereomicroscope used to examine light filth extractions, these downy barbules may appear as a regularly spaced series of black dots with the internodes transparent and difficult to see. The downy barbules of pigeons are larger, usually unpigmented proximally and with triangular shaped nodes (Figure 62).

The pennaceous part of the vane, which is present in the contour feathers of the body and tail and wing feathers, is an intricate interlocking structure that is strong, lightweight and mendable. This is made possible by the adaptations of the pennaceous barbules (Figures 63, 64 and 65), which on the distal side of each barb have small hooklets to grab onto the pennaceous barbules on the proximal side of the next barb (Figure 65). Thus each barb supports the one before and after it giving the pennaceous portion of the vane its unique qualities of lightness and strength, and if separated the hooklets can be easily reconnected by preening.

Pennaceous barbules are basically flat with a base and a pennulum. The base is composed of multiple cells, is bandlike, broader where connected to the rachilla and tapering to the pennulum. The last few cells before the pennulum may have small projections ventrally, called ventral teeth. The pennulum on distal barbules starts with a number of cells each with a ventral hooklet. The remainder of the pennulum of distal barbules and the whole pennulum of proximal barbules is generally filamentous, although thicker and much shorter than that of downy barbules. The pennulum on pennaceous barbules may also be adorned with ventral and occasionally also dorsal prongs or cilia, although this is not common in the three pest species emphasized here. The cellular nature of the pennaceous barbules can usually be discerned at 30× with a stereomicroscope. Individual cells are usually separated by a thin clear area. Pigment is usually present and ranges from light brown to black. Nuclei may be seen as uncolored ovals in the cells of the base. Damage can easily occur to pennaceous barbules, and fragments can be found with missing pennulum or part thereof (Figure 63), or with the base partially or wholly missing. In all cases, understanding of the cellular nature and general morphology should allow recognition of the fragment as a pennaceous barbule.

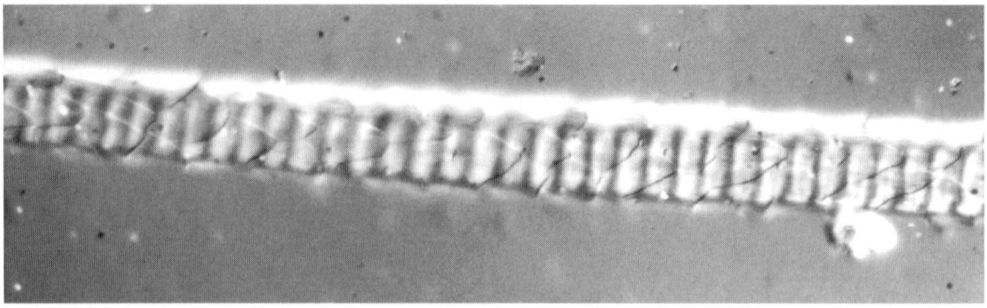

FIGURE 1. *Mus musculus*, albino Webster mouse, petal scale pattern 500× (Differential interference contrast [DIC] microscope).

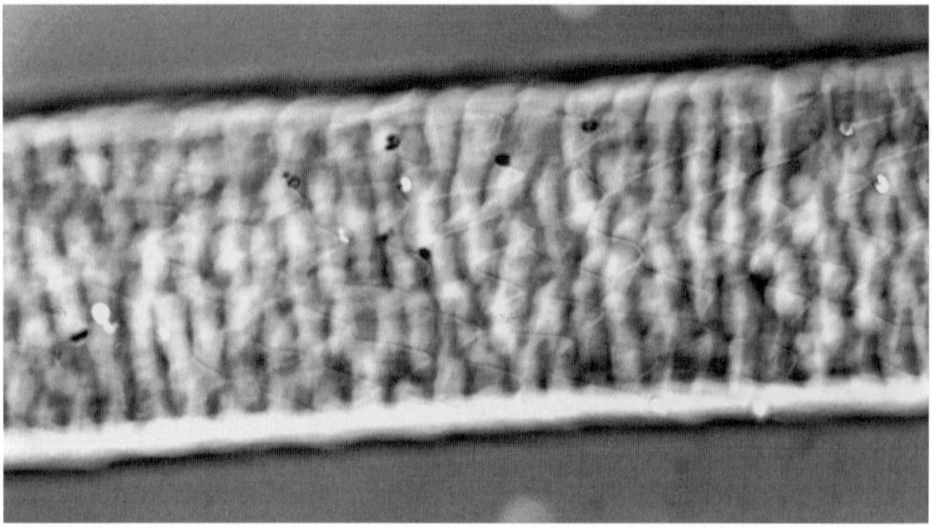

FIGURE 2. *Rattus norvegicus*, Norway rat, guard hair, petal scale pattern, 500× (DIC microscope).

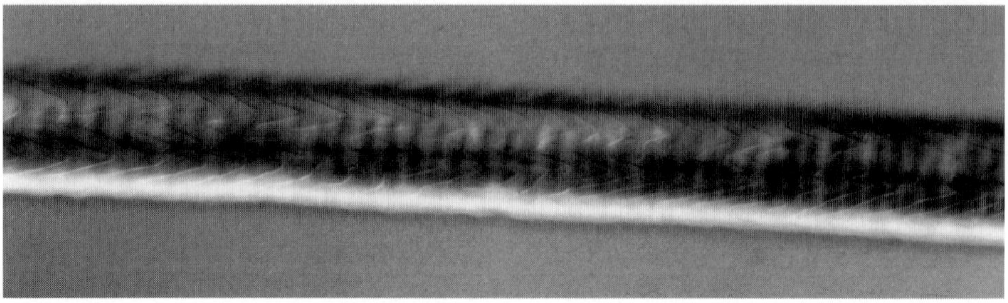

FIGURE 3. Cottontail fur hair, chevron scale pattern, 500× (DIC microscope).

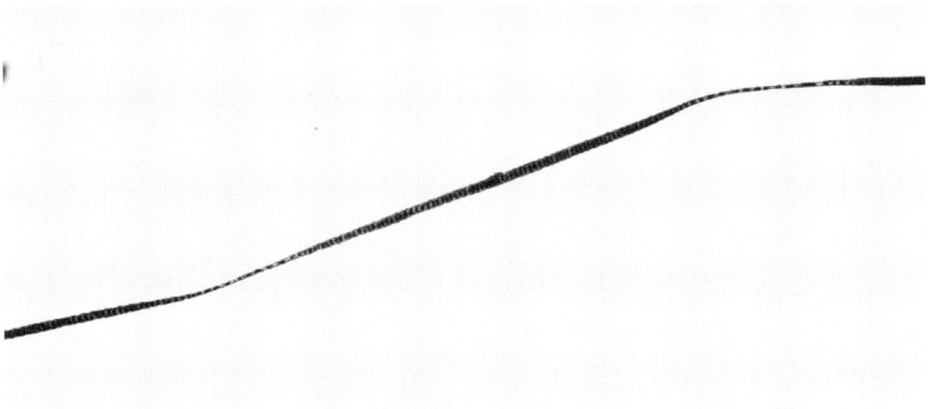

FIGURE 4. *Mus musculus*, house mouse, fur hair, internodes showing zigzag shape, 63×.

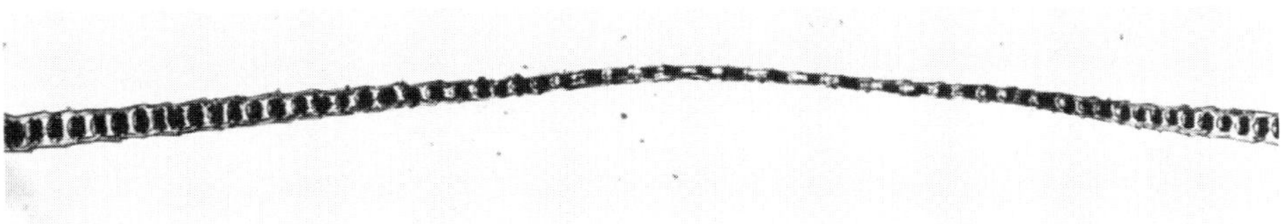

FIGURE 5. *Mus musculus*, house mouse, fur hair, internode, 160×.

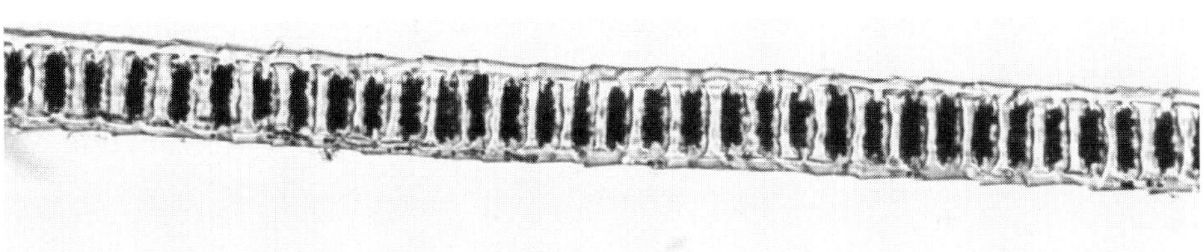

FIGURE 6. *Mus musculus*, house mouse, fur hair, node, 400×.

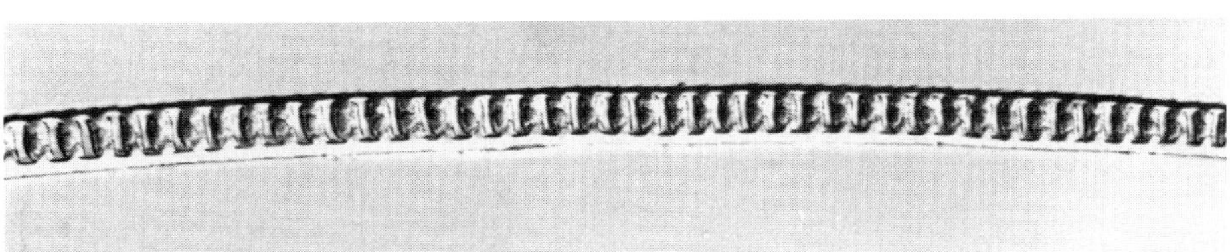

FIGURE 7. *Mus musculus*, house mouse, fur hair, apical node with nuclear ghosts, 500× (DIC microscope).

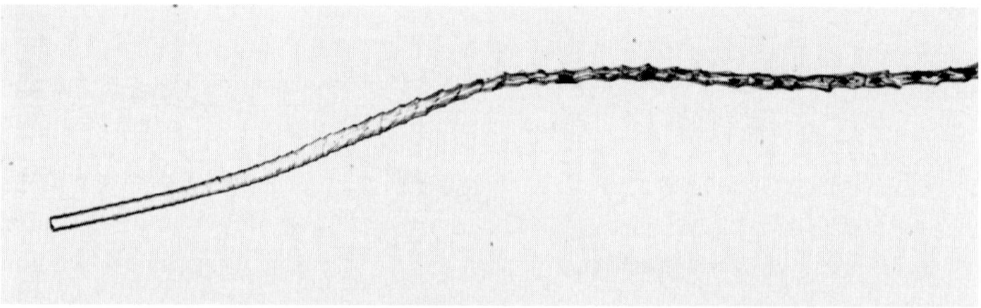

FIGURE 8. *Mus musculus*, house mouse, fur hair, base, 160×.

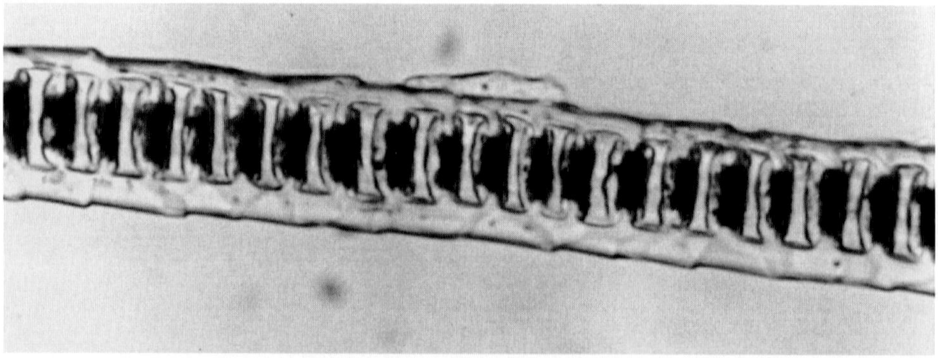

FIGURE 9. *Rattus norvegicus*, Norway rat, fur hair, node, 630× (dry).

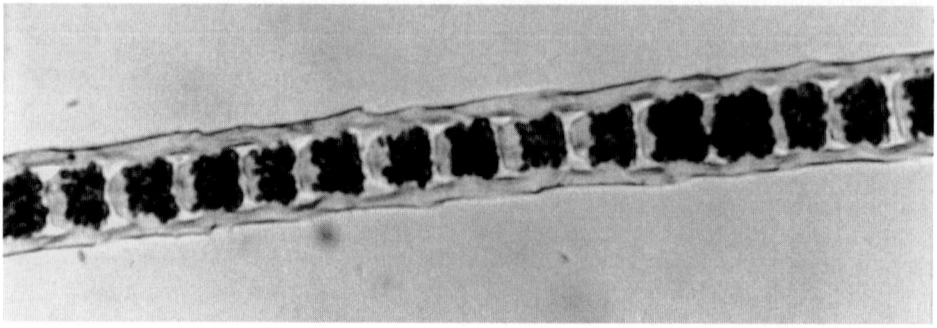

FIGURE 10. *Rattus exulans*, Polynesian rat, fur hair, node, 630× (dry).

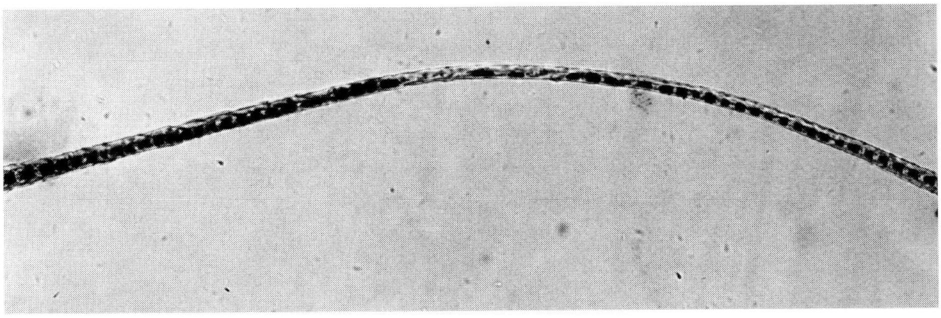

FIGURE 11. *Rattus rattus*, roof rat, fur hair, internode with skipping, 250×.

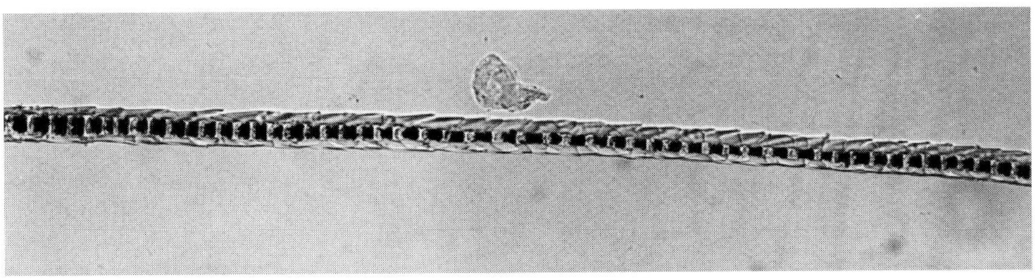

FIGURE 12. *Rattus rattus*, roof rat, fur hair, internode, 250×.

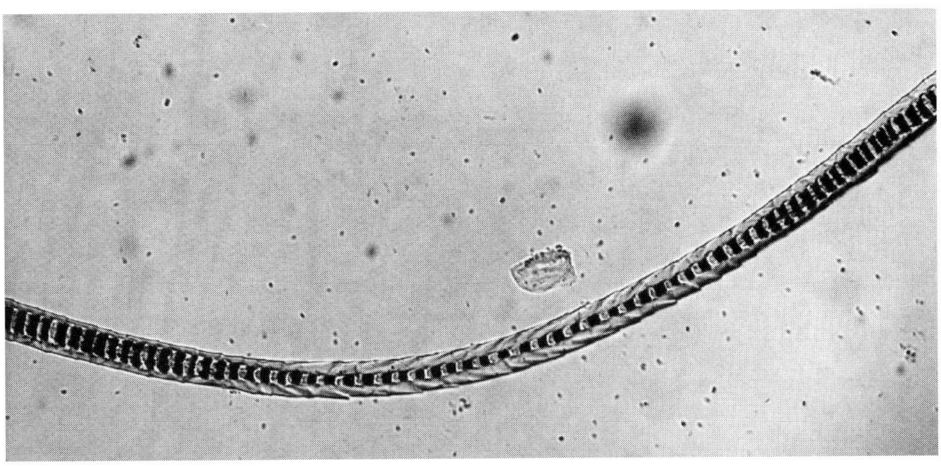

FIGURE 13. *Bandicota bengalensis*, lesser bandicoot rat, fur hair, internode, 250×.

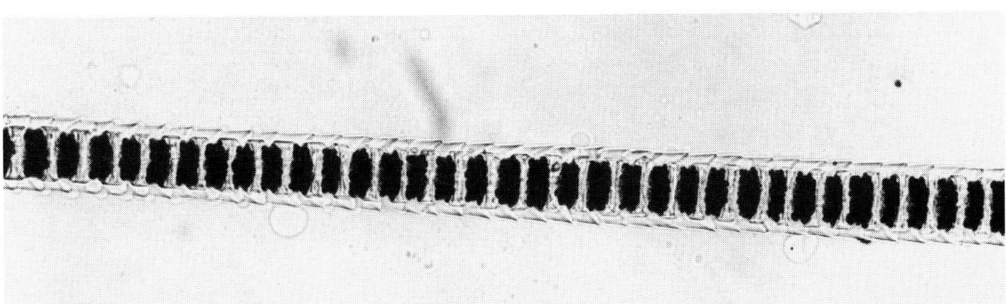

FIGURE 14. *Peromyscus leucopus*, white-footed mouse, fur hair, node, 500×.

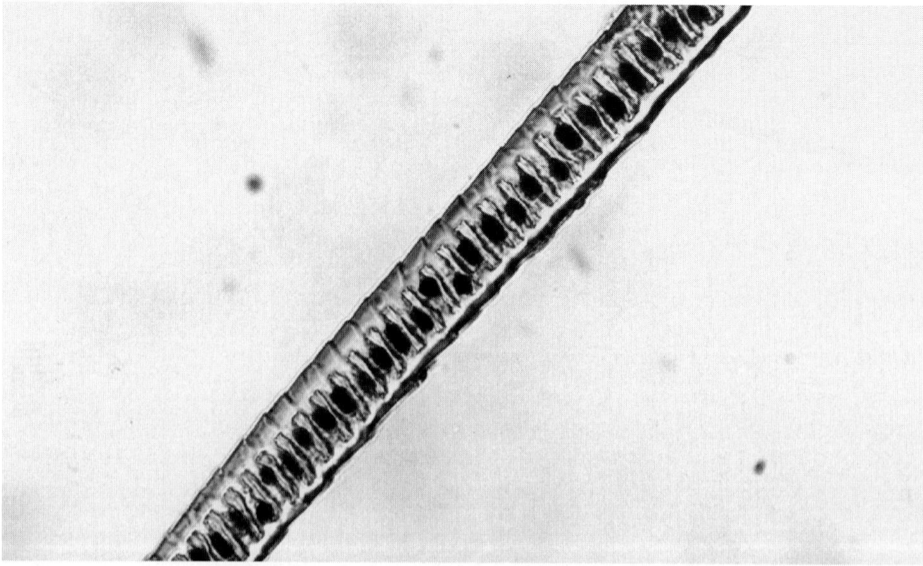

FIGURE 15. *Suncus murinus caerulercens*, Asian musk shrew, apical node, 400×.

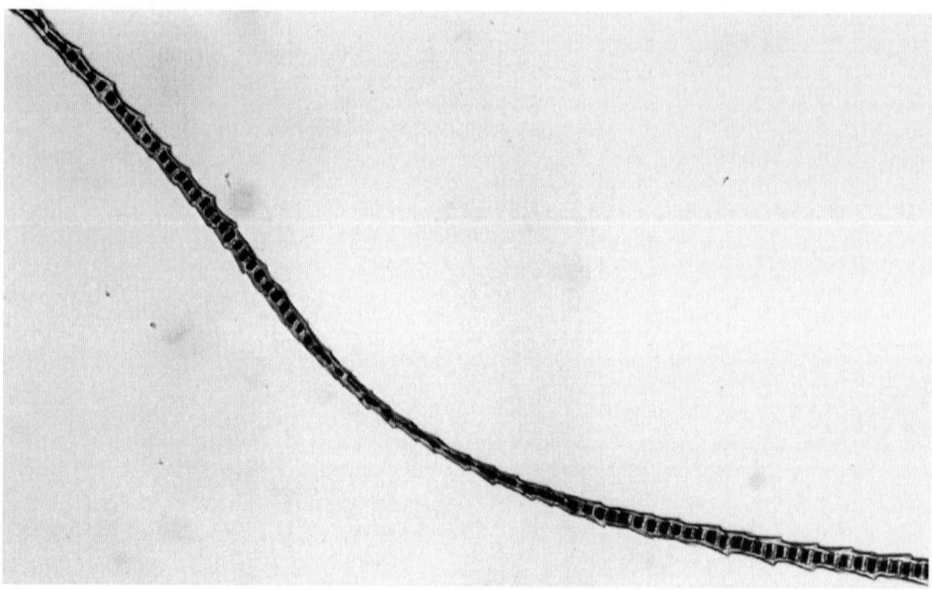

FIGURE 16. *Suncus murinus*, Asian musk shrew, fur hair, internode, 160×.

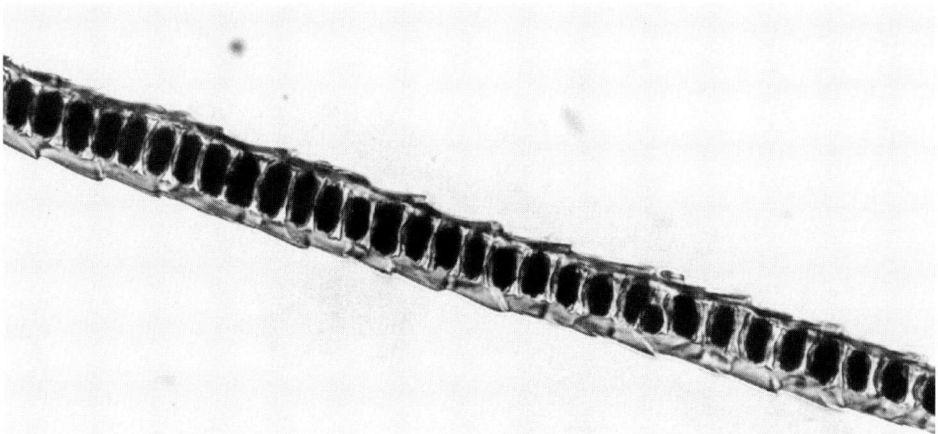

FIGURE 17. *Suncus murinus caerulercens*, Asian musk shrew, fur hair, node, 400×.

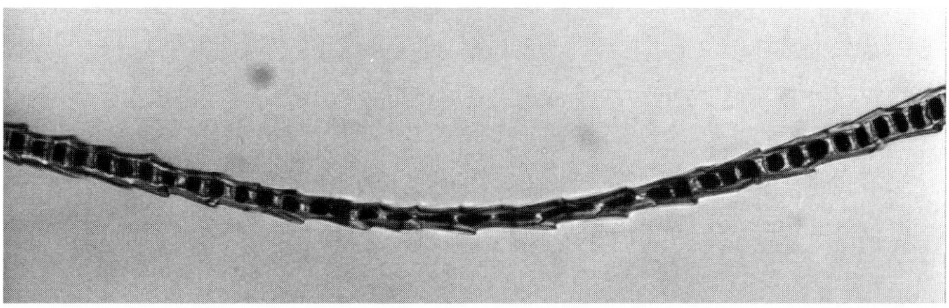

FIGURE 18. *Suncus murinus caerulercens*, Asian musk shrew, fur hair, internode, 250×.

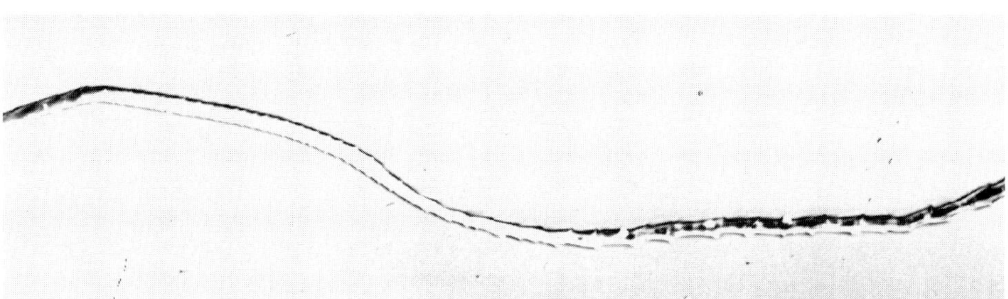

FIGURE 19. *Suncus murinus*, Asian musk shrew, fur hair, base with kink/bulge, 250× (DIC microscope).

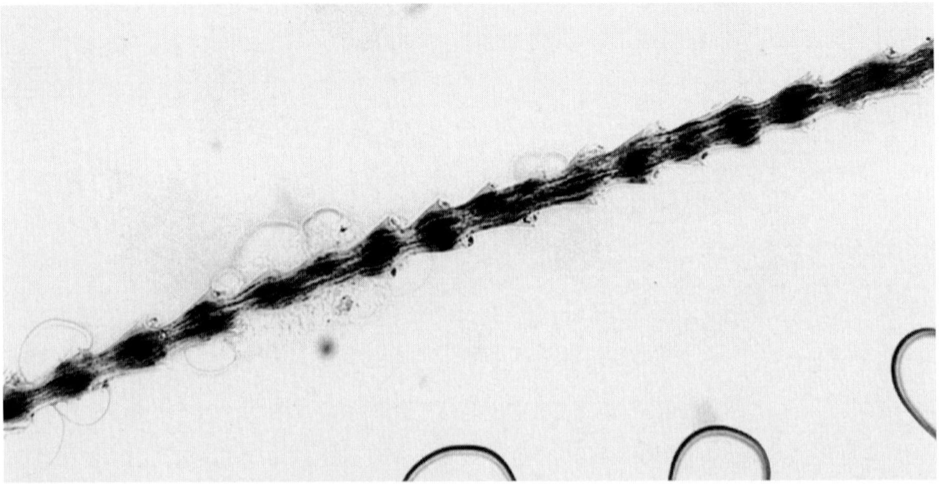

FIGURE 20. *Pipistrellus hesperus hesperus*, pipistrelle bat, fur hair, 400×.

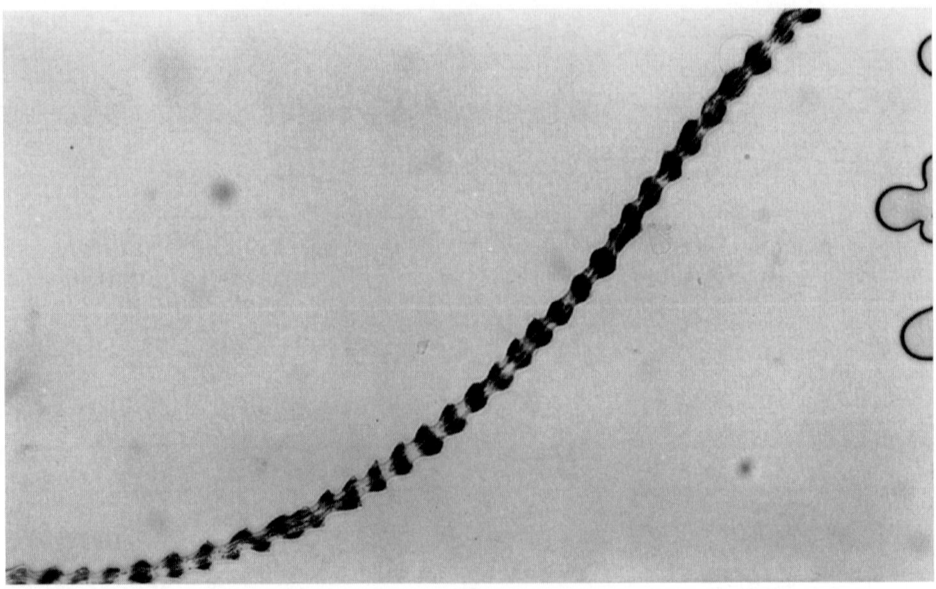

FIGURE 21. *Myotis lucifugus lucifugus*, little brown bat, fur hair, 250×.

FIGURE 22. Squirrel fur hair, 250×.

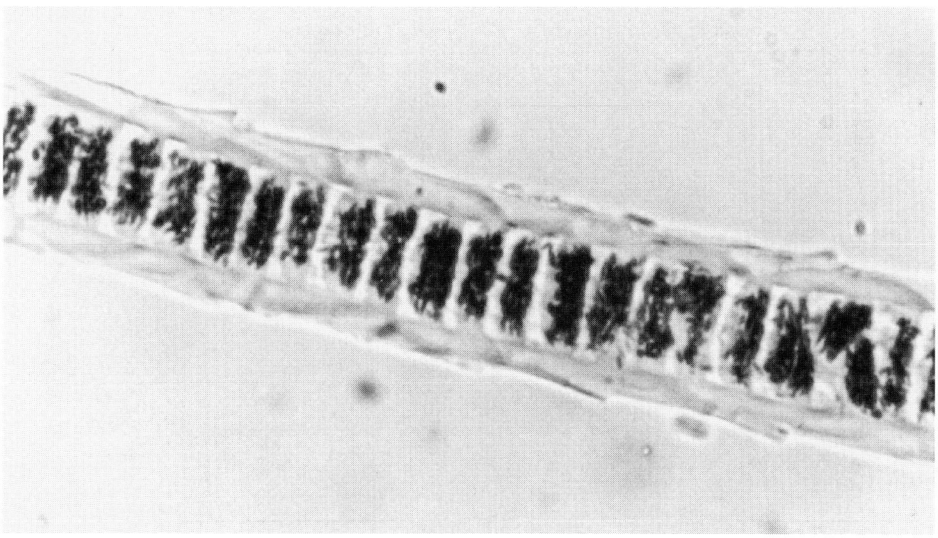

FIGURE 23. *Sciurus* sp., squirrel fur hair, 630× (dry).

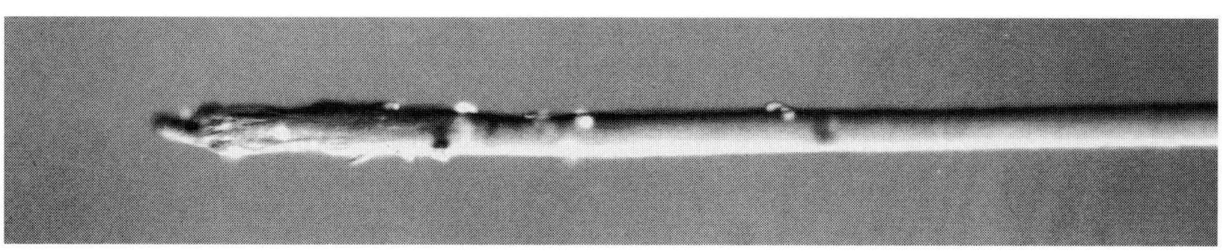

FIGURE 24. Siamese cat fur hair, root, 500× (DIC microscope).

FIGURE 25. Cat (unknown breed) fur hair, near base, 500× (DIC microscope).

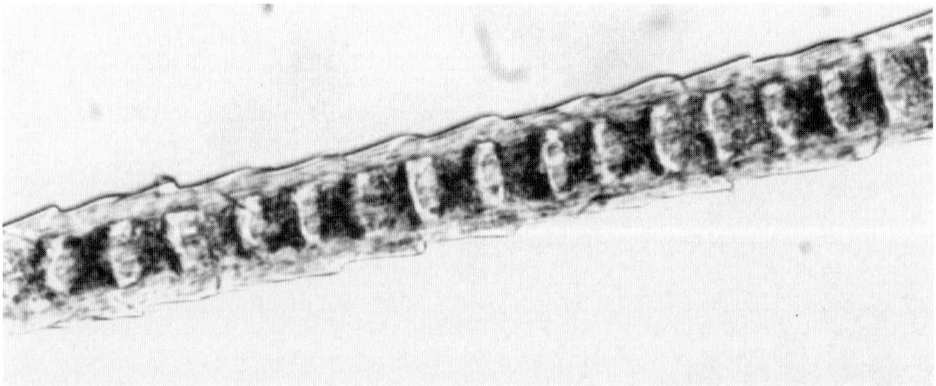

FIGURE 26. Cat (unknown breed) fur hair, 630× (dry).

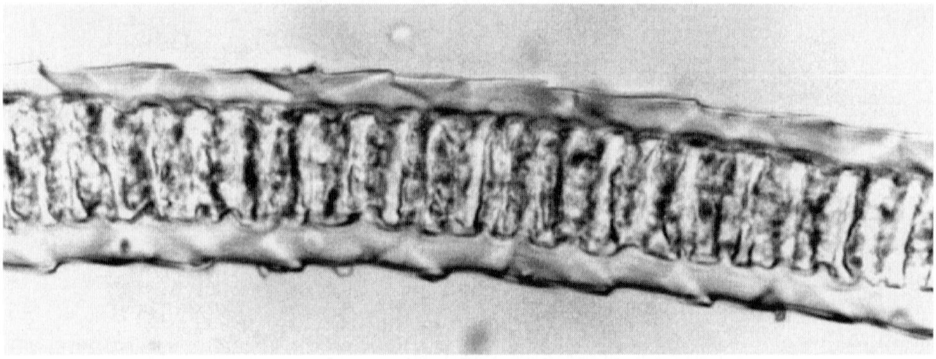

FIGURE 27. Siamese cat fur hair, 630× (dry).

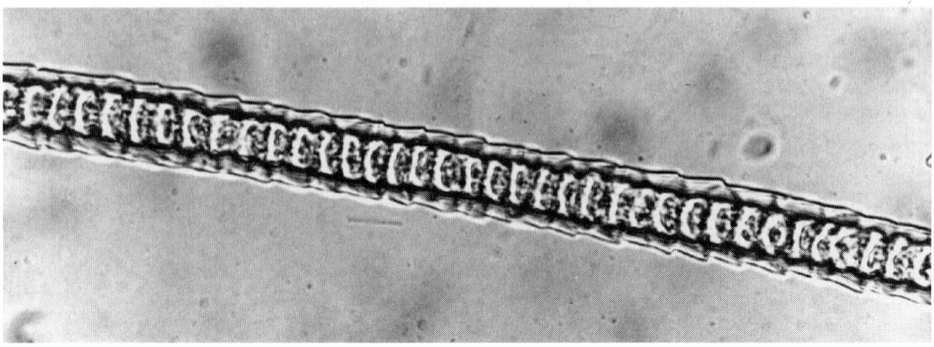

FIGURE 28. Maine Coone cat fur hair, 400×.

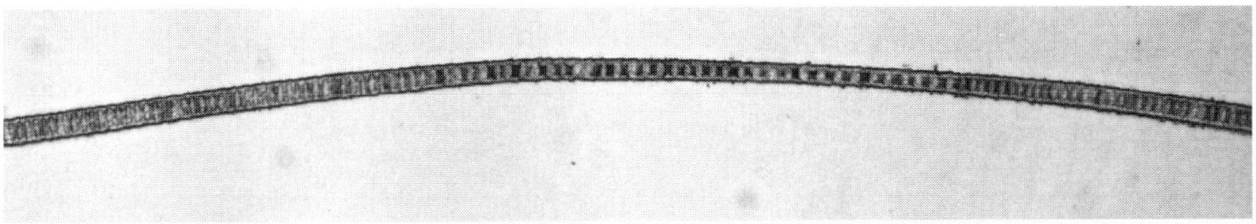

FIGURE 29. Cat (unknown breed) fur hair, internode, 125×.

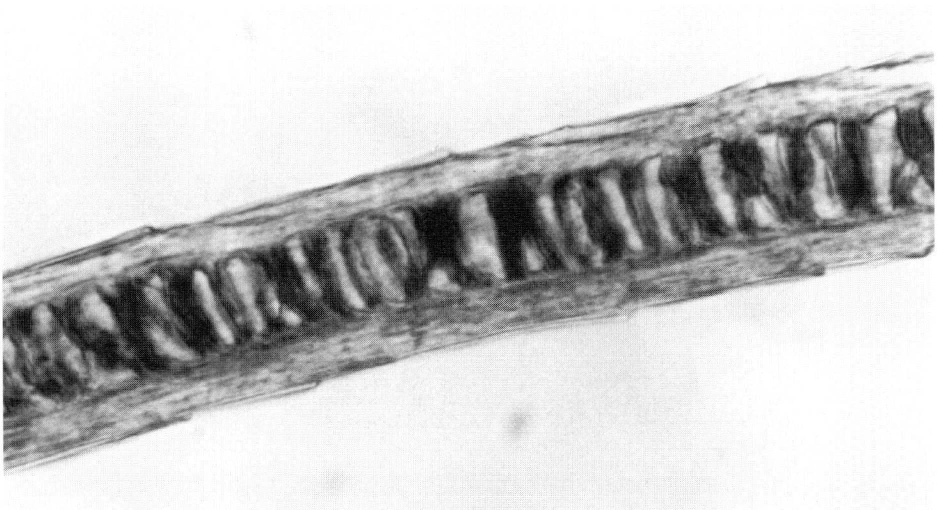

FIGURE 30. Dog (unknown breed) fur hair, 630× (dry).

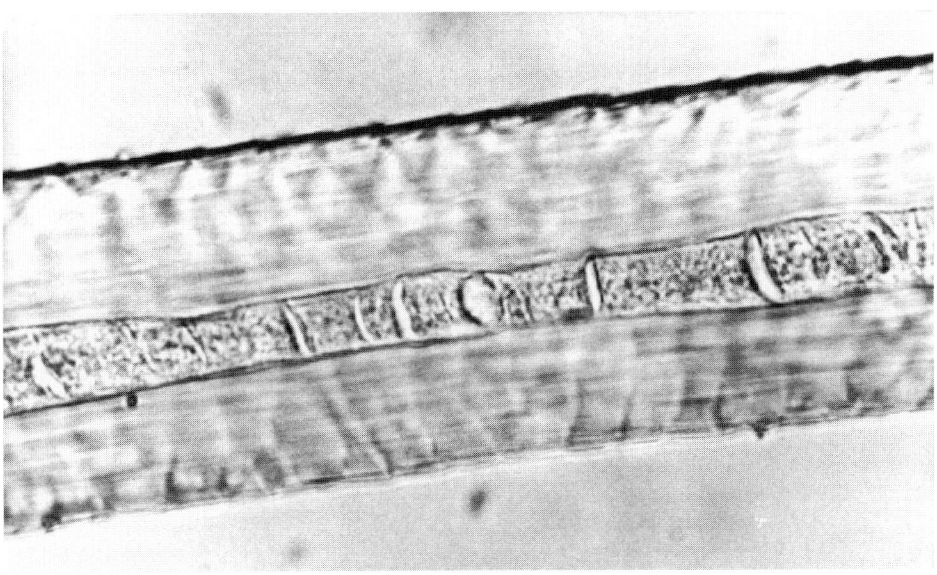

FIGURE 31. Dog (Border collie mix) fur hair, 630× (dry).

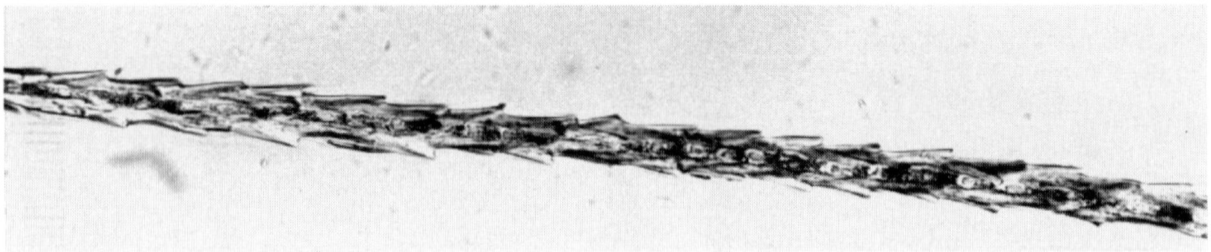

FIGURE 32. *Mustela siberica manchurica*, Siberian weasel fur hair, 250×.

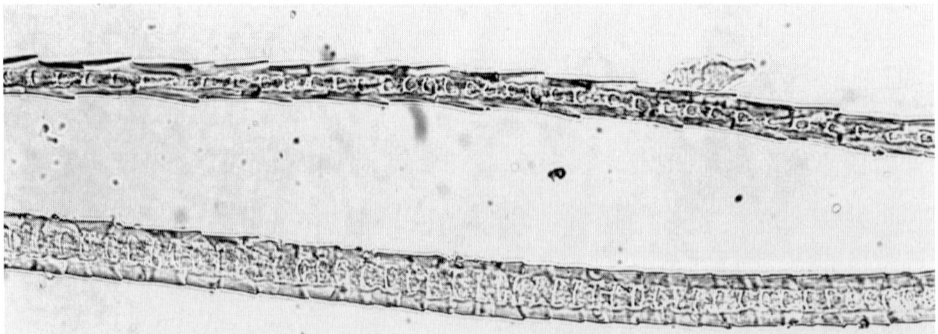

FIGURE 33. *Mustela nivalis*, least weasel, fur hairs, basal (prominent scales) and distal sections, 500×.

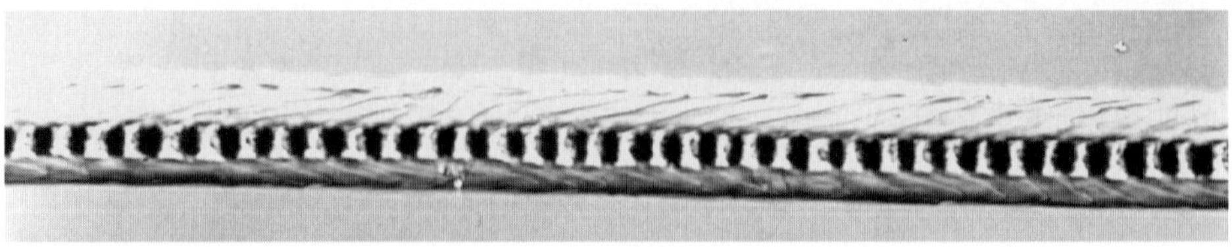

FIGURE 34. Cottontail rabbit fur hair, scales, 500× (DIC microscope).

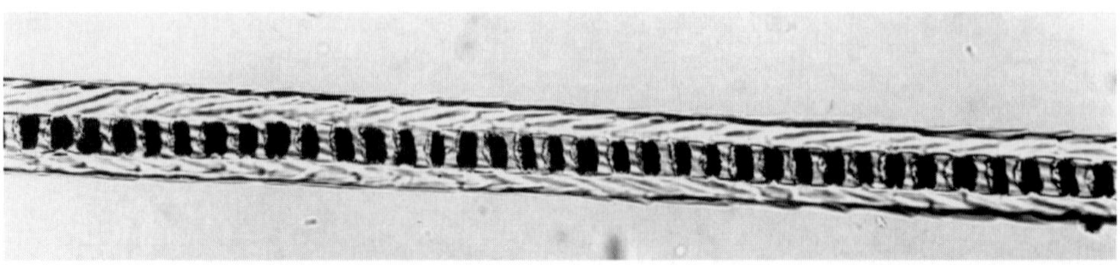

FIGURE 35. Cottontail rabbit fur hair, medulla, 500×.

FIGURE 36. White Angora rabbit fur hair, 500×.

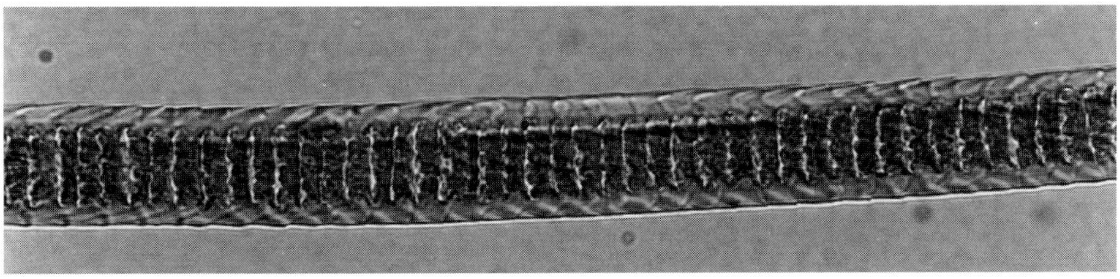

FIGURE 37. Cottontail rabbit fur hair, apical node, 500×.

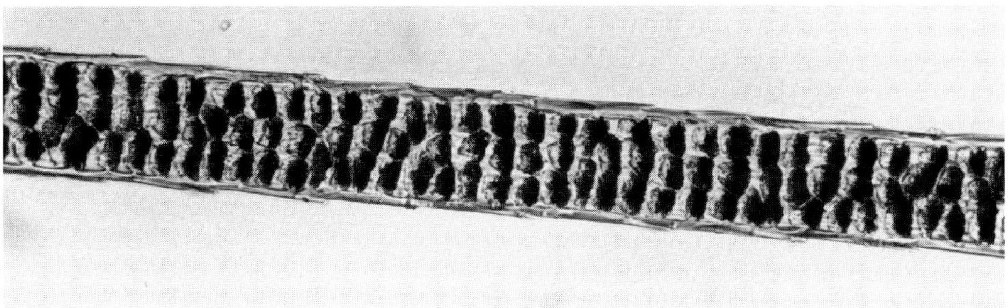

FIGURE 38. *Rattus norvegicus*, Norway rat, guard hair, midsection, 250×.

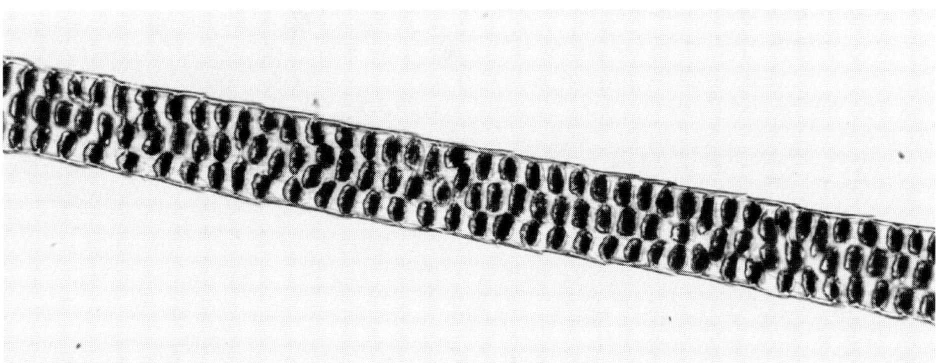

FIGURE 39. *Mus musculus*, house mouse, guard hair, 250×.

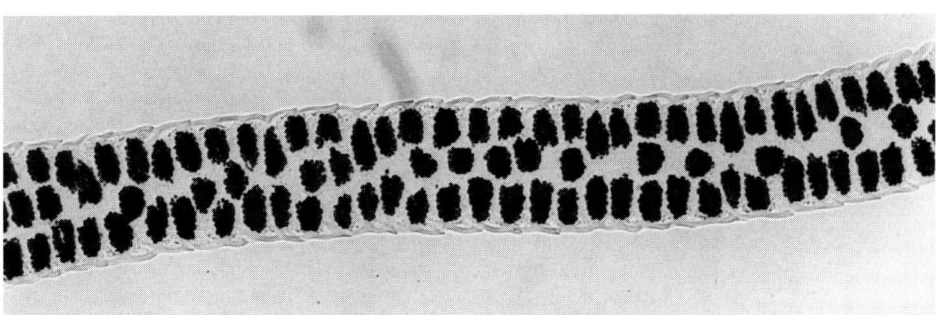

FIGURE 40. *Peromyscus leucopus*, white-footed mouse, guard hair, 500×.

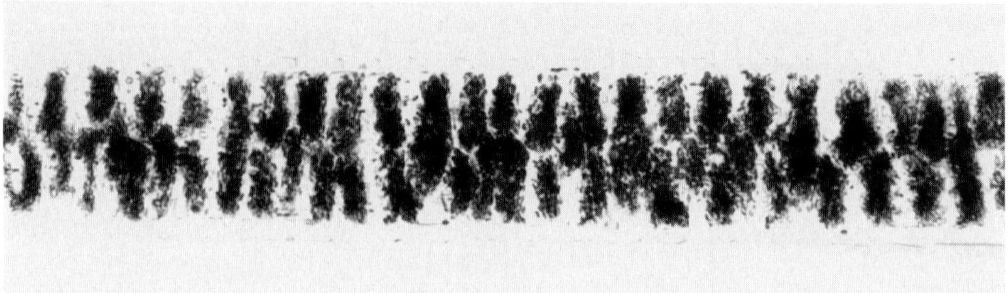

FIGURE 41. *Sciurus* sp., squirrel, guard hair, shaft, 400×.

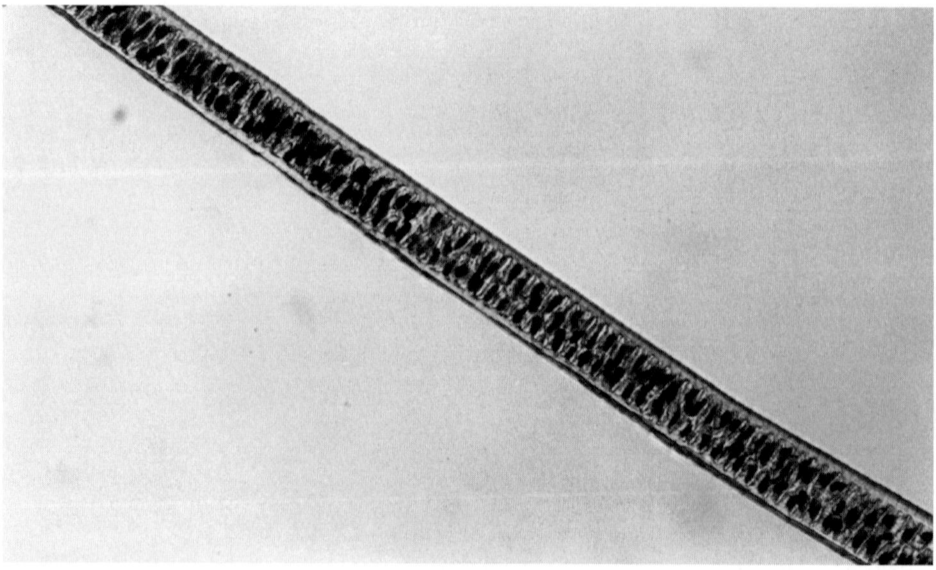

FIGURE 42. *Suncus murinus caerulercens*, Asian musk shrew, guard hair, shield, 160×.

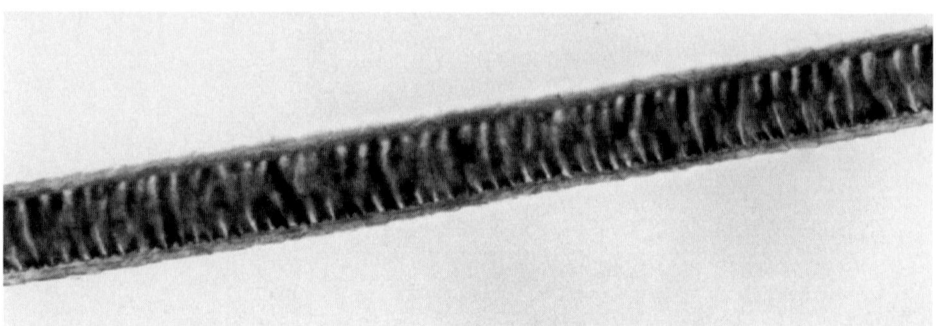

FIGURE 43. Cat (unknown breed) dorsal guard hair, 250×.

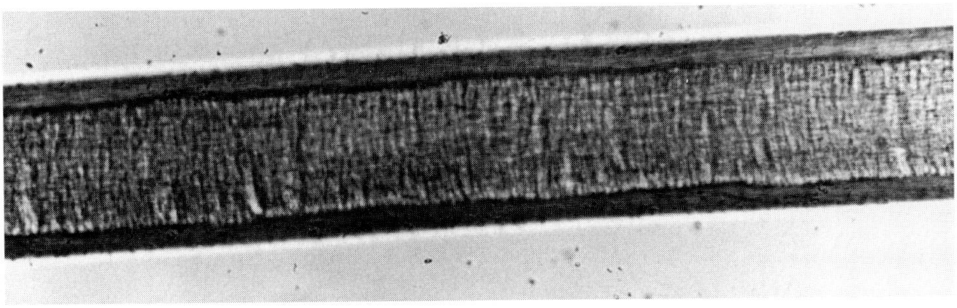

FIGURE 44. Siamese cat guard hair, midsection, 160×.

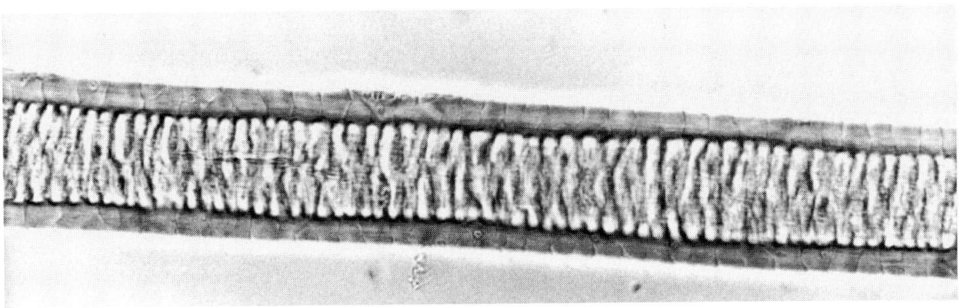

FIGURE 45. Siamese cat guard hair, near base, 160×.

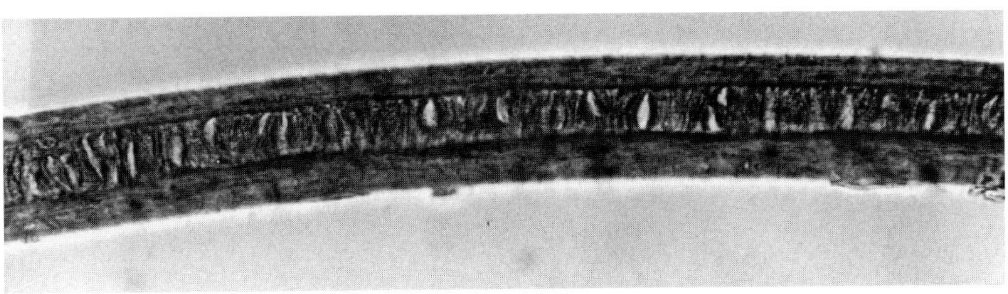

FIGURE 46. Dog (Cocker Spaniel) guard hair, midsection, 250×.

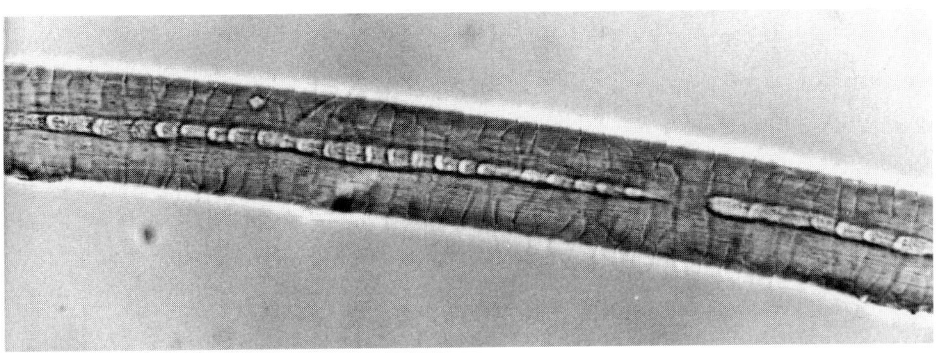

FIGURE 47. Dog (Cocker Spaniel) guard hair, near apical end, 250×.

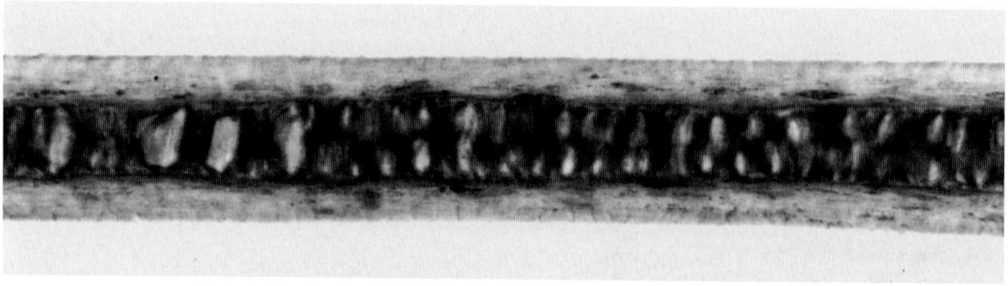

FIGURE 48. Dog (Shih Tzu) guard hair, 250×.

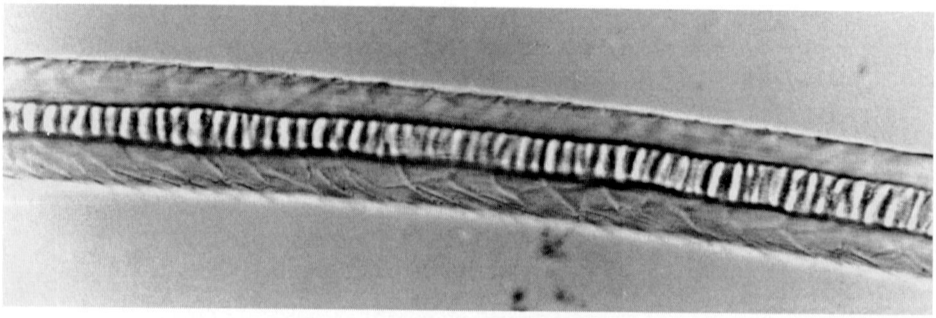

FIGURE 49. Dog (Westhighland terrier/Spitz mix) guard hair, 250×.

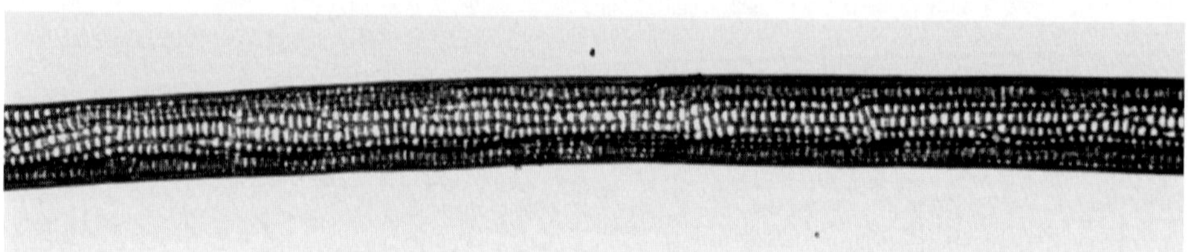

FIGURE 50. French mini lop-eared rabbit guard hair, 63×.

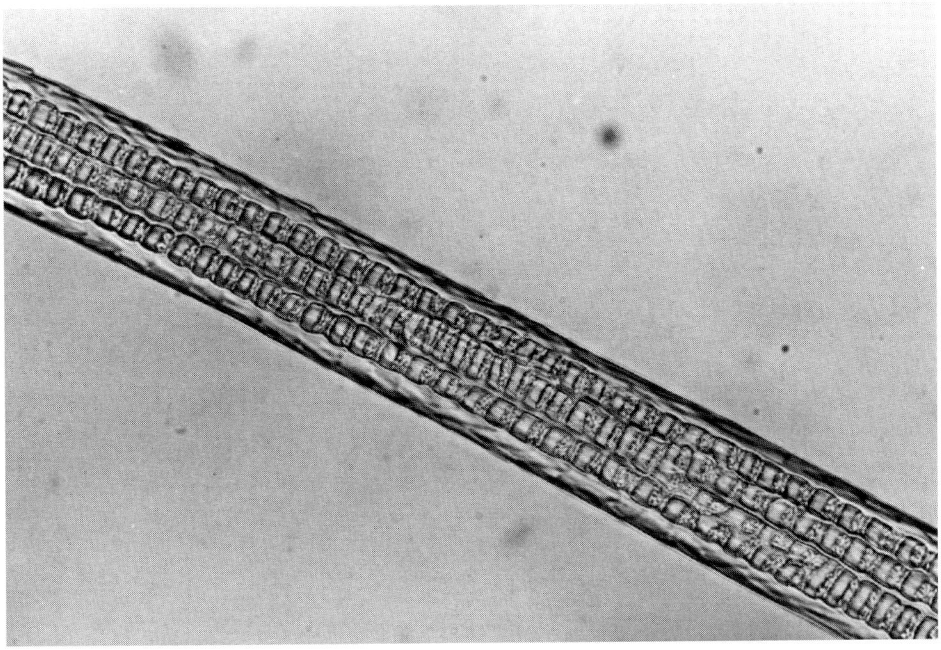

FIGURE 51. French mini lop-eared rabbit guard hair, 250×.

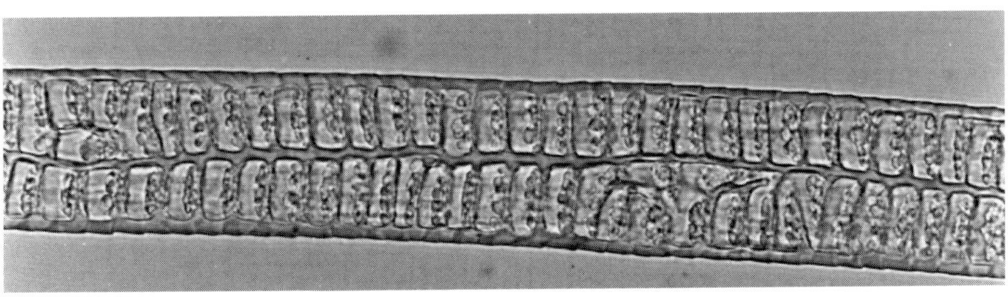

FIGURE 52. White Angora rabbit guard hair, apical node, air spaces, 500×.

FIGURE 53. Human (Hispanic male) arm hair with rounded tip, 63×.

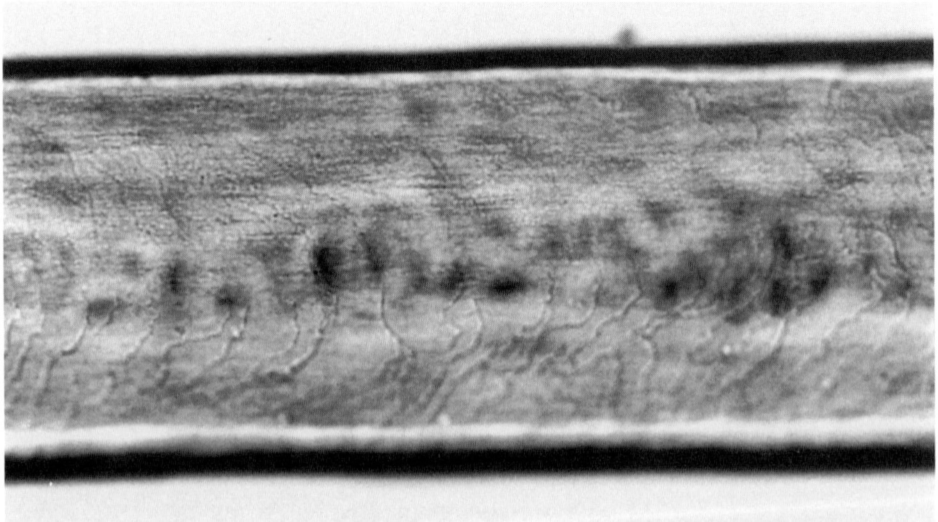

FIGURE 54. Human (Caucasian male) hand hair, cuticular scales, 400×.

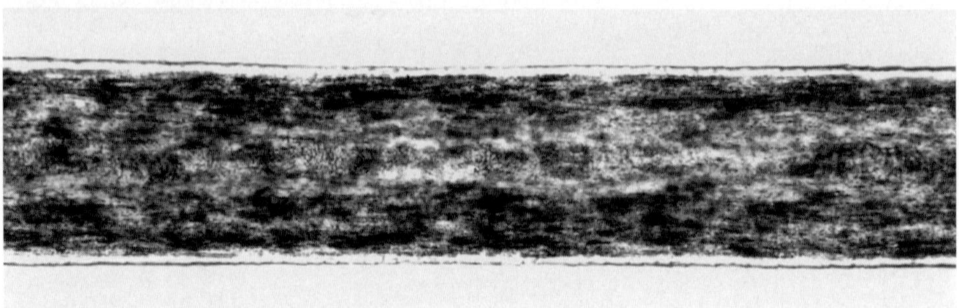

FIGURE 55. Human (Hispanic male) arm hair, 250×.

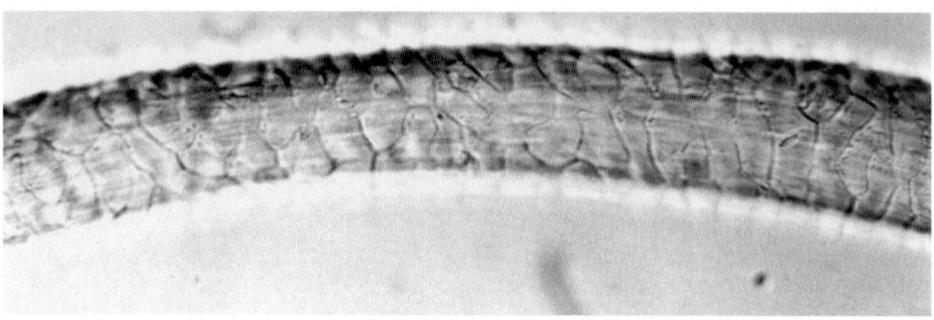

FIGURE 56. Sheep hair (from wool sweater), 250×.

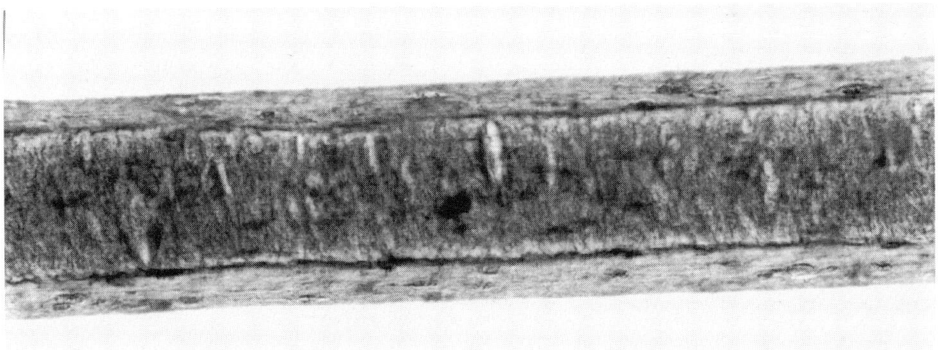

FIGURE 57. Cow hair, 160×.

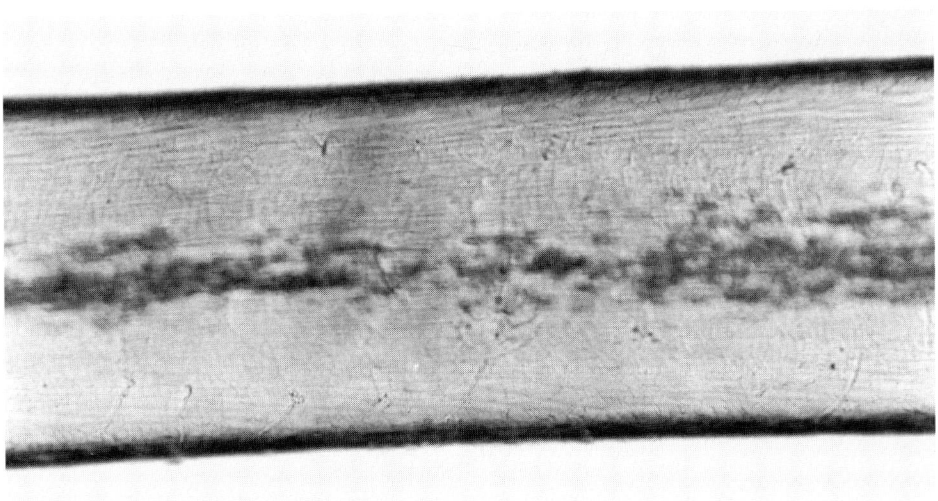

FIGURE 58. Swine hair, cuticular scales, 160×.

FIGURE 59. Swine hair, medulla, 160×.

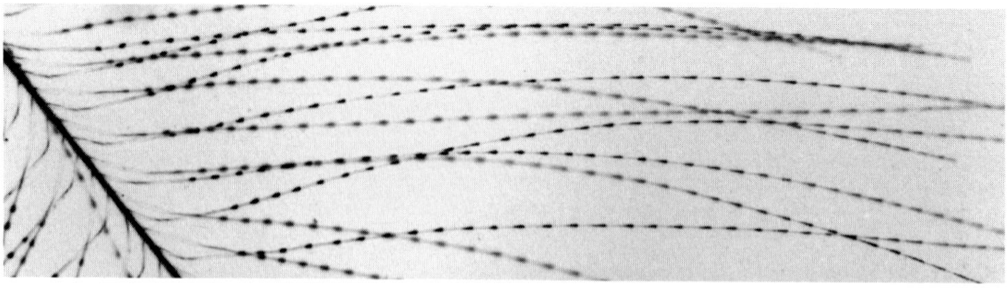

FIGURE 60. English sparrow, section of downy barb showing barbules, 100×.

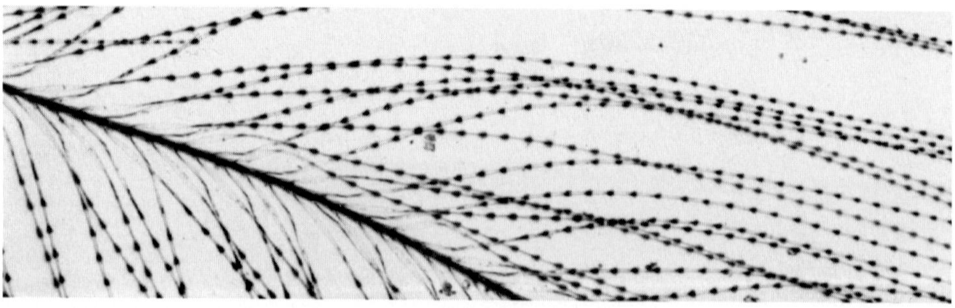

FIGURE 61. Starling, section of downy barb showing barbules, 100×.

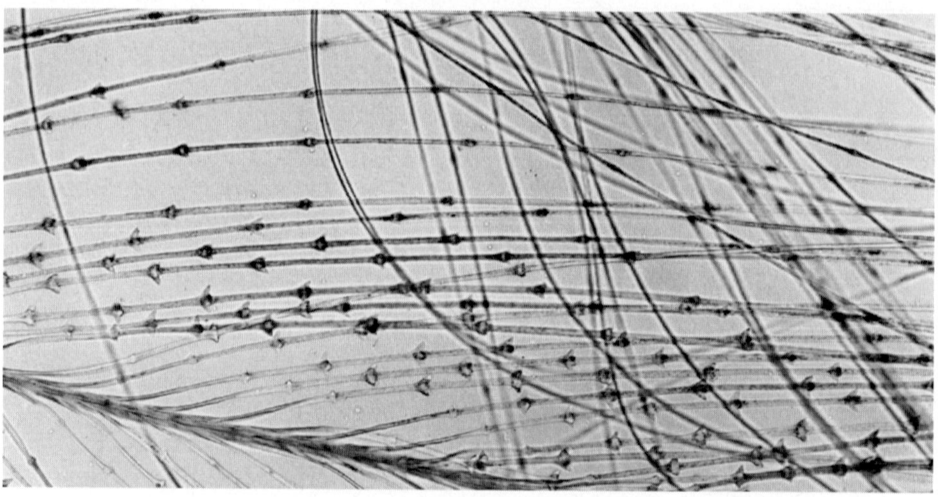

FIGURE 62. Pigeon, section of downy barb showing barbules, 200×.

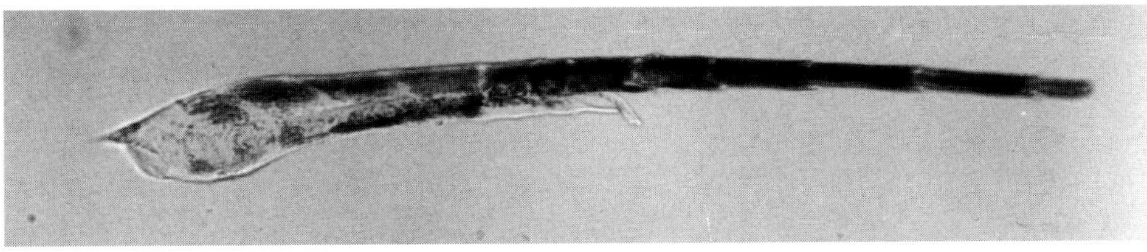

FIGURE 63. English sparrow, pennaceous barbule, 400×.

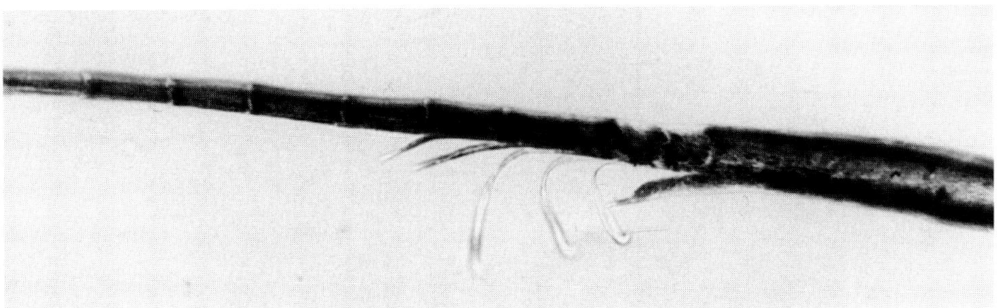

FIGURE 64. Starling, pennaceous barbule, 400×.

FIGURE 65. Pigeon, pennaceous barbules, 100×.

TABLE 1
Diagnostic Characteristics of Striated Fur Hairs of Selected Mammals Found in Foods

CHARACTER	RAT/MOUSE	ASIAN MUSK SHREW *SUNCUS MURINUS*	BAT [1]
GENERAL COMMENTS	Zigzag configuration of hair (Fig. 4) due to abruptly constricted internodes (visible at lower magnification) (Fig. 5).	Zigzag configuration of hair due to abruptly constricted internodes (visible at lower magnification). In *Suncus murinus*, one side of the apical node (tip) appears to be smooth, while the other side appears rougher, having a groove which gives it a double-edged appearance (Fig. 15).	Very fine. May superficially resemble a feather barbule. Unbanded.
CUTICULAR SCALES	Fairly symmetrical, not prominently protruding on one side. Scale pattern diamond petal or lanceolate (scales overlap in a diamond pattern; looks like a pine cone) (Fig. 1).	Asymmetrical - scales protrude prominently on one side ("scalloping") at nodes, crossover to opposite side at internode (more visible at lower magnification) (Fig. 16). This scale pattern distinguishes it from rodents. Occurs at all constrictions except subapical one. Apical node does not have asymmetrical pattern, but side where pattern would normally occur appears to have a double edge (see general comments above).	Coronal [1,2]; divergent scales; can be simple to ornate. May appear as stacked fluted cups with fringes on edges. Scales with fine to large pigment granules in parallel streaks; usually arranged in transverse bands (Fig. 20, 21), sometimes longitudinally; pigment evenly distributed throughout length of hair [3] but decreases at tip and base.
CORTEX	Thin (usually thinner than shrew and other rodents), hyaline, no striations. Prominent cortical intrusion may appear pinkish; pinches medullary cell at proximal end of pigment packet giving cell an hourglass shape (Fig. 6, 9).	Thin, usually hyaline. Cortical intrusion doesn't pinch medullary cell, therefore, no hourglass shaped cell (Fig. 17).	May contain varying degrees of pigmentation. *Myotis* cortex appears dappled with fusiform areas with differing degrees of pigmentation.
CORTICAL PIGMENT	If present, consists of a few fine to large, dark, round or oval granules, usually sparsely scattered or in dense rows (especially near the tip). Subapical node has the greatest amount of cortical pigment. Apical node usually lightly colored (diffuse stain-like pigmentation) with fine pigment in tip. Medulla may be similarly pigmented.	Few large, dark, granules. Apical portion has finer and more profuse pigment.	Variable depending on family. Usually fine, spherical granules which can be arranged in longitudinal streaks or transverse groups.
MEDULLA	Single row, discontinuous (Fig. 6). Nuclear ghosts may appear near tip (Fig. 7). [4]	Single row, discontinuous (Fig. 17).	Absent (except in two families which are not typically found in foods) because the hairs are so fine.

SHAPE OF MEDULLARY CELL	Hourglass shaped. If width > length, then cortex usually relatively thin (Fig. 6); if length > width, then cortex usually relatively wide (Fig. 10). Pigmented distal area 1/2 or more of cell; distal end of pigment packet wider than proximal end. Proximal clear area 1/2 or < cell length.	Variable, not hourglass shaped. Distal pigmented area compact. Small proximal clear area (Fig. 17). At internode, cells appear to elongate with pigment running together (Fig. 18).	None.
MEDULLARY PIGMENT	Large, dark, round or oval granules in dense, compact packet which is irregular in outline (Fig. 6, 9, 10).	Large, dark granules in oval or bean-shaped compact packet located at distal end of cell (Fig. 17).	None.
AIR SPACE	I-beam with distal margin fairly straight, may curve slightly (Fig. 6, 9, 10). Width of airspace much > length. Near tip and base, I-beam may be distorted or be more crescent-shaped. In narrow hairs, I-beam may be short and thick.	I-beam, but main axis is irregular, and sloppy, not as smooth, straight and neat as rat or mouse. Occasionally, clear portion of cell protrudes into air space as a bulge 1/3 or > the height of the air space.	None.
INTERNODES	Very abrupt, 3-6 (usually 4) constrictions; more prominent than in _Peromyscus_ (Fig. 4, 5). Second internode from tip most prominent; remaining internodes less abrupt as move towards base. Medullary cells remain discrete, but are narrower and may become longer. Two most distal internodes may show sections of medullary cells missing (skipping) (Fig. 11).	Abrupt, 1 or 2 constrictions; medullary cells appear to elongate with pigment running together (Fig. 16, 18).	None.
ROOT	Enlarges to bulb. Nonmedullated area near base straight, may have a slight swelling but no asymmetrical scale pattern. Medullary cells adjacent to this clear area remain as discrete units (Fig. 8).	Nonmedullated area near base may have an enlargement on one side which corresponds to where the asymmetrical scale pattern would have occurred; distally, this narrows down to an apparent internode and continues as an asymmetrical scale pattern on opposite side. Medullary cells distad of bulge may run together similar to that at internodes (Fig. 19).	Base narrows in diameter distad of the bulbous root (as compared to the base of a bird feather barbule which is flat).

1 It is difficult to distinguish between fur and guard hairs for bats since they vary only in length and coarseness.[43] Many families are non-medullated because the hairs are so fine,[31] and only 2 families of bats (Pteropidae and Megadermidae) have medullated hairs.[42] In families with coarser hairs, the scales may also be imbricate (these bats are not typically found in foods). Vespertiliontidae also lack medulla and the center of the hair is uniformly unpigmented; however, pigment may be in cortex in apical portion of hair. Pigment occurs in the scales (seen by moving microscope focal plane through hair and observing color changes).[9]

2 Coronal scales are those which encircle the entire hair shaft.[9, 32] Hairs of food-contaminating bats have hastate coronal scales - characteristic of most Vespertiliontidae. The hastate scales may be unequal and cause the hair to look unsymmetrical.[9] These types of scales are characteristic of bats; rarely do they occur in other orders, and then only occasionally in very thin hairs.[41]

3 In _Myotis_ (Little brown bat), the pigment occurs in the distal portion of the scales.

4 Nuclear ghost is remnant of medullary cell nucleus, occurs in tip or apical shield of hairs with finer pigmentation. Appears as oval shaped objects between adjacent air spaces (Fig. 7). Common in some mammal hairs; for example, rat, mouse, rabbit and cat.

TABLE 2
Diagnostic Characteristics of Striated Fur Hairs of Selected Commensal Rodents Found in Foods

CHARACTER	RATTUS SPP. NORWAY, ROOF AND POLYNESIAN RATS	MUS MUSCULUS HOUSE MOUSE	BANDICOTA BENGALENSIS LESSER BANDICOOT RAT
GENERAL COMMENTS	Zigzag configuration of hair due to abruptly constricted internodes (visible at lower magnification). Root enlarges to a bulb.	Zigzag configuration of hair due to abruptly constricted internodes (visible at lower magnification) (Fig. 4). Root enlarges to a bulb.	Zigzag configuration of hair due to abruptly constricted internodes (visible at lower magnification). Similar to Mus and Rattus except that apical node is usually not expanded into a shield shape - remains slender and fairly long; tip unmedullated. Root enlarges to a bulb.
CUTICULAR SCALES	Fairly symmetrical, not prominently protruding on one side. Scale pattern diamond petal or lanceolate (scales overlap in a diamond pattern; looks like a pine cone) (Fig. 1).	Fairly symmetrical, not prominently protruding on one side. Scale pattern diamond petal or lanceolate (scales overlap in a diamond pattern; looks like a pine cone) (Fig. 1).	Fairly symmetrical, not prominently protruding on one side. Scale pattern diamond petal or lanceolate (scales overlap in a diamond pattern; looks like a pine cone) (Fig. 1).
CORTEX	Thin (usually thinner than shrew and other rodents except Mus), hyaline. Prominent cortical intrusion may appear pinkish; pinches medullary cell at proximal end of pigment packet giving cell an hourglass shape (Fig. 9, 10).	Very thin (generally thinner than Rattus or Bandicota), < or = half of medulla width, hyaline. Very prominent cortical intrusion may appear pinkish; pinches medullary cell at proximal end of pigment packet giving distinct hourglass shape (Fig. 6).	Hyaline, thick - about 1/2 - 2/3 the overall width of hair (relatively thicker than Rattus). Prominent cortical intrusion pinches medullary cell giving it an hourglass shape (Fig. 13).
CORTICAL PIGMENT	If present, consists of a few fine to large, dark, round or oval granules, usually sparsely scattered or in dense rows (especially near the tip). Subapical node has the greatest amount of cortical pigment. Apical node usually lightly colored (diffuse stain-like pigmentation). Medulla may be similarly pigmented.	If present, consists of fine to large, dark, round or oval granules, usually sparsely scattered in rows, particularly in the shield. Subapical node has more cortical pigment than other nodes, in short streaks (not as profuse as in Peromyscus). Apical node usually lightly colored (diffuse stain-like pigmentation). Medulla may be similarly pigmented (Fig. 7).	Large granules, scattered, in streaks near tip. Apical node usually lightly colored (diffuse stain-like pigmentation). Medulla may be similarly pigmented.
MEDULLA	Single row, discontinuous. Nuclear ghosts may appear near tip.[1]	Single row, discontinuous. Nuclear ghosts may appear near the tip (Fig. 7). All but apical node have very dark pigment.	Single row, discontinuous. Nuclear ghosts may appear near tip.

SHAPE OF MEDULLARY CELL	Hourglass shape. If width > length, then cortex usually relatively thin; if length > width, then cortex usually relatively wide. Pigmented distal area = 1/2 or more of cell; distal area of pigment packet wider than proximal area (Fig. 9, 10). Proximal clear area < or = 1/2 cell length, square-cut along proximal border. Cells near base of hair more elongate.	Hourglass shape. If width > length, then cortex usually relatively thin (Fig. 6); if length > width, then cortex usually relatively wide. Pigmented distal area = or > 1/2 cell; distal portion of packet wider than proximal portion; proximal area hyaline, <1/2 cell length, rounded along proximal margin, part extends up along both sides of pigmented area. Cells near base of hair more elongate.	Hourglass shape. If width > length, then cortex usually relatively thin; if length > width, then cortex usually relatively wide. Pigment up to 3/4 of cell length (pigmented area usually relatively longer than Rattus). Clear area makes up circa 1/4 to 1/2 cell. In basal portion of hair, proximal corners of clear area rounded. In distal portion of hair, proximal corners of clear area angled.
MEDULLARY PIGMENT	Large, dark, round or oval granules in dense, compact packet; irregular in outline (Fig. 9, 10).	Large, dark, round or oval granules in dense, compact packet; irregular in outline (Fig. 6).	Smaller granules than Rattus, in compact packet which may appear as dense as Rattus or Mus; irregular in outline.
AIR SPACE	I-Beam shape with narrow crossbar, distal margin fairly straight and corners of crossbar angled (Fig. 9, 10). Width of air-space much > length. Near tip and base, may be distorted I-beam or more crescent-shaped. In narrow hairs, shaft of I-beam will be shorter and relatively thicker. I-beam shape may appear bowed or air spaces not obvious.	I-Beam shape clearly outlined. Main axis of I very straight, although may curve slightly on proximal side (Fig. 6); height of I much greater than width. In narrow hairs, shaft of I-beam will be shorter and relatively thicker. I-beam shape may appear bowed or air spaces not obvious.	I-beam shaped; main shaft thick and proximal portion of crossbars may be absent. Distal margin of I-beam may be irregular. In narrow hairs, shaft of I-beam will be shorter and relatively thicker. I-beam shape may appear bowed or air spaces not obvious (Fig. 13). Can be linear with concave distal edge.
INTERNODES	Very abrupt, up to 6, but usually 3 to 4 constrictions; more prominent than Peromyscus. Second internode from tip most prominent; remaining internodes less abrupt as move towards base. Medullary cells remain discrete but are narrower and may become longer. Two most distal internodes may show sections of medullary cells missing (Fig. 11) (also occurs in Mus, but Mus has much thinner cortex). In some cases, overall hair diameter at internode may not decrease abruptly, but the medullary channel may narrow to about 1/2 of its width at the node (Fig. 12, 13).	Very abrupt (more distinctive than Rattus or Peromyscus), usually 3 to 4 and up to 6 constrictions (Fig. 4, 5). Subapical node most pronounced and very thin. Medullary cells remain discrete but are narrower and may become longer. Two most distal internodes may show sections of medullary cells missing). In some cases, overall hair diameter at internode may not decrease abruptly, but the medullary channel may narrow to about 1/2 of its width at the node.	Usually very abrupt. Internodes may not decrease that much in diameter, but the medullary channel narrows to about 1/2 the width of the medulla in the node. Internode at subapical node is most prominent. Medullary cells remain discrete, but are narrower and may become longer. In some cases, overall hair diameter at internode may not decrease abruptly, but the medullary channel may narrow to about 1/2 of its width at the node (Fig. 13).

[1] Nuclear ghost is remnant of medullary cell nucleus; occurs in tip or apical shield of hairs with finer pigmentation. Appears as oval between adjacent air spaces. Common in some mammal hairs; for example, rat, mouse, rabbit and cat.

TABLE 3
Diagnostic Characteristics of Striated Fur Hairs of Selected Noncommensal Rodents Found in Foods

CHARACTER	SQUIRREL	MICROTUS SPP. VOLES, MEADOW MICE	PEROMYSCUS LEUCOPUS WHITE-FOOTED MOUSE
GENERAL COMMENTS	Wavy. Tip short and dark with subapical portion banded yellow. Agouti pattern. Root enlarges to a bulb.	Last node (shield) may have yellowish appearance, medullary pigment looks yellow, darkens near the tip. Root enlarges to a bulb.	Apical node usually yellow brown with very dark tip. The remaining nodes have very dark pigment. Root enlarges to a bulb.
CUTICULAR SCALES	Variable. Chevron scale pattern (non-overlapping scales side to side in wave pattern with valleys and peaks; narrow and V-shaped).	Variable. Chevron scale pattern (non-overlapping scales in wave pattern with either valleys or peaks but not both; narrow and V-shaped).	Variable. Scales at base prominent, like stacked, fluted cups.
CORTEX	Moderately thin (thicker than Rattus), not as clear as rat/mouse (Fig. 22). No cortical - intrusion. May be longitudinally striate (characteristic of squirrel).	Thin, hyaline. Pigmenting near tip may be heavy. Cortical intrusion not prominent, do not pinch into medulla.	Thin, hyaline. Similar to Microtus. Cortical intrusion pinches medullary packet (Fig. 14). Hourglass shape not as prominent as in Rattus or Mus.
CORTICAL PIGMENT	May appear as fine staining in non-medullated body hairs. Can vary from fine to coarse. Large rod shaped pigment granules (more elliptical proximally) may be present; prominent, patchily distributed throughout hair.	Fine to large, dark and round. Present near distal end of hair, may be heavy, in streaks.	Large round to oval dark granules, scattered. Tip has fine granules in short streaks. More profuse than in Mus.
MEDULLA	Discontinuous. Irregular border between medulla and cortex. Usually two cells wide or interspersed with one cell wide (uniserial) (Fig. 23).	Single row, discontinuous. Nuclear ghosts in apical node.	Single row, discontinuous. Nuclear ghosts in apical node.
SHAPE OF MEDULLARY CELL	Distinctive distal end which is concave or flat; proximal end rounded or nearly pointed. Distal portion of the cell is clear, indistinct; may be barely visible (< or = 1/4 of cell). No hourglass shape (Fig. 22).	Cell square to rectangular in outline with rounded corners. Width equals 1 1/2 to 3 length, except at internodes where width less than or equal to length. Pigment packets very closely spaced, occupy 2/3 or more of cell. 1/3 or < of cell is clear.	Cell square to rectangular in outline (Fig. 14). Pigment packet occupies almost the entire cell. Very closely spaced. Clear area approximately 1/4 (up to 2/5) of cell length, not as clear as Mus/Rattus. Distal corners of cell may be very clear.

MEDULLARY PIGMENT	If present, dark and rod shaped or more often elliptical shaped. Large; loosely packed (Fig. 23). Near distal end, may be absent (in finer hairs), or appear as a non-granular to granular stain. Towards very tip may reappear as dark, large granules again. May obscure proximal rounded end of air space, giving it an I-beam appearance but air space never actually I-beam shaped.	Granules large, round, dark, form a dense packet which is square or rectangular in outline with smooth sides and rounded corners.	Granules large, dark (more so than Mus) round to oval; very dark in all nodes except apical node, which is yellow-brown. Form a dense packet which is square or rectangular in outline with smooth sides and rounded corners, cortical intrusion pinches the packet in the middle (Fig. 14); width of packet above and below intrusion is approximately equal.
AIR SPACE	Variable. Large, oblong, concave-convex or bowl shaped. Lateral parallel portion of bowl usually less than the distal side and approximately equal to base of bowl (Fig. 22). Never I-beam shaped, although may appear so due to obscuring medullary pigment.	I-Beam shaped with very straight, narrow main shaft of I.	I-Beam shaped with bowed main shaft of I; cross bars are short. I-beam shaped from straight (Fig. 14) to bowed main shaft. Thickness of main shaft may vary, often very thin.
INTERNODES	None.	Gradual. Medullary cell width decreases at internode to 1/2 to equal to length, but length stays the same.	Gradual, usually 3-4 but up to 5 constrictions; less prominent as move towards base of hair. Medullary cells increase slightly in length, squarish. Width of medulla is approximately 1/3 to 1/2 of the width at the nodes. Clear area about 1/3 to 1/2 of cell length. The internode before the shield is more pronounced, but not like in Rattus or Mus which are abrupt.

Note: Fur hairs of *Mus*, *Rattus* and *Peromyscus* are very similar and are not easily distinguished without familiarity with authentic specimens.[2] However, according to Nowak, *Peromyscus* is found only in North America, Mexico and Central America south to Panama and the adjacent islands but not the West Indies.[10]

TABLE 4
Diagnostic Characteristics of Striated Fur Hairs of Selected Mammals Found in Foods

CHARACTER	CAT	DOG	WEASEL	RABBIT
GENERAL COMMENTS	May be banded. When medulla is pigmented, air spaces are very thin. When unpigmented, bowl-shaped air space looks similar to squirrel hair but more irregular in shape (Fig. 28). Root elongate, no distinct shape; fibrils frayed at base (Fig. 24).	May be banded. Wavy; may have slight changes in diameter over length of hair. Root may be spade-shaped or elongate.	Root similar to cat - thin long base.	Hair may be wavy. Tips long and tapering. Root bulbous.
CUTICULAR SCALES	Spinous, usually very prominent. Margin smooth. But may vary. Slightly elongated near base (Fig. 25).	Easily seen but not prominent. Variable (Fig. 30).	Fingerlike, prominent near basal end. Scales angular in profile (Fig. 32). Near distal end scales are simple and flattened (Fig. 33).	Streaky chevron pattern - scales have oblique margins forming a uniform, parallel, diagonal scale pattern (Fig. 34) (focus up to outer plane of hair).
CORTEX	Translucent, not as clear as rat or mouse. Relatively thick compared to medulla; 1/4 - 1/3 overall hair diameter (Fig. 26, 27). Pigment in cortex, corresponds to color banding of medulla. Cortical intrusion prominent, pinches medullary cell into hourglass shape (Fig. 26).	Thick relative to medulla, usually up to 1/3 hair diameter or rarely more (Fig. 31); can comprise whole shaft when unmedullated. Lightly striate, stained or cloudy. Cortical - intrusion prominent.	May contain fine pigmenting, scattered or in fine parallel rows. Cortical intrusion prominent (Fig. 33).	Hyaline, clearer than rat or mouse; usually very thin, as narrow as or narrower than mouse cortex. Cortical intrusion slightly pinch along the middle of the lateral margins of cell, but no pinching or hourglass shape occurs (Fig. 35).
CORTICAL PIGMENT	Granules small, elliptical, often in faint to dense, parallel rows (Fig. 26). Can appear as a fine stain.	Fine (not as fine as cat) and elliptical, can occur in dark patches. Cortical fusi. in some hairs, may be coarse and extend into root.	Fine brown granules. May be lightly stained.	Few small, dark, round granules. Scattered except at tip where it may be heavier..
MEDULLA	Single row, discontinuous. Column of cylindrical plates; often fragmentary at tip.	Variable. Single row discontinuous, may be interrupted. Rarely < 40% of shaft diameter.	Single row, discontinuous. Makes up more than half of shaft diameter at basal end.	Single row, discontinuous.

SHAPE OF MEDULLARY CELL	Variable, usually length > wide. Concave distally, convex proximally. Frequently hourglass or X-shaped (Fig. 25, 26, 29). "Ball-and-socket articulation" between adjacent cells [2](Fig. 26). No clear area, or if present, appears granular and not hyaline like rat/mouse. Bands of pigmented cells 2-8 cells long. Medullary units smaller in unpigmented sections of hair.	Squarish to somewhat oblong; usually width > length. Concave distally, convex proximally. Can have hourglass shape. Shape of cell less defined than in cat. No clear area.	Varies with hair diameter. > 1/2 distal end pigmented. < 1/2 proximal end clear. Flattened, extends across column. pigment may be in x-shape.	Variable, flat, usually width is 1/3 or more the length. Not hourglass-shaped. Compact, pigmented portion distal, clear proximal area <1/4 cell (Fig. 35). Cloudy. May be short or stout with very small clear area.
MEDULLARY PIGMENT	Light to dark brown, fine and elliptical (smaller than rat or mouse but larger than human). Loosely scattered; more concentrated at distal end (Fig. 25, 26). More defined than in dog.	Fine, elliptical, brown granules, loosely scattered. Usually with large pigment globules.	Light to dark brown, fine, loosely scattered.	Granular, large, very dark, and round in compact packet. May have slight pinching along middle of lateral margins (Fig. 35). Clear area granular or cloudy.
AIR SPACE	Generally bowl or crescent-shaped or more linear; concave distally, convex proximally (Fig. 26, 27, 28).	Generally bowl or crescent-shaped or linear and bowed with distal concave end. Concave end with sharp corners (Fig. 30, 31).	Rectangular or bowl-shaped with distal end concave and ends extending distally (Fig. 33).	Wide. Rectangular with distal end concave (Fig. 36), biconcave with sides slightly bulged or fat I-beam (Fig. 35). Length of air space approximately equal to length of medullary cell. In distal portion of hair - may be flattened with series of semicircular indentations along distal margin (Fig. 37).
INTERNODES	Much less prominent than rat or mouse. If fine hair, often wavy with moderately constricted internodes (Fig. 29).	Absent.	Gradual.	Usually none, may appear as very gradual.

1 According to Vazquez,[2] hair of the different rabbit genera (wild and domestic) have very uniform structure and cannot practically be differentiated.

2 A straight line can be drawn across tips of the concave (distal) end of one segment that would touch the convex (proximal) end of the next one.

Nuclear ghost is remnant of medullary cell nucleus; occurs in tip or apical shield of hairs with finer pigmentation. Appears as oval between adjacent air spaces. Common in some mammal hairs; for example, rat, mouse, rabbit and cat.

TABLE 5
Diagnostic Characteristics of Striated Guard Hairs of Selected Mammals Found in Foods

CHARACTER	RATTUS SPP. NORWAY, ROOF, POLYNESIAN	MUS MUSCULUS HOUSE MOUSE	PEROMYSCUS LEUCOPUS WHITE-FOOTED MOUSE	SQUIRREL	SUNCUS MURINUS ASIAN MUSK SHREW
GENERAL COMMENTS	Broad flat hairs, dumbbell-shaped cross-section. Cuticular scales ovate to elongate to flattened. Diamond petal, several scales across (Fig. 2).	Similar to Rattus but smaller. Cuticular scales ovate to crenate. Appear to point proximally (backwards) near base. Distal 1/3 of shaft brown. Rarely more than 7mm long.	Very tip often heavily pigmented and unmedullated.	Obvious agouti pattern. Large shield - spatulate. Tip more pigmented than rest of hair. Scales irregular at base.	Somewhat H-shaped in cross-section. Usually < 8 mm. Enlarged shield in distal 1/3 - 1/2 of hair. Tapers to a narrower basal portion or stalk 2/3 the width of the shield, one cell wide.
CORTEX	Narrow, very clear (Fig. 38), may have a few scattered granules. Tip may be densely pigmentated with light brown granules. Cortical intrusion projects into medullary cells but if cells are pigmented, intrusion is not discernable.	Very narrow, very clear (Fig. 39), may have a few scattered granules. Tip which may have moderate to dense pigmentation of dark granules. Cortical intrusion projects into medullary cells but if cells are pigmented, intrusion is not discernable. Proportionally narrower than Rattus.	Narrow (Fig. 40).	Longitudinally striate (very characteristic). Similar in proportion to Rattus. Cortex in shaft may have very abundant pigment granules of various sizes.	Hyaline, sometimes with few dark granules; tip more heavily pigmented with finer pigment.
CORTICAL PIGMENT	Few large, round, dark granules when present. Tip may also have moderate amount of smaller, dark granules; in row but rarely as parallel as cat. Apical node usually lightly colored (diffuse stain-like pigmentation. If present, medulla may be similarly pigmented. tip may also have small dark granules.	Few large, round, dark granules when present. Tip may also have moderate amount of smaller, dark granules; in row but rarely as parallel as cat. Apical node usually lightly colored (diffuse stain-like pigmentation). If present, medulla may be similarly pigmented. tip may also have small dark granules.	Little except at distal end in longitudinal streaks.	Granules fine to larger rod shaped, very prominent near the tip and base. At tip very dense black, followed by yellow area, and then subterminal black band (heavy pigment in cortex and medulla). Proximally, granules become more elliptical.	Large, dark granules. Concentrated at distal end of hair.

MEDULLA	Discontinuous, 2-12 transverse rows, staggered (Fig. 38). In the distal approximate 1/3 of hair, medullary cells may be less organized with the demarcation between the clear and pigmented areas becoming indistinct and eventually with cells becoming confluent with each other. Corresponds to area of increased cortical pigmentation.	Discontinuous. Up to 5 (usually 3 or 4) transverse rows, staggered (Fig. 39). Distinct dark and clear areas with distinctive borders. Cells stay discrete near distal portion (unlike Rattus).	3-5 transverse rows (usually 3) (Fig. 40).	Variable. Lattice-like. Border of medullary channel irregular; no clear border between medulla and cortex (Fig. 41). 4-6 transverse rows in shield. Shaft area 2 - 3 rows wide. Rows and cells curved in irregular matrix. In area between the shield and the shaft, small round granules occur among the medullary cells.	Discontinuous. 1-3 rows of cells.
SHAPE OF MEDULLARY CELL	Usually wider than long. Polyhedron with sharper corners than Mus (Fig. 38).	Oval or polyhedron with rounded corners (Fig. 39). Usually wider than long and oblong. Width of cell 1/3 to 2 times its length. Lateral cell pigment packets pinched by cortical - intrusion. Lighter cells (yellow) are shorter and not as well delineated as those of the darker banded area.	Moderately large, clear area - may surround up to 200-300° of proximal part of pigment packet (Fig. 40).	Irregular polyhedron with well angled corners. Width at least 2 times > length. Distal end may be concave.	If 2 rows, parabolic or wedge-shaped with pointed medial end; cells stacked with blunt end alternating from side to side (Fig. 42). If 3 rows, then middle cells are fusiform. Pigment packet at distal end, wider than long, rounded laterally and pointed medially - teardrop shaped.
MEDULLARY PIGMENT	Large, coarse, usually dark, spherical or ovoid; usually at distal end of cell. Proximal end clear but not hyaline as in mouse. In broad part of hair, cells along lateral margins are more heavily pigmented.	Large, dark, spherical or ovoid; usually at distal end of cell. Proximal end about 1/2 of cell length, very clear (clearer than rat).	Large, dark, round granules.	Rod or elliptical shaped (Fig. 41).	Large, dark, round granules.
AIR SPACE	I-Beam	I-Beam	I-Beam.	Fairly large, oblong or concavo-convex.	In double rows, connected in Y-shape or wishbone (Fig. 42); in triple rows or narrow sections, zigzag.

Note: One notable difference separating rodents relates to scale type of the guard hairs. Rats and mice (*Rattus*, *Bandicota*, *Mus*, *Peromyscus*) have diamond petal lanceolate scale patterns; i.e., individual scales make up a diamond pattern that looks like a pine cone (Figure 2). Other rodents (such as *Microtus* and squirrel) have a chevron scale pattern (appears to be made up of a series of irregular waves rather than individual scales) (Figure 3).[31]

TABLE 6
Diagnostic Characteristics of Striated Guard Hairs of Mammals Found in Foods

CHARACTER	CAT	DOG	RABBIT	WEASEL
GENERAL COMMENTS	Shield area absent. Scales flattened, prominent. hair may be wavy; may have transverse banding (a few medullary. units long). Root elongate, frayed fibers at base.	Varies greatly for different breeds. May be banded. Diameter variable. Abrupt tapering at base.	Very distinctive. Flattened; dumbbell-shaped in cross-section. Conspicuous shield (4-5X as wide as shaft), tips may be long and pointed. May be banded. May be in various colors. Distal part of shield and tip less pigmented than rest of hair. No internodes.	Rarely with agouti pattern; if present, tip is light or barely dark. Various colors, but usually brown and white. Shield present.
CUTICULAR SCALES	Flattened, crenate, edges prominent (edges serrate with shallow indentations). Scales Divergent.	Scales not prominent like cat. Difficult to see. Slightly elongate; crenate margins. Basal scales prominent.	Scale outlines or shadows usually inconspicuous. Streaky chevron pattern - a diagonal scale pattern, visible as move through different focal planes under microscope.	In basal portion, scales acuminate, pointing distally. Rest of scales flattened.
CORTEX	Thin to moderately thick (generally thicker than rodent), may be thicker on one side than the other. Translucent, cloudy or pigmented; can be stained (not hyaline like rat/mouse). Cortical intrusion present.	Fairly thick; sometimes one side thicker than the other (Fig. 46). Cloudy, can be striate. Stained or patchily distributed in fusiform aggregates (Fig. 48). Cortex may comprise whole shaft in some areas where medulla is absent. Can occupy most of diameter.	Usually narrow and clear, greenish; forms septa which divides medullary cells along length and width into compartments (very diagnostic). Usually clearer than rat/mouse. No cortical - intrusion apparent.	May be unpigmented, stained, or stained and pigmented
CORTICAL PIGMENT	Very fine, may be much smaller than squirrel; granules become larger, elongate (elliptical), and denser especially at tip. In short, broken parallel rows. Rarely with large round or elliptical dark brown pigment masses. Irregular, may be distributed in patches that correspond to pigmented areas in medulla.	Can be a stain or larger, round granules forming dark patches and not diffuse.	Few large, round, dark granules except at tip, which may range from clear to darkly pigmented and where granules may be finer and occur in short, dense streaks.	when pigmented, Evenly distributed throughout. Granules small and scattered.

MEDULLA	Continuous, smooth or erose medullary channel. Breaks up into indistinctly separated thin discs or plates which don't extend across full width of medulla (Fig. 43, 44, 45). May be vacuolated. Intercellular spaces are narrow and hard to see. May have infrequent lenticular spaces (not as common or as distinctive as dog).	Variable, discontinuous or continuous, vacuolated or amorphous (Fig. 46, 47, 48, 49). Occasionally very broad; 1/3 - 3/5 total width. Tapers sharply at base. Can also form a series of round compartments. May appear as large lenticular spaces interspersed occasionally between closely appressed groups of medullary units (Fig. 46, 48). May appear similar to cows.	Discontinuous. Usually 4-6 (up to 15 in very large hairs) transverse rows, arranged in longitudinal columns like corn-on-the-cob (Fig. 50).	May occupy more than 3/4 shaft diameter. Not as highly vacuolated as in cats and dogs.
SHAPE OF MEDULLARY CELL	Thin flat discs or partial discs. No clear areas. More discrete plates than dog or weasel.	Flat cells skewed or oriented perpendicular to hair axis. Not as closely packed as cat.	Cells small, squarish, rectangular or may be disc-like (Fig. 51, 52). Distal end pigmented, proximal end small and not as clear as rat or mouse.	Thin flat discs or partial discs.
MEDULLARY PIGMENT	Fine and elliptical. May be arranged in a more defined X- or chromosome shape than dog or weasel.	Stain or fine, round granules. Usually heaviest at cell periphery.	Granules large, dark and round. May be absent.	If non-colored, granular to fine; translucent appearance.
AIR SPACE	Usually curved I-Beams, flattened disc or bowl shape or lenticular. Air spaces may appear as plates or Y shapes connected in a zigzag arrangement. Coarser hairs may appear vacuolated. Flattened, thin irregular (not a straight line).	Large, bowl or lens-shaped; randomly occurring between cells. Coarse hair may appear vacuolated.	Rectangular with distal end concave (Fig. 51). May be flattened, sometimes with proximal end convex and distal end concave or irregular with a series of semicircular indentations (caused by pigment granules) (Fig. 52).	Flattened. Can be lenticular.

TABLE 7
Diagnostic Characteristics of Nonstriated Hairs Commonly Found in Foods

CHARACTER	HUMAN	SHEEP (WOOL)	COW	SWINE	GOAT
GENERAL COMMENTS	No banding (unless person diseased or has nutritional deficiency or as a result of dyeing). Coloration usually uniform over its length. Tip may be rounded due to wear (Fig. 53). May have ovoid bodies (dark, solid, oval shaped structures of different sizes). Uniform in general appearance. Area near base unpigmented and unmedullated. Area near base unpigmented and unmedullated.	Similar to human hair. Fine and curly.	Unbanded. No shield. Elongated. Medulla continuous into root area. Characteristic root shape. Slightly tapered.	Texture stiff, brittle. Hair tips frayed into filaments. No shield. Somewhat like a very coarse human hair.	More band-shaped than cow, flat. No shield.
DIAMETER	Varies, fine or wider depending on part of the body it came from.	Overall diameter approximately uniform, no noticeable taper at base or tip. However, small microscopic changes (narrowed and bulged areas) over the length of the hair.	Guard hairs thick and coarse. Coarse and uniformly tapered from root to tip.	Guard hairs thick and coarse. Coarse and uniformly tapered from root to tip.	Guard hairs thick and coarse. Coarse and uniformly tapered from root to tip.

CUTICULAR SCALES	Cuticle can comprise up to 25% of diameter. Heavily imbricated, flattened (Fig. 54). Small, edges irregular with little serration, change from smooth to crenate from base to tip. As hair diameter decreases, scales become spaced further apart. More tightly overlapping near base, widely spaced apart near tip. Normally 4-10 layers thick. Overlaps 5/6 or more of length of scale proximal to it. Scales of small, fine hairs may be more pronounced and more widely spaced due to wear.	Cuticular margins very serrated. Larger and less imbricated than human (Fig. 56). Scale has distinctive pattern, more of scales show. Usually 1-2 layers thick. Scales protrude (elevated more than other domestic animals). Scales overlap 1/6 or less of length.	Edges divergent.	Thick. Do not protrude from shaft. Diagnostic pattern - scales consist of delicate series of wide, short crenelated edges (Fig. 58).	No characteristic scale shapes or patterns. Smooth, scales don't protrude.
CORTEX	Color consistent throughout its length. Cortical fusi may be present.	Makes up majority of hair (90%). Clear to pigmented brown.	Variable. May appear fibrous with longitudinal striations. Moderately thick. May contain ovoid bodies - ovoid or fusiform clumps composed of tightly compressed pigment granules or large pigment granules. Infrequently, huge cortical fusi may be present. These air pockets are unique to cows (dogs and goats may have pigment masses). On the other hand, there may not be any structures in the cortex. Cortical intrusion present. Thicker than goat.	Usually very thick with medulla fairly narrow to absent (Fig. 59).	Moderately thick, longitudinally striate. May have occasional large pigment masses but rounder than dog or pig. Cortical - intrusion present.

TABLE 7 *(Continued)*
Diagnostic Characteristics of Nonstriated Hairs Commonly Found in Foods

CHARACTER	HUMAN	SHEEP (WOOL)	COW	SWINE	GOAT
CORTICAL PIGMENT	Evenly distributed or more concentrated towards periphery of cortex versus animal hair, where pigment is more concentrated around medulla. Size and shape of granules and density is variable. Fine to coarse, ovate to elliptical, may be in clumps (Fig. 50).	Clear to brown. If red, appears as stain. If brown or black, made up of granules.	Coarse. Ovoid bodies abundant. Small pigment granules, uniformly dispersed; may be in strings or large clumps.	Fine pigment granules, uniformly dispersed; may be in strings. Lighter hairs may appear stained.	Cortical globules of pigment present. In dark hairs, heavily pigmented, granules in dense strings.
MEDULLA	Absent, fragmented, or continuous. Usually 1/3 or < of shaft diameter. More commonly present in coarse hairs. Amorphous; fine granular or frothy appearance. Broken into rounded or spindle-shaped compartments with no distinct divisions between cells. If continuous, diameter along hair length is irregular. Non-pigmented.	Absent.	Variable. Absent or continuous. Amorphous or vacuolated. Width varies from narrow to broad, usually greater than 1/3 of hair diameter; irregular over length of hair. No discrete units with clear areas and air spaces. Walls of medullary channel smooth. Appears as closely appressed granular disk shaped objects, occasionally interspersed with large vacuoles (Fig. 57).	Absent to continuous; may be interrupted. Amorphous (Fig. 59), may be irregular with areas protruding into cortex. Fine grained and spongy often with rounded units. Usually broad to very broad; greater than 1/3 diameter. No thin, disk shaped units. More globular and irregular. Narrow at base, widens distally.	Greater than 1/3 of hair diameter (usually 1/2 or >). Unbroken lattice with honeycomb texture or disk or lens-shaped units. May occupy almost entire shaft width. Similar to cow but walls of medullary channel serrated, may not be obvious. No air vacuoles. Pattern more regular than other domestic ungulates.

Note: In nonstriated animal hairs, it is difficult if not impossible to distinguish between animals using the finer diameter hairs. These fur hairs are nondescript, very fine, with medulla absent or fragmentary. Therefore, they are not covered in this table. The thicker hairs must be used because they exhibit greater variation among species. In general, the nonstriated hairs of ungulates or hoofed animals have medullae that consist of a continuous channel; there are no discrete units with clear areas as in striated hairs. The medullary channel contains globules of pigments and air vacuoles. The cortex may appear fibrous.

REFERENCES

1. Vazquez, A.W., Structure and identification of common food-contaminating hairs, *J. Assoc. Off. Agric. Chem.*, 44(4), 754-779, 1961.
2. Vazquez, A.W., Hairs, in *Principles of Food Analysis for Filth, Decomposition, and Foreign Matter*, 2nd ed., Gorham, J.R., Ed., U.S. Food and Drug Admin. Tech. Bull. No. 1, Washington, DC, 125-170, 1981.
3. Scott, D.B., Identification of rodent fur hairs, *J. Assoc. Off. Agric. Chem.*, 38(2), 503-506, 1955.
4. Eisenberg, W.V., Sources of Food Contaminants, in *Principles of Food Analysis for Filth, Decomposition, and Foreign Matter*, 2nd ed., Gorham, J.R., Ed., U.S. Food and Drug Admin. Tech. Bull. No. 1, Washington, DC, 11-25, 1981.
5. Mills, R. and Pederson, J., *A Flour Mill Sanitation Manual*, eagan press, St. Paul, MN, 1-164, 1990.
6. Frantz, S.C. and Davis, D.E., Bionomics and Integrated Pest Management of Commensal Rodents, in *Ecology and Management of Food-Industry Pests*, Gorham, J.R., Ed., U.S. Food and Drug Admin. Tech. Bull. No. 4, Assoc. Off. Anal. Chem., Arlington, VA, 243-311, 1991.
7. Gorham, J.R., Food Pests as Disease Vectors, in *Ecology and Management of Food-Industry Pests*, Gorham, J.R., Ed., U.S. Food and Drug Admin. Tech. Bull. No. 4, Assoc. Off. Anal. Chem., Arlington, VA, 477-482, 1991.
8. Constantine, D.G., Ecology and Management of Bats as Food-Industry Pests, in *Ecology and Management of Food-Industry Pests*, Gorham, J.R., Ed., U.S. Food and Drug Admin. Tech. Bull. No. 4, Assoc. Off. Anal. Chem., Arlington, VA, 237-241, 1991.
9. Olsen, A.R., Distinguishing common food-contaminating bat hairs from certain feather barbules, *J. Assoc. Off. Anal. Chem.*, 64(4), 786-791, 1981.
10. Nowak, R.M., *Walker's Mammals of the World*, Vols. I & II, 5th ed., Johns Hopkins University Press, Baltimore, 1-1629, 1991.
11. Longrée, K., *Quantity Food Sanitation*, Wiley Interscience, NY, 95-110, 1967.
12. Imholte, T.J., *Engineering for Food Safety and Sanitation: A Guide to the Sanitary Design of Food Plants and Food Plant Equipment*, TIFS—Technical Institute of Food Safety, Crystal, MN, 1-283, 1984.
13. Thrasher, J.J., Detection of Metabolic Products, in *Principles of Food Analysis for Filth, Decomposition, and Foreign Matter*, 2nd ed., Gorham, J.R., Ed., U.S. Food and Drug Admin. Tech. Bull. No. 1, Washington, DC, 201-216, 1981.
14. Sinha, R.N., Storage Ecosystems, in *Ecology and Management of Food-Industry Pests*, Gorham, J.R., Ed., U.S. Food and Drug Admin. Tech. Bull. No. 4, Assoc. Off. Anal. Chem., Arlington, VA, 17-30, 1991.
15. Canby, T.Y., The rat, lapdog of the devil, *Natl. Geograph.*, 152, 60-87, 1977.
16. Churchfield, S., *The Natural History of Shrews*, Cornell Univ. Press, Ithaca, NY, 1-178, 1990.
17. Olsen, A.R., Fecal pellets from a commensal shrew (*Suncus murinus*) and a house gecko (Gekkonidae), *J. Assoc. Off. Anal. Chem.*, 67(6), 1035-1036, 1984.
18. Chenchittikul, M., Daengpium, S., Hasegawa, M., Itoh, T. and Phanthumachinda, B., A study of commensal rodents and shrews with reference to the parasites of medical importance in Chanthaburi Province, Thailand, *Southeast Asian J. Trop. Med. Pub. Health*, 14(2), 255-259, 1983.
19. Walton, D.W. and Tun, U.M.M., Fleas of small mammals from Rangoon, Burma, *Southeast Asian J. Trop. Med. Pub. Health*, 9(3), 369-377, 1978.
20. Brooks, J.E., Naing, U.H., Walton, D.W., Myint, D.S., Tun, U.M.M., Thaung, U., and Kyi, D.O., Plague in small mammals and humans in Rangoon, Burma, *Southeast Asian J. Trop. Med. Pub. Health*, 8(3), 335-344, 1977.
21. Singh, S.P., Sethi, M.S., and Sharma, V.D., The occurrence of salmonellae in rodent, shrew, cockroach and ant, *Int. J. Zoonoses*, 7(1), 58-61, 1980.
22. Schmidt, G.D. and Roberts, L.S., Phylum Apicomplexa: Gregarines, Coccidia, and Related Organisms, in *Foundation of Parasitology*, 4th ed., Times Mirror/Mosby Coll. Publ., St. Louis, MO, 123-129, 1989.
23. World Health Organization, *Ecology and Control of Public Health Importance*, Tech. Rep. Ser., No. 553, World Health Organization, Geneva, 1-42, 1974.
24. Lyne, A.G., The development of hair follicles, *Aust. J. Sci.*, 28(10), 370-377, 1966.
25. Wildman, A.B., *The Microscopy of Animal Textile Fibres*, Wool Industries Research Association, Leeds, 3-10, 49-59, 1954.
26. Ling, J.K., Pelage and molting in wild mammals with special reference to aquatic forms, *Quart. Rev. Biol.*, 45, 16-54, 1970.
27. Hausman, L.A., A comparative racial study of the structural elements of human head-hair, *Am. Naturalist*, 59(665), 529-538, 1925.
28. Hausman, L. A., The relationships of the microscopic structural characters of human head-hair, *Am. J. Phys. Anthrop.*, 8(2), 173-177, 1925.
29. Hausman, L.A., The pigmentation of human head-hair, *Am. Naturalist*, 61(677), 545-554, 1927.
30. Teerink, B.J., *Hair of West-European Mammals*, Cambridge Univ. Press, Cambridge, 1-224, 1991.

31. Day, M.G., Identification of hair and feather remains in the gut and faeces of stoats and weasels, *J. Zool.*, 148, 201-217, 1966.
32. Moore, T.D., Spence, L.E., and Dugnolle, C.E., *Identification of the Dorsal Guard Hairs of Some Mammals of Wyoming*, Hepworth, W.G., Ed., Wyoming Game and Fish Department, Bull. No. 14, Cheyenne, WY, 1-177, 1974.
33. Bisbing, R.E., The Forensic Identification and Association of Human Hair, in *Forensic Science Handbook*, Saferstein, R., Ed., Prentice Hall, Englewood Cliffs, NJ, 184-220, 1982.
34. Myers, R.J., The structure and identification of hair, *J. Assoc. Off. Agric. Chem.*, 41(2), 368-376, 1958.
35. Robbins, C.R., The Morphology and Chemistry of Human Hair, in *Proceedings of the International Symposium on Forensic Hair Comparisons*, Laboratory Division, Federal Bureau of Investigation, Washington, DC, 3-21, 1985.
36. Eddy, M.W., Hair classification, *Proc. Penn. Acad. Sci.*, 12, 19-26, 1938.
37. Hausman, L.A., Applied microscopy of hair, *Scientific Monthly*, 59, 195-202, 1944.
38. Hicks, J. W., *Microscopy of Hairs: A Practical Guide and Manual*, Issue 2, Federal Bureau of Investigation, Washington, DC, 1-53, 1977.
39. Kaszynski, E., Hair Growth: Mechanism and Regulation, in *Proceedings of the International Symposium on Forensic Hair Comparisons*, Laboratory Division, Federal Bureau of Investigation, Washington, DC, 23-33, 1985.
40. Mathiak, H.A., A key to hairs of the mammals of Southern Michigan, *J. Wildl. Management*, 2(4), 251-268, 1938.
41. Brown, F.M., The microscopy of mammalian hair for anthropologists, *Proc. Am. Philos. Soc.*, 85(3), 250-274, 1942.
42. Cole, H.I., Taxonomic value of hair in Chiroptera, *Philipp. J. Sci.*, 24(1), 117-121 and Plates 1 & 2, 1924.
43. Mayer, W. V., The hair of California mammals with keys to the dorsal guard hairs of California mammals, *Am. Midland Naturalist*, 48(2), 480-512, 1952.
44. Williams, C.S., Aids to the identification of mole and shrew hairs with general comments on hair structure and hair determination, *J. Wildl. Management*, 2(4), 239-250, 1938.
45. Bryce, J.R., Identification of hairs of three asian commensal mammals: *Suncus murinus*, *Bandicota bengalensis*, and *Rattus exulans*, *J. Assoc. Off. Anal. Chem.*, 77(2), 403-410, 1994.
46. Stains, H.J., Field key to guard hair of Middle Western furbearers, *J. Wildl. Management*, 22(1), 95-97, 1958.
47. Hausman, L.A., Structural characteristics of the hair of mammals, *Am. Naturalist*, 54(635), 496-523, 1920.
48. Metcalf, Z.P., The construction of keys, *Syst. Zool.*, 3(1), 38-45, 1954.
49. Weber, W.J., *Health Hazards from Pigeons, Starlings and English Sparrows: Diseases and Parasites Associated with Pigeons, Starlings and English Sparrows which Affect Man and Domestic Animals*, Thomson Publications, Fresno, CA, 1-138, 1979.
50. Baur, F.J. and Jackson, W.B., Eds., *Bird Control in Food Plants—It's a Flying Shame!*, Am. Assoc. Cereal Chem., St. Paul, MN, 1-90, 1982.
51. Fenlon, D.R., Wild birds and silage as reservoirs of *Listeria* in the agricultural environment, *J. Appl. Bacteriol.*, 59, 537-543, 1985.
52. Ogg, J.E., Ryder, R.A., and Smith, H.L., Jr., Isolation of *Vibrio cholerae* from aquatic birds in Colorado and Utah, *Appl. Environ. Microbiol.*, 55(1), 95-99, 1989.
53. Chandler, A.C., A study of the structure of feathers, with reference to their taxonomic significance, *Univ. Calif. Publ. Zool.*, 13(11), 243-446, plates 13-37, 1916.
54. Voitkevich, A.A., *The Feathers and Plumage of Birds*, Sidgwick and Jackson, London, 1-335, 1966.
55. Davies, A., Micromorphology of feathers using the scanning electron microscope, *Forensic Sci. Soc. J.*, 10(3), 165-174, 1970.
56. Messenger, N.G., Methods used for identification of feather remains from Wetherill Mesa, *Am. Antiquity*, 31(2) Pt. 2, 206-215, 1965.
57. Laybourne, R.C., Deedrick, D.W., and Hueber, F.M., Feather in amber is earliest New World fossil of Picidae, *Wilson Bull.*, 106(1), 18-25, 1994.

Chapter 13

ETHNOENTOMOLOGY: INSECTS FOR HUMAN CONSUMPTION

Alan R. Olsen and Jeremy M. Olsen

CONTENTS

I. Introduction 217

II. Historical Perspective 217

III. Major Food Insect Commodities 218
A. Scale Insects 218
B. Bee Products 218

IV. Ethnic Items 219
A. Lepidoptera Larvae 219
B. Giant Water Bugs 221
C. Ants 222

V. Future Trends 222

VI. Regulation of Food Insects 224

References 225

I. INTRODUCTION

Insects and other terrestrial arthropods figure in the arts, religion, folklore, recreation and diet of many cultures. Ethnoentomology is the study of the interactions between humans and arthropods in these spheres of culture. Traditionally, ethnoentomology, has dealt with the roles that arthropods play in nonindustrial or so-called primitive societies.[1] In the modern world, however, there are few truly primitive societies that are so isolated that their customs and traditions are immune from exportation to other countries and vice versa. Cultural cross-fertilization is accelerated by today's global communications and transportation networks and, in this global village, food is a major exportable item of trade. A minor, but persistent, component of this trade consists of insects that are packed and sold for human consumption. [2] One result is that ethnoentomologies of the use of food insects in the diets of other cultures find applications that reach into the societies of even the most industrialized countries.

II. HISTORICAL PERSPECTIVE

Entomophagy, the eating of insects, is an ancient practice dating from prehistoric times.[3,4] The manna of the Exodus is similar to the sweet exudate of the tamarisk manna scale, *Trabutina manipara* (Ehrenberg) (Homoptera: Eriococcidae),[5,6] and the locust-like insects permitted under levitical food laws[7] undoubtedly included the migratory locust, *Locusta migratoria* L., and the desert locust, *Schistocerca gregaria* Forskal (Orthoptera: Acrididae).

These insects still occur in the Middle East and are used as food by local residents.[8] Ethnoentomologists find that entomophagy is practiced among cultures throughout the modern world and is not merely a relic of a primitive past.[9]

It is often assumed that entomophagy in developing countries is a last resort to avoid starvation, and this may be true in some cases. In many indigenous hunting/gathering and subsistence farming societies, however, arthropods are a planned component of the diet, either seasonally or throughout the year.[10] There are more than 400 known species of such edible insects.[11] Examples range from the bugong worm, *Agrotis infusa* (Boisduval) (Lepidoptera: Noctuidae), and the giant wood spider, *Nephila maculata* (F.) (Araneae: Tetragnathidae), which are delicacies in Australia and Papua New Guinea, to the kungu cake of Africa made from phantom midge larvae, *Chaoborus edulis* Edwards (Diptera: Chaoboridae), that grow in Lake Nyasa.[5,6,12-14] These and other arthropods are harvested as prized delicacies to be enjoyed and shared with family or friends. As a society emerges from relative isolation and begins to participate in the global economy, entrepreneurs devise techniques for manufacturing and packing cultural food items, such as arthropods, for sale to affluent members who retain their ethnic diet preferences but no longer forage or farm. Inevitably, some ethnic food insects become exportable, either to supply the needs of emigrant communities or as a general commodity in international trade. The ethnoentomologies for groups that traditionally consume a particular insect become important sources of information for determining how to regulate and control the safety of the commercial food insect product once it is offered for trade to other countries.

III. MAJOR FOOD INSECT COMMODITIES

A. SCALE INSECTS

Citizens of industrialized societies consume insect products as food, but the food is usually refined to where its arthropod component is not obvious. Cochineal, for example, is a color additive used in processed foods that is extracted from a scale insect, *Dactylopius coccus* Costa (Homoptera: Dactylopidae). The Indian lac insect, *Laccifer lacca* (Kern) (Homoptera: Lacciferidae), is another scale insect that produces an exudate that is refined to make shellac. Among its many uses, shellac serves as a stabilizer or thickener in certain foods and as a surface-finishing agent in other products. It is also used as an adhesive for food packages and as an adjuvant for pesticide chemicals that are applied to food crops and raw agricultural commodities after harvest.[15] Although cochineal and shellac are intrinsically harmless, their production is regulated to prevent adulteration. Cochineal extract must be pasteurized or otherwise treated to destroy viable *Salmonella* microorganisms and both products must be of food-grade quality that is wholesome and free of harmful or deleterious substances. In addition, their use must be declared in the ingredients statement of mandatory labeling that is required in many countries.

Ethnoentomologists have found that the cochineal insect and Indian lac insect were not used for food in the cultures where they were first cultivated. Native Americans used cochineal as a dye and it was originally exported by 16th century Spaniards as a fabric dye. In India, shellac is used as a dye and as an ingredient in paints and varnishes.[5]

B. BEE PRODUCTS

Another widely used insect product, honey, has always been prized as a food by nearly every culture that has honey-producing bees in its traditional territory. The product of the honeybee, *Apis mellifera* L. (Hymenoptera: Apidae), is commercially produced throughout the world and there is a plethora of regulations and laws governing its quality and standardization of its components. Honey houses, where honey is extracted from the comb, and packing plants must be operated under sanitary conditions to avoid infestations by rodents, roaches and other pests.

Microanalytical examinations of honey that was produced under insanitary conditions may detect rodent hairs, excreta and insect fragments, including fragments of honeybee workers that were attracted to the extracted product as a convenient means of resupplying the hive. Honeybee cultivation has been widely exported from Europe and India where it originated, but not all honey comes from *A. mellifera*. In many tropical countries, for example, the honey from small stingless bees, *Trigona* spp., is valued for its sweetness despite the fact that stingless bees do not separate honey combs from brood combs and bee larvae are distributed throughout the sticky mass of dark honey. In most cases, the larvae are picked out and discarded rather than consumed with the honey. Conversely, some cultures consider bee brood from *A. mellifera* to be food and separately harvest honeybee larvae and pupae for consumption.[(8)]

IV. ETHNIC ITEMS

Direct consumption of whole insects is practiced in many cultures. In most cases, this entomophagy remains a local phenomenon that is limited by the natural distribution or abundance of the particular food insect. The majority of ethnoentomologies that report insects in a culture's diet involve situations where insects are foraged from the surrounding environment and consumed by the foragers and their immediate social group.[(16-18)] This type of food resource is rarely abundant enough to support sustained commercial exploitation beyond a local level. The few exceptions mostly involve insects that are associated with another indigenous industry that experiences a transition to wider commercial success. Entomophagy that exists in companion with an industry in its cottage or subsistence stage may persist as a tradition among workers and captains of the industry through its growth phases simply because the means of producing the food insect is related to the industrial processes and thus keeps pace with the growth of the companion industry. Eventually, the production and processing of this type of food insect becomes sufficiently sophisticated and profitable to invite exportation.

A. LEPIDOPTERA LARVAE

Two excellent examples of this process are the silkworm and the maguey or tequila worm. In 2697 B.C. (or so), sericulture began as a cottage industry, first in China and later spreading to the rest of Asia and Europe.[(8)] Over time, various Asian cultures recognized the food value of the naked silkworm pupa, *Bombyx mori* (L.) (Lepidoptera: Bombycidae), that remained after its silk cocoon was reeled off. Silkworm pupae are a popular, seasonally abundant food in the silk-producing areas of Asia.

With the expansion and industrialization of the silk trade in the 20th century, modern food-processing techniques were applied to the silkworm pupa, whose abundance had similarly grown. Today, pupae in hermetically sealed cans are mass produced and exported throughout the world (Figure 1). Dried silkworm larvae are also commercially produced and exported. There have been instances where pupa-eating is introduced to new territories along with the exportation of sericulture technology.[(9)]

The maguey worm or caterpillar, *Aegiale hesperianus* Kirby (Lepidoptera: Megathymidae), is a borer of the agave plant that is cultivated in Mexico for the production of tequila. The larvae are relished in local subcultures where tequila production is an economic mainstay. Early distillers began a tradition of placing a single larva in each bottle of distillation batches that were of exceptional quality. The tradition apparently assured lasting popularity for the maguey worm as a food insect. The larvae are served in posh restaurants in Mexico and commercially canned and dried products are exported.[(6,9)] The larva in the tequila

FIGURE 1. Commercially canned silkworm pupae, *Bombyx mori*, packed by an establishment in Korea for local sale and exportation. In the silk-producing areas of Asia, fresh (still alive) pupae are often stir-fried. (Photo taken by D. Ludwig.)

originally causes concern in cultures where tequila is newly imported, but as consumers gain familiarity with the product the insect component becomes accepted as a part of the product.[2]

Mopane caterpillars, *Gonimbrasia belina* Westwood (Lepidoptera: Saturniidae), are a food item of commerce in Africa. The caterpillars are gathered in season and dried for the export trade among central and southern African countries.[9] The mopane achieved recent notoriety as the object of poachers in Malawi who felled trees in the Kasungu National Park in order to swiftly harvest the caterpillars without being apprehended.[19] The World Wildlife Federation responded by funding a project to promote harvesting techniques that are less devastating to the ecology of the region. The park reportedly yields two tons of caterpillars during the harvest season from October through December.

The pallid emporer moth larva, *Cirina forda* Westwood, is another saturniid caterpillar that is a common food item in subsaharan Africa. The dried caterpillar is commercially marketed in Nigeria (Kwara state), where it is an essential ingredient in recipes for a popular vegetable soup.[20] The pallid emporer occurs in tropical Africa wherever its host plant, the sheabutter tree, is found and the caterpillar is generally known to be edible throughout its range.[21,22] Such is the importance of saturniid larvae and other caterpillars as a food

FIGURE 2. Giant water bug, family Belostomatidae, as marketed in fresh or frozen condition. (Photo taken by D. Ludwig.)

resource that researchers in Afria are developing forest management techniques for protecting and increasing the annual harvest of these insects.[23]

B. GIANT WATER BUGS

A more recent entrant into international commerce is the Thai giant water bug, *Lethocerus indicus* Stal (Hemiptera: Belostomatidae) (Figure 2). This traditional food insect of Southeast

Asia is exported to many countries including the United States.[24] Although pastes and extracts of this aquatic insect are manufactured, the most common product is whole, frozen insects that are packed in trays with a clear polyfilm cover, much as ground beef and other meats are packed for supermarkets and butcher shops. Reportedly, the insects grow in or near artificial ponds designed to cultivate shrimp or fish and their general availability has increased along with the recent growth in the shrimp and fish farming industry in Southeast Asia. The construction of refrigeration and freezer facilities for the seafood farming industry has probably influenced the developmental direction of the companion water bug industry toward the production of freshly frozen water bugs.

C. ANTS

Ants are another recent entrant into the world trade in food insects. Hardly the chocolate-covered novelty items that were marketed in the 1950s, weaver ants, *Oecophylla smaragdina* F. (Hymenoptera: Formicidae) (Figure 3), from Southeast Asia are steamed and packed in plastic bags for export as a genuine food item. The commercially packed product is usually kept frozen and consists of larvae and pupae with a few adults including alates. They are also used to manufacture a condimental paste.[3] In the tropics, *O. smaragdina* builds large aerial nests in trees and a single colony may contain tens of thousands of individuals.[25] Another ant, *Liometopum apiculatum* Mayr, is an occasional export from Mexico. The nests of both ants are easily harvested, partially explaining the commercial exploitation of these particular species.

V. FUTURE TRENDS

Food insects presently comprise an unimpressive proportion of the world trade in food commodities but this may not forever be the case. The mopane sequence of events in Malawi may represent a new avenue for ethnic food insects to expand their markets through media exposure and outside support, although not necessarily with such a criminal undercurrent. A small, though vocal, advocacy for expanding the food use of insects has developed, and among the more serious endeavors of these advocates are ethnoentomologies that inventory the species of insects that are dietary items in particular cultures. These inventories are an important resource for establishing identity standards for future entrants into the commercial trade in processed food insects. A focal point of this movement is the University of Wisconsin's Food Insects Research and Development Project, established in 1986, which maintains a global inventory of edible insects and publishes a newsletter of anecdotal reports concerning the food use of insects. Of perhaps greater impact was the first Ethnobiology Congress, convened in Belem, Brazil, in July 1988. The chairperson of this congress was D.A. Posey, an ethnoentomologist who inventoried entomophagy among Brazilian native peoples.[26] The congress issued a policy known as "The Declaration of Belem" that calls for increased developmental aid for native peoples to use to preserve their cultures and global biodiversity. An underlying theme of this declaration is support for the development of markets for indigenous products and renewable natural resources as a means of increasing the intrinsic value of natural biodiversity as a capital asset for native peoples. Food insects fit this concept well and, if successful, the Declaration of Belem may prove a turning point in the mass marketing of food insects.

Future research needs include conducting additional ethnoentomological surveys to catalog, among other information, species of edible insects that could enter the marketplace in the future. This is a pressing need in Amazonia and other tropical regions where the marketing of food insects may be developed as a component of programs to preserve the biodiversity of tropical rain forests.[26] Another need is to develop taxonomic keys, reference collections and other reliable aids for identifying food insects. Research on the systematics of food insect

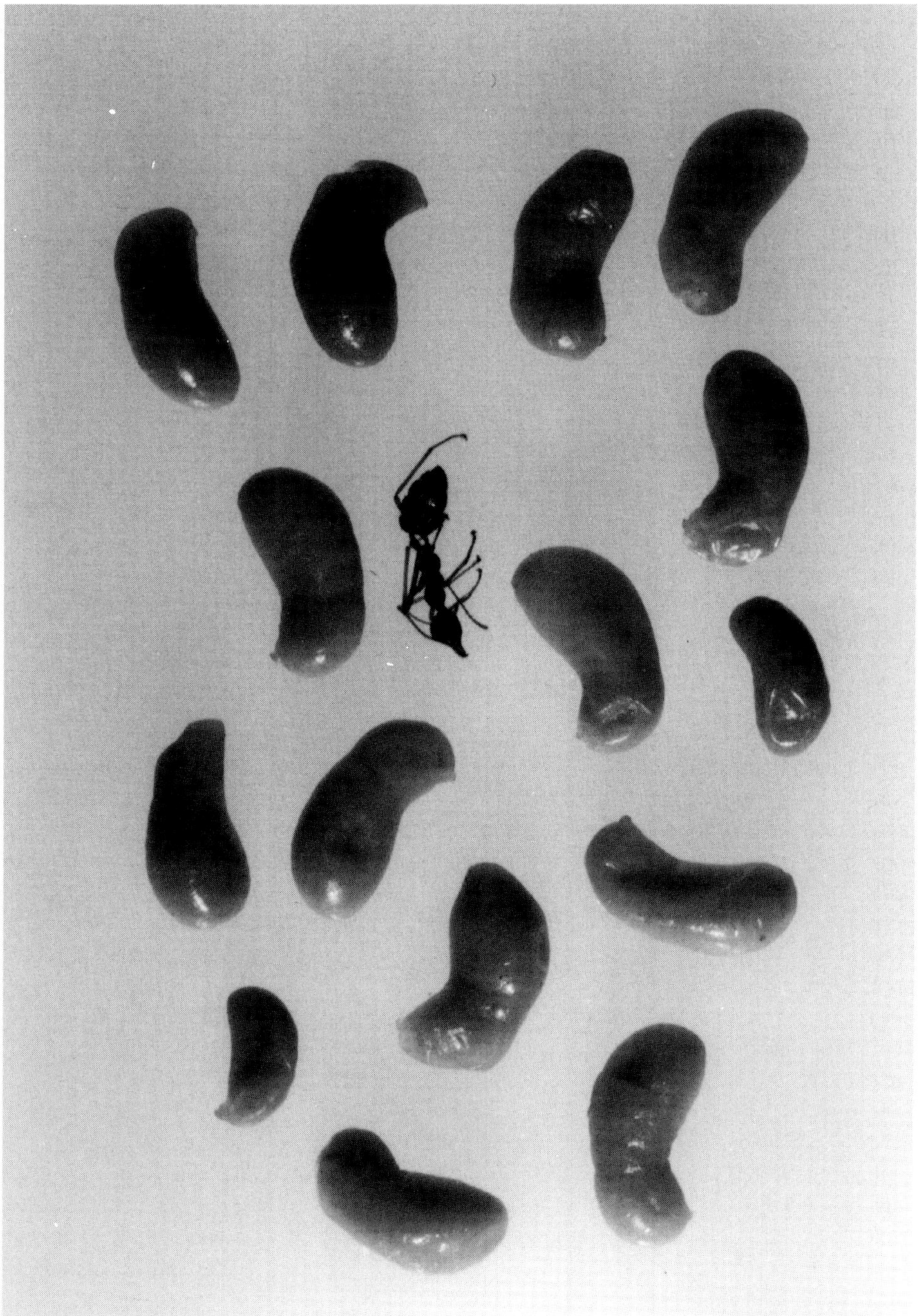

FIGURE 3. Various life stages of the weaver ant, *Oecophylla smaragdina*, that are steamed, then packed in plastic bags and tubs to be sold locally in Thailand or shipped under refrigeration to other countries. (Photo taken by D. Ludwig.)

groups is especially important for those groups that have economic potential that has been recognized but not yet realized in the marketplace. Examples of potential entrants into the market include the palm weevil larvae, *Rhynchophorus phoenicis* (F.) (Coleoptera: Curculionidae); silk spiders *Nephila* spp. (Araneae: Tetragnathidae); and tropical crickets, locusts and related orthopterans.[12,27,28] Identification keys are needed not only to identify food insects but also to identify poisonous or toxic species that could be mistaken for edible species.[29] Eventually, analyses must be conducted to determine the nutritional content of the various food insect species in order to comply with the nutritional labeling requirements of the United States, other countries and international codices.

VI. REGULATION OF FOOD INSECTS

From a regulatory standpoint, food insects must comply with the same regulations that apply to any other food. They must be clean, wholesome, free from harmful or deleterious substances and correctly labeled. They must have been manufactured, packaged, stored and transported under sanitary conditions and there must be no filth in the product. In the United States, food insects must also comply with all applicable standards of the Food, Drug and Cosmetic Act including, but not limited to, the above standards. The fact that a food product consists largely or entirely of insects intentionally processed and packaged for use as human food does not automatically bar it from commercial distribution in most countries, including the United States.[2] As with any food product, handlers of food insects must operate under protocols that will prevent the proliferation and persistence of pathogens and toxins. Many insects contain toxic substances, allergens and anatomical hazards that can harm the consumer.[3,30] Otherwise harmless insects may contain pesticide residues. This has been a concern in the Middle East where locusts that migrate from Africa to Kuwait (where they are gathered for food) are occasionally found to contain pesticide residues. Insect products are also susceptible to contamination with food-borne pathogens, especially aquatic insects and insects that live on or in soils where pathogens are widely distributed. Precautions are essential in these respects because food insect products that were mishandled have caused deaths of consumers.[31,32]

The task of the microanalytical entomologist with regard to food insects is twofold. First, the food insect must be accurately and precisely identified. Second, the product must be examined for the presence of filth and other extraneous materials. The reasons for accurately identifying a food insect are many, not the least of which is truth in labeling. Identification is also necessary to determine if the insect may contain toxins or other harmful substances. Some insects have a substantial folklore history of unsubstantiated medical claims, in which case they would be more appropriately regulated as pharmaceutical drugs, to which much more stringent laws and regulations apply. Identification also reveals the life history of the insect. Life history information can prove valuable in determining the etiology of adulterants that may subsequently be detected in the product.

Finally, identification enables comparison with ethnoentomological information to determine if an insect is a genuine food article, if it has a folkloric history of medicinal use, if it may contain a hazardous or deleterious component, or if it may have been grown or harvested under conditions whereby it may become adulterated. It is equally important for ethnoentomologists to supply full, accurate and precise identifications of the insects they study for the benefit of microanalytical entomologists and other scientists who rely on their scientific reports.

Historically, food insects have proven vulnerable to adulteration from the same types of stored-product and commensal pests as other foods (see chapter 18). Analyses of most of the products discussed in this chapter have, on occasion, detected rodent hairs, feather barbules, insect fragments (other than the food species), mites and other filth. Fresh and refrigerated or frozen products are also susceptible to decomposition and mold adulteration as a result of mishandling. Dried products may become adulterated with mold, live insects and mites. Highly processed or comminuted products may contain various types of microscopic filth. In summary, although the food insect may be an exotic and unusual item, the filth that is detected will not be unfamiliar to the experienced microanalytical entomologist.

REFERENCES

1. Hogue, C.L., Cultural entomology, *Annu. Rev. Entomol.*, 32, 181-199, 1986.
2. Brickey, P.M., Jr. and Gorham, J.R., Preliminary comments on federal regulations pertaining to insects as food, *Food Insects Newsl.*, 2(1), 1-7, 1989.
3. Bequaert, J., Insects as food. How they have augmented the food supply of mankind in early and recent times, *J. Am. Mus. Natur. Hist.*, 21, 191-200, 1921.
4. Ramos-Elordy, J. and Moreno, J.M.P., *Los Insectos Comestibles en el Mexico antiguo (estudio ethnoentomologico)*, Mexico, D.F., 1-108, 1988.
5. Essig, E.O., *College Entomology*, Macmillan, NY, 1-900, 1942.
6. Borror, D.J., Triplehorn, C.A., and Johnson, N.F., *An Introduction to the Study of Insects*, 6th ed., Saunders Coll. Publ., Philadelphia, PA, 1-875, 1989.
7. Leviticus 11:21-23.
8. Metcalf, C.L., Flint, W.P., and Metcalf, R.L., *Destructive and Useful Insects*, McGraw-Hill, NY, 1-1087, 1962.
9. DeFoliart, G.R., The human use of insects as food and as animal feed, *Bull. Entomol. Soc. Am.*, 34(2), 22-35, 1989.
10. Defoliart, G.R., Insects as human food, *Crop Prot.*, 11, 395-399, 1992.
11. Ramos-Elordy, J., *Los Insectos como fuente de Proteinas en el Futoro*, Editorial Limusa, Mexico, D.F., 1-148, 1991.
12. Bristowe, W.S., Insects and other invertebrates for human consumption in Siam, *Trans. Entomol. Soc. Lond.*, 80, 387-404, 1932.
13. Robinson, M.H. and Robinson, B., Comparative studies of the courtship and mating behaviors of tropical araneid spiders, *Pac. Insect Monogr.*, 36, 1-218, 1980.
14. Oldroyd, H., *The Natural History of Flies*, Norton, NY, 1-324, 1964.
15. *Code of Federal Regulations, Title 21*, USGPO, Washington, DC, 1994; secs. 73.1100, 175.105, 184,1706.
16. Hitchcock, S.W., Insects and the Indians of America, *Bull. Entomol. Soc. Am.*, 8, 181-187, 1962.
17. Bourne, G.H., The food of the Australian aboriginal, *Proc. Nutr. Soc.*, 12, 58-65, 1953.
18. Ruddle, K., The human use of insects: Examples from the Yukpa, *Biotropica*, 5, 94-101, 1973.
19. Anon., Saving a threatened African tree, *Focus*, 15, 5, 1993.
20. Fasoranti, J.O. and Ajiboye, D.O., Some edible insects of Kwara state, Nigeria, *Am. Entomol.*, 39, 113-116, 1993.
21. Pinhey, E.C.G., *Emporer Moths of South and South Central Africa*, C. Struik, Cape Town, 1-150, 1972.
22. Pinhey, E.C.G., *Moths of Southern Africa*, Tafelberg Publ., Cape Town, 1-273, 1975.
23. DeFoliart, G.R., Forest management for the protection of edible caterpillars in Africa, *Food Insects Newsl.*, 4(2), 1-2, 1991.
24. Pemberton, R.W., The use of the Thai giant waterbug, *Lethocerus indicus* (Hemiptera: Belostomatidae), as human food in California, *Pan-Pacif. Entomol.*, 64, 81-82, 1988.
25. Holdobler, B. and Wilson, E.O., *The Ants*, Belknap, Cambridge, 1-732, 1990.
26. Posey, D.A., Ethnoentomological survey of Brazilian Indians, *Entomol. Gener.*, 12, 191-202, 1987.
27. DeFoliart, G.R., Food insects, *Principes*, 37, 42-47, 1993.
28. Robinson, M.H. and Robinson, B., The biology and behavior of *Nephila maculata*: A supplement, *Smithson. Contrib. Zool.*, 218, 1-22, 1976.
29. Gorham, J.R., The significance for human health of insects in food, *Annu. Rev. Entomol.*, 24, 209-224, 1979.
30. Gorham, J.R., Insects as food, *Bull. Soc. Vector Ecol.*, 3, 11-16, 1976.
31. Nightingale, K.W. and Aym, E.N., Outbreak of botulism in Kenya after ingestion of white ants, *Br. Med. J.*, 281, 1682-1683, 1980.
32. Steyn, D.G., Grasshopper (*Phymateus leprosus* Fabr.) poisoning in a Bantu child, *S. Afr. Med. J.*, 36, 822-823, 1962.

Chapter 14

REFERENCE COLLECTIONS AND MATERIALS

Howard G. Semey

CONTENTS

I. Introduction 227

II. Basic Equipment Needs 228

III. Techniques 229
A. Preserving Whole Insect Specimens 229
1. Pin Mounts 230
2. Labels 230
3. Relaxing Specimens 231
4. Minutens and Point Mounts 231
5. Glue Boards 232
6. Preserving Insects in Liquids 232
7. Specimens for Training 233
B. Microscope Slide Mounts of Insects and Fragmentary Items 233
1. Mounting Media and Mounting Techniques 234
2. Clearing and Staining Reagents 235
3. Preserving and Storing Microscope Slide Mounts 235
C. Miscellaneous Techniques 236
1. Vertebrate Specimens 236
2. Product Specimens 236
3. Genitalia Specimens 236

IV. Collection Management 236

V. Sources and Resources 237

References 238

I. INTRODUCTION

Although a whole insect may be found during microanalysis of a food product, a whole rodent, bird or bottle is not. What is possible is to find fragmentary evidence of contamination—evidence as small as 0.5 mm. A portion of a single hair, a feather barbule or a shard of glass may be used by the microanalytical entomologist to determine what has contaminated the product, and where along the route from producer to consumer the contamination could have occurred. To do this requires an extensive amount of knowledge and experience.[1] But even with years of training and practice, there are simply too many and varied items that can be seen in a food sample to give one the confidence to stand up and testify in a court of law based on memory alone. Just think about a food product, possibly containing flour, spices, fruits and nuts, or beans and peas, with some Oriental herbs thrown in for appearance. Digest this in a mild acid, transferring the residue to a lined disk of filter paper for microscopic analysis. Each of these items of plant material could leave more than a little residue on the filter paper. Now add to this mass a number of insect fragments, say from four species of stored-product and field pests, possibly a few hairs, and even an insect's head, without antennae, of course. You begin to see how memory alone may begin to be strained trying to remember the characteristics for all the potential insects and animals that could have gotten into the product based just on these remnants. You need something more concrete to confirm what you think you have found.[1,2] Probably the most accurate and definitive means of identifying unknown specimens and confirming identities of those items you are not so sure about is to establish and maintain a reference collection of authenticated specimens.

II. BASIC EQUIPMENT NEEDS

Although budgetary considerations have an impact on the available equipment and storage facilities you can purchase for the reference collection, it is still possible to have a very efficient, usable collection at a low price. Storage units may range from one of the more formal cabinets, such as Cornell, California Academy, or U.S. National Museum, to simpler designs, such as glass-topped display boxes, wooden field boxes, or even homemade boxes with a styrofoam bottom. What is essential is that your storage units protect the collection from damage, promote ease of use of any specimen in the collection, and take up as little space as possible. Consideration must also be given for proper storage of slide-mounted specimens. Again, storage units may range from slide units for the type of formal cabinet you have, slide folders or slide boxes (holding from 25 to 100 slides), or even the boxes that slides are packed in when purchased.[3,4]

Mounting equipment procurement is a simpler task. The first item on any list should be insect pins (which range in size from 000, 00, 0, and 1 to size 7) in sufficient quantities. Size 0 to size 3 seem to be the most commonly used, especially for stored-product insects, so more of these sizes would be advisable. Ensure that your pins are stainless steel, to prevent rust from becoming a problem with your specimens. Never use ordinary straight pins for anything in a reference collection, not even for holding labels. They are too short, too thick, and rust very quickly and are therefore considered unacceptable. The size of the specimen you are mounting will determine the size of the insect pin you use. In some cases, the specimen you are trying to mount will be too small to pin directly, so another method of pinning must be used. There are basically three methods from which to choose: the double mount, the point and the glue board. It would be a good idea to have a supply of each type of mount on hand to start with. Use of these will be discussed later in this chapter. Regardless of specimen size, you will need a pinning block, which ensures that all specimens and labels are at a uniform height, and at least one pair of fine-tipped forceps (either straight or curved) for handling your specimens (a pair of pinning forceps is a wise investment because they help prevent fumbling fingers from damaging the fragile specimens). Dissection needles may be purchased or made from dowels and insect pins and/or minutens. You will need a spreading board, which is used to spread the wings of lepidoptera adults. Additionally, various sizes of vials will be needed, including genitalia vials. And let's not forget the labels, which may be preprinted, computer generated or individually handwritten.[3-6]

Another aspect of mounting specimens is the use of microscope slides for mounting fragments and mammalian hairs. Generally the standard 1″ × 3″ clear glass microscope slide is used, although larger ones—ones with depressions and ones with a wedge-shaped cavity—may also be used. Clear glass cover slips may be square or round and of various sizes. Stick-on labels for slides should be used.

There is also a variety of mounting media, clearing and preservation fluids to be considered when doing microscopic mounts. Some stains may also be desired to accent certain features. Many of the formulations that may be used in making microanalytical mounting fluids can be

TABLE 1
Suggested Formulae for Preservatives, Clearing Fluids and Mounting Media Used in Maintaining Reference Collections

Item	Use	Formula
AGA	Preservative	8 parts 95% ethanol, 5 parts water, 1 part glacial acetic acid, 1 part glycerin
Barber's fluid[a]	Relaxant	53 parts 95% ethanol, 7 parts benzene, 49 parts water, 19 parts ethyl acetate
Glycerin jelly	Mounting	60 ml water, 10 g gelatin, 70 ml glycerin, 1 g phenol
Hood's solution	Preservative	95 ml 75% ethanol, 5 ml glycerin[b]
Hoyer's solution[a]	Mounting	200 g chloral hydrate, 50 ml water, 20 ml glycerin, 30 g gum arabic[c]
Lactophenol solution[a]	Clearing	50 parts lactic acid, 25 parts phenol crystals, 25 parts water
Nesbitt's fluid[a]	Clearing	40 g chloral hydrate, 25–50 ml water, 2.5 ml 1*N* hydrochloric acid
Pempel's fluid[a,d]	Larval preservative	15 parts 95% ethanol, 30 parts water, 4 parts glacial acetic acid, 6 parts formalin[e]
Polyvinyl lactophenol	Clearing and mounting	2.5 g polyvinyl alcohol, 30 ml lactophenol solution,[f] 10–15 ml water (optional)

[a] Available commercially.
[b] Some prefer a 1:1 mixture.
[c] Use crystal or powder only.
[d] May also see spelled "Pampel's."
[e] Contains 37–40% formaldehyde.
[f] 45 g phenol in 45 ml lactic acid.

found in the references listed in the bibliography. Many of these can also be purchased commercially. Basic needs for any type of mounting should include 70–80% ethyl alcohol (or isopropyl alcohol), glycerin jelly, Hoyer's solution, phenol, moth balls or crystals, and nail polish.[(4)] These are discussed later in this chapter. Table 1 lists formulations for the most commonly used preservatives, clearing fluids and mounting media. The formulae in Table 1 are offered as basic needs for a reference collection. Amounts of individual ingredients may vary in the other literature. Those available commercially are cited. Note that all water used in formulas should be distilled.

To aid in the identification of your specimens, you will need a good quality stereomicroscope (also called a dissecting microscope) and a compound microscope (see chapter 3). It cannot be overemphasized that the quality of the scopes is paramount. Many of the structures you will be looking at are extremely small and the better the optics, the better are the chances to see and identify the structure. The scopes must also have the proper lighting attachments included.[(1,7,8)]

III. TECHNIQUES

A. PRESERVING WHOLE INSECT SPECIMENS

Proper preservation mounting of your insect specimens is the keynote of a good reference collection. Even the most unique specimen is worthless, unless it is mounted or preserved in such a way as to display its key features easily. After all, two or more insects can look alike, except for one minute feature. If you have never pinned an insect before, it is suggested that you do a bit of practicing before getting to your real reference specimens. It is common to break off legs or antennae during pinning, but in time you will be less prone to damage your specimens. You should pin a specimen as soon as possible after death, because insects tend to dry quickly and become brittle, causing more appendages to break off readily.[(4,9)]

1. Pin Mounts

The primary method of mounting adult, hard-bodied insects is by direct pinning. The appropriate size insect pin is pushed through the insect itself, to a predetermined height, with the aid of a pinning block. Usually the insect can be held between the index finger and thumb of one hand, and the pin pushed through the body with the other hand. Then the mount is put in the top-most hole of the pinning block to set the proper height. Where the pin is inserted is determined by the order to which the insect belongs. We will be discussing only those types of insects that will commonly be encountered in food products and are able to be pinned directly at this time. Beetles (Coleoptera) are pinned through the right elytron, halfway between the ends (Figure 1). You should avoid damaging the coxa of the leg when the pin is pushed through. With practice you will be able to pin beetles without damage to the surrounding structures. Flies (Diptera) are pinned just to the right of center of the thorax (Figure 2), whereas moths (Lepidoptera) are pinned exactly in the center of the thorax (Figure 3). Roaches and other orthopterous insects are pinned through the shieldlike pronotum (Figure 4). True bugs (Hemiptera) should be pinned to the right of the scutellum's midline. All specimens must be positioned parallel to the substrate, with no canting to the left or right nor head higher or lower than the body. The pinning block will set the proper height. The legs should be positioned in a natural pose, allowing examination of the underside of the insect. This can be achieved by placing the pin with the mounted insect into a piece of styrofoam until the specimen is at its natural height and using a dissecting pin to move the appendages. It may be helpful to have a disk of filter paper between your specimen and the styrofoam to prevent the insect's tarsal claws from getting caught in the nooks and crannies of the styrofoam. They may be secured with other insect pins. In a day or two, the specimen should be dried in the set position, ready to be labeled.[4]

2. Labels

Labeling is an important part of mounting. These labels carry all the identification information, such as genus and species, date and place collected, collector and who identified the specimen. You may also include information about the type of product the animal was found in, if it was taken from an actual infestation. Generally, there are two labels required on a specimen. The upper label is the collection information and the lower label is the identification. The collection label contains the sample number or other identification number; the

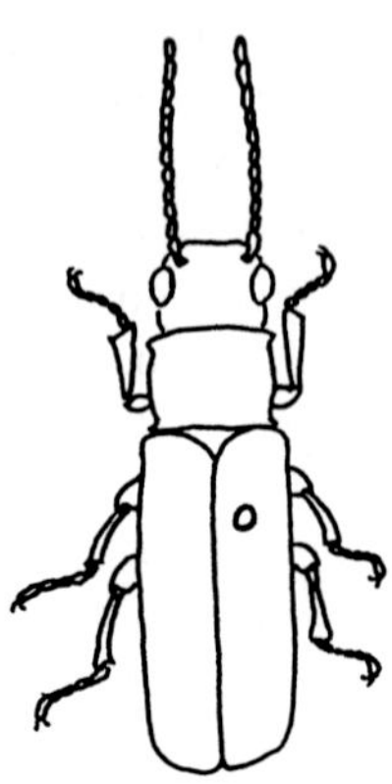
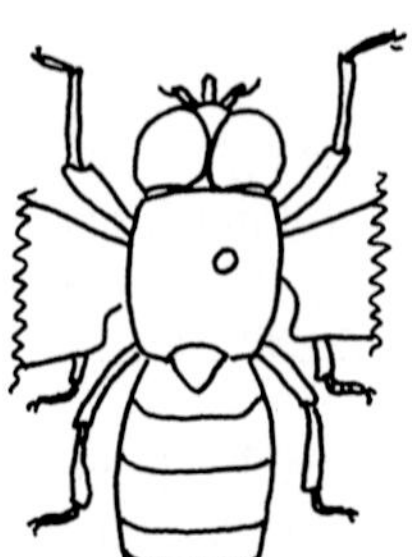
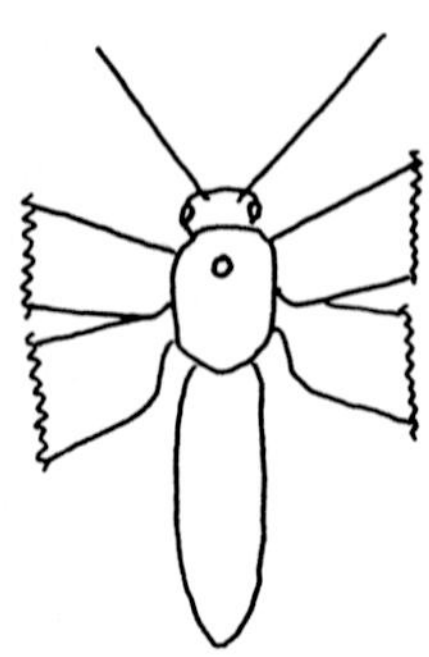
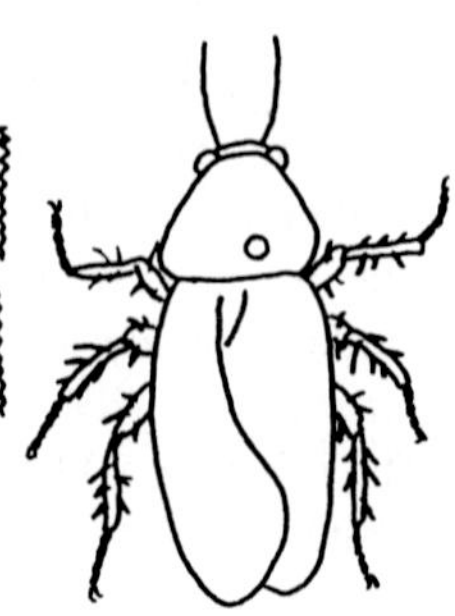

FIGURE 1. (Far left) diagram of the pinning position for beetles.
FIGURE 2. (Middle left) diagram of the pinning position for flies.
FIGURE 3. (Middle right) diagram of the pinning position for moths.
FIGURE 4. (Far right) diagram of the pinning position for roaches.

product or substrate where the specimen was found; the locality; the date collected and/or examined; the collector's or analyst's name; and any other information regarding the collection locality or techniques. The identification label contains the complete known identification beginning at the family level; the names(s) of the identifier(s); and the date identified. Additional information such as sex, instar stage, etc., may be included where known and applicable. You must ensure that your identification is correct in order for the specimen to be of any value. If you use common names for organisms, use only those that are recognized.[10] As stated before, these labels can be preprinted, handwritten, or computer generated.

3. Relaxing Specimens

Sometimes an insect has already dried and become brittle. To be pinned, it must first be relaxed. To do this, get a vacuum chamber or a large, wide-mouth jar with cover, and fill the bottom with a couple of inches of sand, sawdust or other porous material. Add water to moisten the sand and a small amount of phenol to prevent mold formation. If phenol is unavailable, paradichlorobenzene, naphthalene or ethyl acetate may be used. Put a substrate over the sand. This could be the porcelain stage that comes with a vacuum chamber or may be a simple circle of cardboard cut to the size of the jar. It is important not to let the insect lay on the moist sand, or damage may result. It may take a day or longer to relax the specimen, so check on it periodically. You will need to maintain the moisture content of the sand by adding water as needed. If left in the relaxing jar too long, specimens may fall apart when handled. Keep the relaxing jar away from windows or heat sources to prevent the buildup of moisture in the chamber or jar. If an appendage is still not supple enough to manage, it may be painted with a fine brush dipped in Barber's fluid or a 30–50% ammonium hydroxide solution.[4]

4. Minutens and Point Mounts

Minute, hard-bodied insects would be destroyed if pinned directly. We can use one of three different methods to mount these specimens. Each method has advantages and disadvantages, but all have their place in a reference collection. As with the direct pinning technique, these methods will also require some practice. Don't hesitate to use the dissection microscope when mounting small specimens.

A common method of mounting a small insect is the point (Figure 5). This is a heavy paper or cardstock mount, cut to the shape of a triangle. You can cut them out yourself, but it is far easier and much more attractive to get a point punch from an entomological supply company and punch out the amount you need. The insect is then glued to the tip of the point, with a clear adhesive (such as Elmer's Glue), shellac or clear nail polish. The tip is generally positioned just barely to the midline of the sternum. Some insects are better mounted with the tip of the point bent down and then glued to the right side of the thorax. A regular insect pin is then pushed through the opposite end of the point. The advantage, as with the double mount, is that just about all areas of the insect will be visible. Unless care is taken, this method can be messy, in as much as a glob of adhesive may obscure a major portion of the insect's ventral side.[3-5]

The minuten nalden (or simply minuten) is a small (12 mm × 0.15 mm to 0.20 mm) pin, sharpened at both ends. One end is pushed into a block of cork, balsa or silicon rubber (commonly known as a double mount), with the other end inserted into the insect (Figure 6). Looking down on the insect, with the head oriented away from the person pinning the insect, the entomologist would usually insert the minuten into the right side of the thorax of the creature. Some curators may prefer to push the minuten into the insect's back, with the legs toward the left. This position affords a better view of the insect without turning it upside down all the time. Others prefer to pin the insect with the minuten above the double mount block, rather than off the tip, with the insect oriented in either manner as before. A regular insect pin

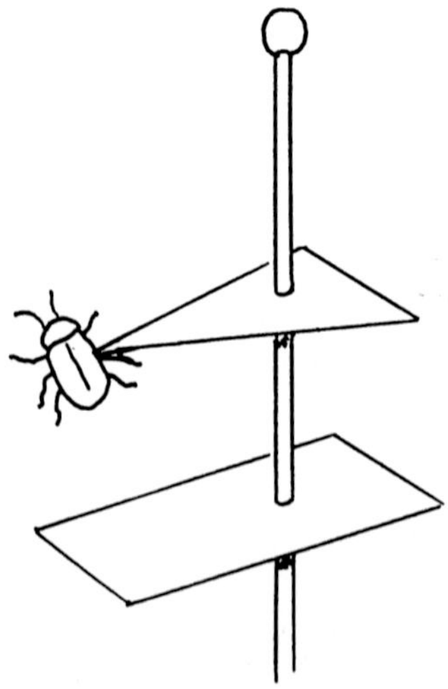

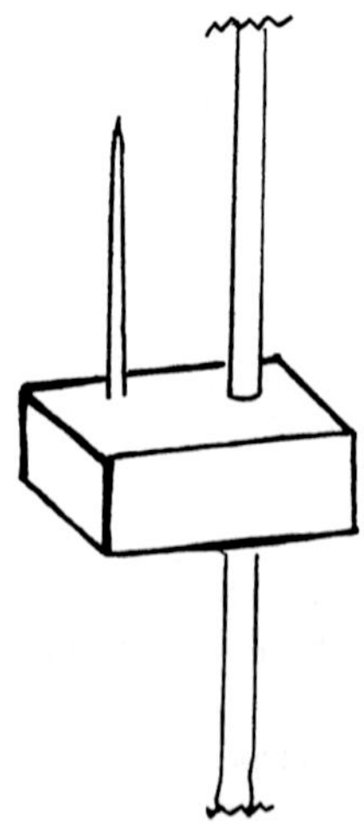

FIGURE 5. (Left) diagram of a point mount for small specimens.
FIGURE 6. (Right) diagram of a minuten assembly for pinning small specimens.

is then inserted through the opposite end of the block, perpendicular to the plane of the minuten. The pinning block is again used to set the proper height of the double mount. The advantage of using the double mount is obvious: with such a small pin, all areas of the insect are visible. The disadvantage is the small size of the pin. Minutens are difficult to handle, whether using forceps or your fingers. It is very easy to stick yourself, so caution is urged whenever handling minutens.[3-5]

5. Glue Boards

The glue board is a heavy paper or cardstock rectangle, with a minute drop of a clear drying adhesive, securing the insect's sternum to the board. The insect is oriented along the length of the board with the legs usually laid along side of the insect's body, facing forward, backward, or both, to present as natural a pose as possible. The regular insect pin is then pushed through the glue board, generally near the posterior end of the board, and the height set with the pinning block. This method works well for long, somewhat soft-bodied adults (such as Staphylinidae) and has also been used by some for mounting larvae. Glue boards vary in size and usually have rounded edges. Use boards just large enough for the insect being mounted. They can be purchased or you can make your own. But be sure to keep your glue boards uniform in appearance. Because this method has the sternum glued to the glue board, the ease of mounting is a definite advantage, but this method does not permit you to examine the sternal portion. You may mount a companion specimen on a separate pin showing the ventral side.[4]

6. Preserving Insects in Liquids

Every reference collection should include a "wet" collection for specimens that require storage in liquid preservatives. Very soft-bodied insects, particularly larvae, are best kept in a vial of preserving fluid. The most common and easy to prepare is a 70–80% ethanol solution. The size of the insect will dictate the size of the vial, but the 1 dram through 3 dram sizes appear to be the most frequently used. If the insect specimen is extremely delicate, you may want to put a wad of absorbent cotton into the vial first, place the specimen on this, then fill the vial to the lip with preservative and cap. The cap may be the neoprene stopper type, or a screw-on type with a plastic liner. (*Note:* Neoprene stoppers must be "bled" to seat properly in the vial.) Labels are written in waterproof and solvent-resistant ink (or pencil, if the ink is not available) on good quality paper. Although this is the most satisfactory way to maintain a larval collection, it can be difficult to remove a specimen from its vial because of the narrow mouth. An alternative for large collections is to store specimens in test tubes or straightsided vials with cotton stoppers. These containers are then placed in wide-mouthed jars (such as the jars used for home canning of jams and jellies) that are also filled with the preservative liquid. If not sealed properly, vials and jars can allow the preservative to evaporate, so a periodic check of fluid levels is necessary.[4,6,11-13]

7. Specimens for Training

In addition to those for formally pinned specimens in the reference collection, each species and life stage should also have a small vial of its kind kept separately. These insects can be used for training purposes, such as whole insect identification, "reasonably" whole insects (insects with major parts missing), and dissected or crushed for fragment identification. These insects should be from an in-house–reared collection or obtained through other mass-rearing sources where their identities are known to be correct.[14] Verification of each identity should be done prior to inclusion in the collection. Each vial should be labeled with the identification data and source of the insect. It is appropriate to have some "old dried up" known specimens (especially larvae) kept for reference, because this is the way many are found in a food item. Seldom will a "museum quality" specimen be encountered in a sample.

B. MICROSCOPE SLIDE MOUNTS OF INSECTS AND FRAGMENTARY ITEMS

The major portion of microanalytical entomology work involves examining foodstuff residues on a circle of lined filter paper with a stereomicroscope set at 30× magnification. This magnification enables an analyst to easily distinguish food residue from a possible contaminant. Now the problem becomes how to precisely identify the contaminant. Sometimes a tiny whole insect may be found, but most often the filth element is only a fragment of the whole insect. With a little training and experience, an analyst can find "a couple of insect mandibles." But, is it from a storage pest, a filth-related insect, a field insect or an incidental pest? Are they larval or adult mandibles, or larval *and* adult mandibles, which could show an ongoing infestation? How about if they were a male and a female of the species, which could have mated and produced a second generation in the product. In other words, can the analyst identify the insect that had been in the product from a lone fragment? Even considering that the analyst is permitted and encouraged to use higher power magnifications, including magnifications incorporated in a compound microscope to aid in the identification of the fragment, this is not the easiest of tasks. This also holds true for hair fragment and feather fragment identification. It is up to the analyst to determine, through microscopic examination, entomological keys and comparison with known fragmentary specimens, if the fragment is from a pest species, if a sanitation or a storage problem exists and if the evidence found will stand up in a court of law.[1]

Many body parts of insects can be identified as belonging to a particular insect by shape, positions of hairs, setae or setal pits and sculpturing (see chapter 3). But because most insect filth elements that are found are from 1 to 3 mm at their longest dimension, positive identification is difficult at best. Although mandibles are probably the most readily identifiable and most common fragments found (because of their ability to survive processing and grinding relatively intact), any body part can be found. Taking an identified, whole insect and fragmenting it by both dissection and random crushing gives you fragments that can be permanently mounted on a microscope slide for ready comparison with fragments of an unknown specimen found in a sample. The more insects you can take apart and mount, the more useful your fragment reference collection will become. But each fragment must be properly mounted to ensure its usefulness. An air bubble in the wrong place can make a slide unusable for identification purposes. Uniformity in mounting and labeling is as important with microscopic mounts as it is with whole mounts.[15]

1. Mounting Media and Mounting Techniques

A myriad of fluids and solutions are used in the clearing, mounting and staining of insect fragments. There are even various products that can be used to "ring" the slide to make it more permanent. All are considered to be hazardous to some degree, so care should be exercised when preparing or using them. The selection of a medium will, in large part, be the preference of the person mounting the specimen on the slide. Some may yield better results, but are more than enough trouble to use to achieve those results. Others are easy to use, but the results leave a little something to be desired. Odor and ventilation may influence which ones are used. A fluid may be bought from a biological supply house or prepared by the person using it. Try each one of them, beginning with the professionally prepared variety. You may find one is good for one type of specimen, something else for another type of specimen.[4,11]

Glycerin jelly is the most commonly used mounting medium because fragment specimens can be transferred directly from aqueous solutions (e.g., glycerin alcohol) to the medium. The formulation for glycerin jelly is 10 g of gelatin, 70 ml of glycerin, 60 ml of water and 1 g of phenol. Pour the gelatin in cold water to soak, then heat over a steam bath to completely dissolve the gelatin. Mix in the glycerin and phenol while still hot and then cool to room temperature. The cooled mixture should have the consistency of jelly and may be stored at room temperature in a tightly capped container for years. An acceptable commercial variety is sold by biological supply houses.

To mount a specimen, place a small lump of jelly on a clean 1″ × 3″ glass microscope slide and gently warm the slide until the jelly melts. Place the specimen in the middle of the melted drop of jelly and push it to the bottom, using a probe or insect pin to orient the specimen. Gently place a glass cover slip on top of the drop. The weight of the cover slip will flatten the drop, filling the area beneath the coverslip with jelly engulfing the centered specimen. If the jelly has set, rewarming the slide with coverslip in place will cause the jelly to flow again to fill the entire area under the cover slip. Rewarming also helps remove any trapped air bubbles. Care must be taken when boiling off trapped air that the specimen does not migrate from its original position.

Hoyer's solution is used for mites and small, soft-bodied insects. Hoyer's solution is formulated with 50 ml of water, 30 g of gum arabic, 200 g of chloral hydrate and 20 ml of glycerin. There are numerous variations on this formula, usually varying the proportions of the ingredients only. The ingredients are mixed in the water in the order given above. One ingredient must be completely dissolved before adding the next. Filter the final product through bolting cloth or glass wool to remove impurities from the gum arabic. *Note:* Extreme care is required when working with chloral hydrate to avoid exposure through contact or breathing the fumes. Use appropriate ventilation and other safety precautions.

Like glycerin jelly, Hoyer's solution has the advantage of accepting specimens directly from aqueous solutions. In addition, the chloral hydrate has a clearing effect on sclerotized specimens, rendering them more transparent over time. Hoyer's solution may be gently heated to remove entrapped air bubbles. Gentle heat also helps penetrate whole specimens and contract their musculature so that the appendages spread out nicely. Again, care is necessary to use proper ventilation and avoid overheating the medium, which releases small amounts of fumes.

Technique in mounting microscopic fragments will vary from curator to curator, but it is as important with slide mounts as with whole mounts to be consistent in the method and manner that the specimens are mounted. For example, some curators prefer to mount both mandibles with the condyle up and teeth facing each other. Others mount them with the teeth facing each other, but with the condyles down. Still others prefer to have two right mandibles mounted on the same slide, one condyle up, the other down. They then do the same for the left mandible. The same can be argued for the mounting of legs. Should they be individually mounted on a slide, should they be mounted in pairs or should all three (individually or in pairs) be mounted on the same slide in proper taxonomic order? Even the age-old argument for how to mount mites (ventral up or ventral down?) is moot, as long as the mounting is consistent. Specimens can be manipulated into position with microprobes prior to placing the cover slip on. It is best performed under a dissection microscope. After the specimen is mounted, appropriate identification labels can be applied to one or both ends of the slide. (*Hint:* Write your labels before placing them on the slide, because trying to write them while they are on the slide could move the cover slip or crush your specimen.)

Hairs present a special problem when mounting. Part of their inner structure (called the medulla) has tiny air spaces that can trap air bubbles and obscure the image. For information on mounting hair specimens see chapter 12 or one of the many literature references on hair identification.[2,13,15-22]

2. Clearing and Staining Reagents

Some insects and mites, and subsequently their body parts, are too dark to clearly see certain identifying features. Sometimes too many internal structures are present. These must be cleared to ensure optimum viewing. The more commonly used clearing agents are lactophenol, a 10–20% potassium hydroxide (KOH) solution, Nesbitt's fluid and Hoyer's solution (actually a mounting medium!). Clearing a specimen can be more of an art than a science. There are numerous agents to use, some alcohol soluble, others dissolve in water. They may work well at room temperature, especially if time is not a factor, or can be heated to accelerate the process. Too much clearing, either by excessive time in the agent or by raising the temperature of the agent too high, can cause connective tissues between sclerites or leg joints to weaken, shrinking or distortion of the specimen, or possible clearing to the point of invisibility. Some specimens may need to be dehydrated in various strengths of ethanol prior to mounting in a particular medium. The degree of sclerotization present (darkness of the plates that make up the exoskeleton) will also affect the clearing time necessary.[4,11]

If a specimen is lightly sclerotized, or was cleared too much, it can be made more visible with a wide variety of biological stains. These stains, like the mounting and clearing fluids, may be alcohol or water soluble. This presents the advantage that if one were to overstain a specimen, the color can be leached out to the desired hue. The biological stains will stain human skin (somewhat permanently), so care is urged.[4]

3. Preserving and Storing Microscope Slide Mounts

Once a specimen is mounted, the cover slip should be ringed to prevent evaporation of mounting fluid. Various types and colors of nail polish and enamel paints are commonly used. Ringed slides will be more or less permanent, regardless of the mounting medium used. Care must be used in ringing a cover slip, especially if your specimen is close to the edge of the slip. Even a clear nail polish will distort your view if it covers the surface of the slip directly over the specimen. Small gaps in the continuity of the ring will allow evaporation over time, so be sure none occur. Have a place for the slide to dry *flat*, where nothing will rest on top of it and it is relatively dust free. Proper drying may take up to a week, so be patient. A drying oven or light bulb in a box can help dry a slide mount better. If slides are not "cured" properly, the cover slip or specimen may shift its position, or even stick to the bottom of the slide box. Slides should be stored in a slide box of some type, and although you put the slides in the box on their side, they should be stored flat, with the cover slip up to prevent any slippage of the cover slip or the specimen. This is especially true if the slide is not going to be used for a long time. Simply turn the box on end for storage. Extremely large collections may have special cabinets with flat drawers to hold the slides in the proper orientation.[4,11]

Mounting media such as glycerin jelly and Hoyer's solution are nonpermanent, but with careful preparation and maintenance, slides mounts using these media can last for years. Slide mounts should be left undisturbed at relatively uniform room temperature for 3–4 weeks to cure. The curing process can be accelerated by holding the slide mount at 45°C for 48 hours, but temperatures above this will dehydrate the medium. After curing, the cover slip edges should be ringed (see below) to provide a moisture barrier against dehydration. Glycerin jelly and Hoyer's solution slide mounts should be stored flat or horizontally to avoid gravity migration of the specimen.

C. MISCELLANEOUS TECHNIQUES

1. Vertebrate Specimens

Some reference collections include whole mounted birds and mammals. Although this chapter is not designed to instruct in taxidermy, you will note a reference in the bibliography relating to this subject.(23) If you are not inclined to perform taxidermy, a qualified taxidermist may be found in the local telephone book.

2. Product Specimens

Vials of dry products, such as grains, seeds, Oriental herbs, spices and flours may be useful with seldom seen products. They should be in whole, crushed and ground forms and should be kept dry. These items are also beneficial in training new personnel to distinguish the micromorphology of plant fragments from insect fragments, as well as being able to see the whole product. When coupled with the whole and fragmentary insect reference collections, the analyst will have a better eye for putting together the "jigsaw puzzle" on the petri dish before him or her. Vials will vary, depending on the physical make-up of the whole product, but generally a 2-dram vial will be sufficient for crushed or ground products. Products such as seaweed and whole Oriental herbs may be kept in a large jar or ziplock type bag. Seaweed may be rehydrated, unfolded, redried and pressed as one would do to a flower. Seaweed may also be rehydrated and preserved in formalin. Specimens may then be kept in a botanical cabinet. (*Note:* Keep green vegetable matter in the dark, because they may tend to lose their color in direct light.)(24,25)

3. Genitalia Specimens

For some groups of insects, including stored-product beetles and moths, it is necessary to examine the genitalia in order to identify the animal. Often the genitalia can be examined *in situ*. If dissection of the genitalia proves necessary, the dissected structures must be preserved in a way that they can be related back to the original specimen. In pinned collections, it is common to use tiny genitalia vials made of clear glass or plastic to store dissected genitalia. The dissected genitalia are placed in a genitalia vial containing some preservative (alcohol and glycerin is suitable) and a label. The vial is stoppered with a tiny cork. Sealing the cork with wax or nail polish helps reduce fluid loss during storage. The corked vial can then be stored in the pinned specimen collection, by stabbing an insect pin through the soft cork, or in the wet collection.

An alternative for stored-product insects, which are generally small to minute in size, is to clear the intact whole insect specimen with one of the clearing reagents discussed above. Clearing should continue to the point that the internal genitalia are visible through the exoskeleton. Cleared specimens can be mounted on glass microscope slides for easy viewing and storage. This technique has proved successful for stored-product beetles.(26)

IV. COLLECTION MANAGEMENT

Collection management is a relatively simple task for the curator of a reference collection if a layout plan is worked out in advance. Because a collection's usefulness is also based on ease of use, a theme must be established and the specimens arranged in keeping with this theme. A large, random collection serves no purpose other than looking impressive. Collections can be arranged by taxonomic groups, by type of product infested (i.e., grains and seeds, fish, sauces, etc.), by country of origin of the product, or even a straight alphabetical arrangement by scientific or common name. It is important that the collection include as many insects as possible, both by species and numbers within species to show variety and possible variations.(5) In building a collection related to microanalytical entomology, such as food sanitation or quality assurance, the curator should concentrate on filth and stored-product pests first. These pests can directly affect the health of the consumer and the wholesomeness of the product. The collection could then move to obtaining field and incidental pests, which can be useful in preproduction contaminant identification.

One of the most important practices for collection management is the regular replacement of mothballs, moth crystals or other fumigant. These products will prevent infestation by dermestids, psocids, moths and other dried animal or vegetable pests. Also regularly check (and refill as necessary) any alcohol or liquid preservatives. Even though all precautions may have been taken to minimize evaporation, some will still occur over time. "Permanent" slides may also need replacing if broken, cracked or drying out. In all cases, ensure not only that the medium is in good condition, but also that there is no damage to the specimen being preserved.

V. SOURCES AND RESOURCES

The reference collection must be more than just an accumulation of known, adult stored-product pest insects. A complete reference collection should include, at a minimum and in addition to the adult stored-product insects: all instars of larval stored-product insects, their pupae and fragments of all these (i.e., dissect each and mount the various appendages and sclerites separately); larva, pupa, adult and fragments of as many field and filth insect pests as possible; mites associated with stored-products, pest birds or pest vertebrates of stored-products; predatory mites; arachnids other than mites; animal fur and guard hairs from various species of rats, mice, shrews, rabbits, squirrels, kolinskies and bats; all varieties and colors of hairs of cattle and oxen, horses, goats, sheep, pigs, monkeys, cats and dogs; hairs from the families of man (to include the female of the species) from the head, wrist, arms and beard; feather fragments, and individual barbules, from domesticated poultry and fowl, exotic birds and wild "pest" species from all over the world; droppings and urine stains from rodents and birds. Additionally, you will need various natural and artificial fibers; shards of various types of glass, plastic and metal; whole, crushed and ground forms of spices, herbs and grains. And let's not forget various types of fish scales. Specimens may be obtained from an actual sample being analyzed, provided that the specimen is in suitable condition. This allows you to have an identified contaminate, the product contaminated and an origination point for the contamination. The product itself may even be used as a known, especially if it is a seldom seen product. Be sure to check with your supervisor or review your agency's policy on sample accountability before procuring any specimens from an actual sample.

A wide variety of filth and field specimens can be obtained by collecting them in nature yourself. Using a basic textbook or field guide, anyone can become proficient in finding, attracting and catching live specimens without damage to the specimen. These can then be killed and mounted, or kept alive, bred and reared through the various life cycle stages. Collecting the specimen yourself can give you more of an insight into the various habits and habitats of the creatures that can contaminate your food supply.(9,25,27-32)

Each of the specimens acquired must be positively identified and labeled. This should be done by two analysts experienced with the type of specimen being identified, by the use of specific keys, or even by "creating" a situation to achieve the desired result, such as rearing a known insect to obtain all life stages, or allowing an animal to gnaw and soil packaging materials to obtain verifiable damage. (*Note:* It is highly advisable not to create your situations in the same area where you analyze samples.) Specimens of known insects may be purchased from biological supply houses, but each one ordered should be verified before including it in the reference collection. This is not to belittle the efforts of these sources, but remember they are a business not a scientific facility. It would seem prudent to "key out" any specimen received from these sources, preferably by two experienced analysts. To avoid obtaining questionable specimens, one can contact the Agricultural Extension Service or the entomology, zoology and biology departments of their state university. All states have departments dealing with foods and agriculture, and local governments have their health department or vector control department, which may be contacted not only for known specimens, but also for scientific and practical expertise. The federal government has entomologists and biologists at the Food and Drug Administration, the Department of Agriculture (including the U.S. Forest Service) and on most military installations of the Army, Navy and Air Force. Museums, research institutions, zoos and entomological societies have both scientists and collections that can be used and specimens borrowed for taxonomic research.[3,14] The curators and staff of these collections can be an invaluable resource when it comes to specimen identification and management of your collection. Before long your reference collection will help all the microanalytical technicians and professionals do their job in a more efficient, accurate manner. The evidence you find in a sample will be better accepted by a court of law because you can back up your finding by comparing it with a known, positively identified specimen, rather than with only your memory.

REFERENCES

1. Boese, J.L. and Bandler, R., Extraneous Materials: Isolation, in *Official Methods of Analysis of the Association of Official Analytical Chemists*, 15th ed., Helrich, K., Ed., Assoc. Off. Anal. Chem., Arlington, VA, 1, 372-373, 1990.
2. Parry, J.W., *Spices*, Vol. II, Chemical Publishing, NY, 1-6, 1969.
3. Borror, D.J., Triplehorn, C.A., and Johnson, N.F., *An Introduction to the Study of Insects*, 6th ed. Saunders Coll. Publ., Philadelphia, PA, 755-770, 1989.
4. Martin, J.E.H., *The Insects and Arachnids of Canada, Part I*, Ministry Supply and Services, Quebec, 1-182, 1970.
5. Oman, P.W. and Cushman, A.D., Collection and preservation of insects, *U.S. Dept. Agric. Misc. Publ.*, 601, 1-25, 1948.
6. Steyskal, G.C., Murphy, W.L., and Hoover, E.M., Insects and mites: Techniques for collection and preservation, *U.S. Dept. Agric. Misc. Publ.*, 1443, 23-44, 1989.
7. Hicks, J.W., *Microscopy of Hairs: A Practical Guide and Manual*, Federal Bureau of Investigation, Washington, DC, 1-14, 1077.
8. McCrone, W.C. and Delly J.G., *The Particle Atlas*, Vol. II, Ann Arbor Sci. Publ., Ann Arbor, MI, 9-27, 1973.
9. Bland, R.G. and Jacques, H.E., *How To Know The Insects*, 3rd ed., Wm. C. Brown, Dubuque, IA, 13-29, 1978.
10. Stoetzel, M.B., *Common Names of Insects and Related Organisms*, Entomol. Soc. Am., Lanham, MD, 1-199, 1989.
11. Krantz, G.W., *A Manual of Acarology*, 2nd ed., Oregon St. Univ., Corvalis, 85-98, 1986.
12. Peterson, A., *Larvae of Insects, Part I*, Edward Brothers, Ann Arbor, MI, 4-10, 1951.
13. Smith, F.R., The effect of natural and laboratory digestion on the microscopic appearance of certain animal hairs, *J. Assoc. Off. Agric. Chem.*, 41, 479-480, 1958.
14. Edwards, D.R., Leppla, N.C., and Dickerson, W.A., *Arthropod Species in Culture*, Entomol. Soc. Am., College Park, MD, 1-49, 1987.
15. Gorham, J.R., Ed., *FDA Training Manual for Analytical Entomology in the Food Industry*, FDA Tech. Bull. No. 2, Assoc. Off. Anal. Chem., Washington, DC, 1-145, 1977.
16. Brickey, P.M., Jr., Gecan, J.S., Thrasher, J.J., and Vazquez, A.W., Notes on microanalytical techniques in the analysis of foods for extraneous materials, Part II, *J. Assoc. Off. Anal. Chem.*, 55, 51-56, 1972.
17. Bryce, J.R., Identification of the hairs of three Asian commensal mammals: *Suncus murinus*, *Bandicota bengalensis*, and *Rattus exulans*, *J. Assoc. Off. Anal. Chem.*, 77, 403-410, 1994.

18. Day, M.G., Identification of hair and feather remains in the gut and faeces of stoats and weasels, *J. Zool.*, 148, 201-217, 1966.
19. Nason, E.S., Morphology of hair of eastern North American bats, *Am. Midland Natur.*, 39, 345-361, 1948.
20. Palenic, S., Light microscopy of medullary micro-structure in hair identification, *Microscope*, 31, 129-137, 1983.
21. Vail, D.J., *Improved Procedure for Liquid Replacement of Entrapped Air in Mammalian Hairs,* Lab. Information Bull. No. 2243, U.S. Food and Drug Administration, Washington, DC, 1978.
22. Vasquez, A.W., Structure and identification of common food-contaminating hairs, *J. Assoc. Off. Anal. Chem.*, 44, 754-779, 1961.
23. Hall, E.R., *Collecting and Preparing Study Specimens of Vertebrates*, Univ. Kansas, Lawrence, KS, 1-46, 1962.
24. Dawson, E.Y. and Foster, M.S., *Seashore Plants of California*, Univ. Calif. Press, Berkeley, 16-18, 1982.
25. Prescott, W.G., *How to Know the Fresh-Water Algae*, Wm. C. Brown, Dubuque, IA, 16-17, 1954.
26. Banks, H.J., Identification of stored product *Cryptolestes* spp. (Coleoptera: Cucujidae): A rapid technique for preparation of suitable mounts, *J. Austr. Entomol. Soc.*, 18, 217-222, 1979.
27. Arnett, R.H., Downie, N.M., and Jaques, H.E., *How to Know the Beetles*, 2nd ed., Wm. C. Brown, Dubuque, IA, 9-18, 1980.
28. Borror, D.J. and White, R.E., *A Field Guide to the Insects of America North of Mexico*, Houghton Mifflin, Boston, MA, 1-404, 1970.
29. Chu, H.F., *How to Know the Immature Insects*, Wm. C. Brown, Dubuque, IA, 19-28, 1949.
30. Helfer, J.R., *How to Know the Grasshoppers, Crickets, Cockroaches and Their Allies*, Dover Publ., NY, 3-4, 1987.
31. Kaston, B.J., *How to Know the Spiders*, Wm. C. Brown, Dubuque, IA, 10-13, 1972.
32. Slater, J.A. and Baranowski, R.M., *How to Know the True Bugs*, Wm. C. Brown, Dubuque, IA, 1-256, 1978.

Chapter 15

LABORATORY PROCEDURES

Edwin C. Washbon and Kenneth R. Halcrow

CONTENTS

I. Introduction 241

II. Filth Extraction Methodology 242
A. Apparatus and Reagents 242
1. Apparatus 243
2. Reagents 245

III. Macroscopic Filth Analysis 247

IV. Microscopic Filth Analysis 247
A. Techniques 247
B. Quantitative Transfers 252
C. Sample Contamination 253
D. Disposal 254

V. Examination of Filth Elements 254
A. General Techniques 254
B. Microscope Specifications 255
C. Care and Use of the Microscope 255
1. Caveats 255
2. Cleaning 256
D. Microscopes and Accessories 256
1. Types of Microscopes 256
2. Accessories 258

VI. Laboratory Information Bulletins 259

References 260

I. INTRODUCTION

Laboratory procedures are standard practices that are followed by laboratory scientists of the same discipline. Many laboratory procedures are concerned with safety, handling of hazardous materials, record-keeping and reporting requirements for results of experiments. These general procedures are beyond the scope of this book and most laboratories have manuals that prescribe procedures for these activities. Every scientific discipline also has a unique set of practices that address issues peculiar to that discipline, the purpose of which is to achieve fundamental uniformity in the conduct of experiments and analyses so that the results from different laboratories can be compared and tested for reproducibility. They are fundamental in that the scientists of the discipline mutually agree that if the procedures are not followed then the validity of the results could be suspect. Much of the material included in this chapter is excerpted from internal government manuals that describe the procedures fundamental to the practice of microanalytical entomology. Most of these procedures were developed by FDA scientists and many are found in the FDA manuals for laboratory procedures and the compendium of analytical methods issued periodically by the Association of Official Analytical Chemists (AOAC).[1-3] The microanalytical entomologist must not only be familiar with these written procedures but also understand how they are executed in actual practice. The focus of this chapter is to elaborate on the "tricks of the trade" that are essential to the successful practice and application of the science of microanalytical entomology.

II. FILTH EXTRACTION METHODOLOGY

METHODOLOGY! What is it and what does it mean to those involved in filth analysis of foods?

Methodology primarily involves the standard analytical procedures by which filth elements are isolated from foodstuffs and the ways that they are subsequently treated so that they can be recognized. It is important for two main reasons: (1) standardization, which assists in establishing reference points, and guidelines and (2) improving the means of recovering and recognizing filth elements under the microscope.

Several reference manuals and compendia are available that describe the specific laboratory procedures used to detect and isolate filth from foods. A major compendium is the *Official Methods of Analysis, 15th ed.*, issued by the AOAC.[3] It devotes a chapter to descriptive recipes for analytical methods to isolate microscopic filth from various food product matrices. Other useful compendia include the FDA's *Macroanalytical Procedures Manual* and its predecessor the *Microanalytical Methods*, which describe mostly visual, or macroscopic, methods of filth detection.[2,4] Many times, laboratories are faced with an analytical problem that is not addressed by the available manuals and so must modify a method or develop a new method to solve the problem. In such a case, a record must be kept of the procedure that was developed and used. If the analytical problem is a recurring one, the record of the original successful procedure eventually is circulated and ultimately the procedure is fully developed to be included in a manual or in the literature. The FDA issues individual *Laboratory Information Bulletins* that describe interim techniques in current use by FDA laboratories. Also helpful are the FDA technical bulletins, *Principles of Food Analysis for Filth, Decomposition and Foreign Matter* and *Training Manual for Analytical Entomology in the Food Industry*.[5,6] Most analytical methods that involve a chemical extraction procedure are found in the AOAC compendium of official methods which is the main source of specific analytical methods for extracting filth from foods. This compendium is referred to as "AOAC" in subsequent discussions.

A. APPARATUS AND REAGENTS

A microanalytical laboratory requires little in the way of elaborate equipment or facilities. A basic chemistry laboratory environment is suitable. The laboratory must have chemical fume hoods and standard benchwork but with a relatively deep sink for sieving operations. Useful utilities include vacuum and ample power outlets for appliances. Vacuum may be supplied by benchtop pumps if a house system is not available. Water utilities must be capable of delivering running water at a variety of controlled temperatures and the water must be free of particulates. This section describes the specific laboratory apparatus and materials that are most often used in present day filth extraction analyses. These items are fairly standardized and use of alternative apparatus is not recommended without validation data to show that equivalent recoveries can be achieved. Some general precautions must be observed in regards to apparatus. Worn or damaged items should not be used because they may reduce recoveries if minute filth elements become lodged in cracks or crevices. Plastic items should never be used in contact with the analytical portion of a product or extracted filth because many filth elements adhere to plastic surfaces.

1. Apparatus

Where appropriate, the AOAC compendium's reference numbers are given in parentheses for the items listed.[3]

Aerator (AOAC 945.75B.[a]). The aerator is an attachment to a sink faucet used to provide a smooth-flowing, forced stream of temperature-controlled water. The aerator required for AOAC extraction methods and most other filth extraction methods is available from Faucet-Queens, Inc., 550 Palwaukee Dr., Wheeling, IL 60090 (item No. 00200). It must have the lower screen removed in order to provide the appropriate forced stream of water. It is usual practice to attach the aerator to the faucet with a flexible hose of sufficient length to provide vertical and horizontal mobility over the sink area.

Autoclave (AOAC 945.75B.[b]). Filth work requires the slow exhaust type autoclave. Set the "slow exhaust" cycle to lower pressure from 15 to 0 in 15–20 min.

Blender (AOAC 945.75B.[c]). A high-speed blender is often used for comminuting products to free internal filth from a product matrix. The standard blender uses a 1-L, 4-lobed blender jar fitted with 4-blade assembly. Two blades are tilted upward approx. 30° with diameter or span of 60 mm, and the other two blades are tilted downward approx. 25° with diameter or span of 55 mm. Operating speeds specified in a particular method may be controlled using a variable transformer. Use a tachometer to calibrate the speed of the blender's blade shaft (with blades removed).

Extraction vessels (AOAC 945.75B.[h]):

Kilborn funnel (AOAC 945.75B.[h][1]). The Kilborn funnel is a glass settling vessel that can be drained from the bottom. Filth methods use a 1-L vessel, 3.5 in. outside diameter by 9.5 in. high, with an 8 mm opening at tip. Rubber tubing, 3/8 in. inside diameter, fitted with a pinch clamp provides convenient control of the drain flow.

Percolator (AOAC 945.75B.[h][2]). Similar in general form to the Kilborn funnel, the standard percolator is the 2-L, Corning Glass Works No. 7040, or an equivalent piece of glassware that conforms to the following general size and shape: 115 mm inside diameter × 400 mm long, approx. 90 mm inside diameter at 200 mm down from top, with 8–9-mm bore tip, with a clamp as described above. Use a stirring rod 370 × 10 mm diameter, when specified, to prevent compacting of sample in drain opening.

Trap flask (AOAC 945.75B.[h][4]). The Wildman trap flask is the "workhorse" piece of glassware for filth extractions (Figure 1). It is a 1- or 2-L Erlenmeyer flask into which is inserted a close-fitting rubber stopper or wafer stopper (Entomological Supply Co., Inc., 2411 S. Harbor City Blvd., Melbourne, FL 32901) supported on a stiff metal rod. The rod is 5 mm (3/16 in.) in diameter and at least 10 cm longer than the height of the flask. The stopper or wafer must be slightly larger in diameter than the diameter of the neck of the flask. An acceptable rod can be constructed by drilling a 5-mm center hole in a rubber or neoprene stopper and inserting a rod with one threaded end through the hole, using nuts and washers to hold the rod in place on the stopper. Countersink the lower nut and washer in the rubber stopper to prevent metal from striking the flask bottom.

Filter paper (AOAC 945.75B.[i]). For filth work use smooth, high wet-strength, rapid-acting filter paper ruled with oil-, alcohol-and water-proof lines 5 mm apart. S&S No. 8 (9 cm diameter) is the most widely used.

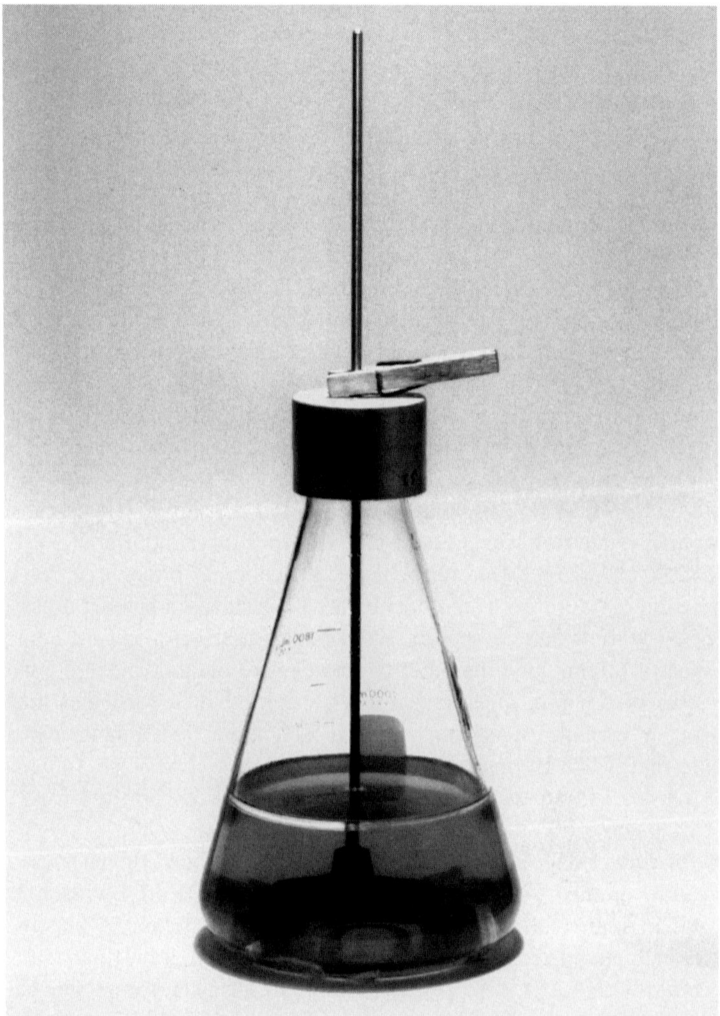

FIGURE 1. Trap flask with aqueous phase. (Photo taken by W.E. Savary.)

Funnels (AOAC 945.75B.[k]). For filtration with suction, use a Hirsch or Buchner funnel whose size allows the filter paper to be cupped up on the funnel sides to ensure quantitative transfer and prevent loss of solids. A typical filtration apparatus is pictured in Figure 2.

Magnetic stirring bar and stirrer-hot plate (AOAC 945.75B.[n]). When stirring analytes use a teflon-covered magnetic bar, about 47 mm long × 9 mm outside diameter, in combination with a hot plate equipped with independent, continuously variable heat and speed controls.

Sieves. For both wet sieving and dry sieving work, standard 8-in. diameter metal testing sieves are required. Either brass or stainless steel are acceptable. The sieve body should provide at least 2 in. from rim to fabric surface. A good selection will include sieves with fabric weaves of U.S. standard numbers 8, 10, 20, 40, 140 and 230 mesh. The sieve fabric must be regular weave, not twill. A #8 or 10 mesh sieve is useful for retaining dry or wet bulk product, separating loose insects and other filth in the "fines" or "throughs." A #20 mesh sieve retains most adult whole stored-product insects found in dry products. A #40 mesh sieve is used in wet sieving to retain product residue while washing minute filth particles through onto a finer mesh sieve nested below the #40 mesh one. Fine-meshed #140 and 230 mesh sieves are used in wet sieving to retain microscopic filth. The choice between mesh #140 and 230

FIGURE 2. Filtration apparatus. (Photo taken by W.E. Savary.)

depends largely on the degree of comminution of the product, which affects the size of any insect fragments or other filth in that product.

Additional standard apparatus and equipment needed for filth analyses include analytical balances; petri dishes; glass beakers (an assortment with 50 ml through 2000 ml capacities); graduated glass cylinders; glass rods; rubber tips (for rubber policemen); a 4-L plastic carboy with a spigot, used to dispense solutions; and 4000-ml Erlenmeyer flasks, used to make dilutions of alcohol and other solutions. Most methods that require nonstandard or specialized apparatus not listed above describe these items in the method.

2. Reagents

The extraction of microscopic filth elements from food products requires the use of chemical reagents, most of which are listed in the AOAC compendium and available from chemical supply houses.[3] These reagents are useful in developing methods for nonstandard and unfamiliar products. Insect, hair and feather filth are not destroyed by most of the reagents listed below. There are three basic types of reagents used in filth extraction methods: digestion solutions, saturation solutions and flotation liquids. Digestion solutions are usually dilute acids that are used to break down, or digest, the components of a food product to release any filth contained in the product matrix. Saturation solutions are usually aqueous solutions that are used to soak the food product in preparation for separation of the filth. Most filth elements are resistant to penetration by water-based solutions but most food product matrices are hydrophilic, readily absorbing aqueous solutions. Once saturated by soaking and relatively denser than the filth, the food material tends to sink to the bottom of the extraction vessel. Filth elements on the other hand are often lipophilic and easily penetrated and saturated by petroleum solvents. These latter reagents are used as flotation liquids by selecting a solvent that is less dense than and immiscible with the aqueous phase (saturation solution) of a method. When mixed with the aqueous phase, the flotation liquid differentially penetrates the filth in the product matrix. As the system is allowed to rest and reach equilibrium, the flotation liquid and lipophilic filth float to the top of the system while the saturated food product remains in the lower aqueous phase. This separation allows the filth to be easily skimmed off the top of the system. Several reagents are also used to enhance the lipophilic flotation of filth elements while increasing the hydrophilic nature of the food particles, aiding in the sinking of food particles to the center and bottom of the aqueous phase.

The listing below updates common reagents used, per AOAC section 945.75C, and includes the purpose of each.(3) In most cases, the reagents specified by a method must be used in the specified amounts, dilutions and quality specifications provided by the method. Substitutions or variations should be validated by recovery comparisons.

Acid alcohol solution. Acidified alcohol is sometimes used as the aqueous phase in filth flotation procedures. The AOAC official formulation is a 1+9 dilution of hydrochloric acid with 60% ethanol or 40% isopropyl alcohol. Dilutions using 22–24% isopropyl alcohol may be substituted in unofficial methods using percolator, Kilborn funnel or trap flask operations.

95% ethyl alcohol and 100% isopropyl alcohol. These reagents are necessary for cleaning glassware. Dilute solutions are used for the aqueous phase of filth flotations and also as the liquids in the wash bottles used for transferring material from one vessel to another. If the wash bottle liquid is not specified in a particular method, care must be taken to match the formulation of the wash bottle liquid with the formulation of the aqueous phase of that method.

Calcium chloride. $CaCl_2$ is used in the aqueous phase liquids for the analysis of nut products to enhance the saturation of the nut tissue. It may be effective at lower concentrations than called for by AOAC methods. Some experimentation with concentrations may be in order in laboratories that analyze nut products.

Antifoam A solution. Antifoam is a surfactant used to keep materials from overflowing their vessels during autoclaving or boiling on a hot plate. Only small amounts are used.

Sodium lauryl sulfate. This is a strong detergent used in solution to strip oils from vegetable product matrices. It is sometimes used as a final rinse at the very end of an extraction procedure to loosen any filth adhering to the vessel from which extractant is poured onto the filter paper for examination under a microscope. Its dust may severely irritate the nasal passages and safety precautions are necessary when formulating solutions from the pure powdered form of this reagent.

Emulsifiers. The nonionic (in water) surfactants Igepal DM 710 and CO 730 are most commonly used to wash away fats and oils during wet sieving. They make particles wetter, resulting in less clumping of product matrices during digestion or during soaking in an aqueous phase solution.

Flotation liquids. Most methods use either *n*-heptane or mineral oil as flotation liquids. Commercial grades of heptane are usually acceptable. Mineral oil flotation liquids have exact specifications in terms of viscosities and specific gravities. Paraffin oil and mineral oil are sometimes used interchangeably as long as the viscosity and specific gravity characteristics are acceptable.

Pancreatin. Pancreatin solution is a digestion reagent that is sometimes used for insect filth in grain products. Pancreatin digestion has been found to alter animal hairs in samples, and therefore methods using pancreatin are rapidly falling into disuse.[7] A dilute pancreatin solution is effective for cleaning fine mesh sieves after use.

Tween 80. This is a polysorbate that inhibits plant particles from rising during extractions by providing a layer between the aqueous and flotation phase while filling up the trap flask. This prevents the layers from mixing. Tween 80 appears to be even more effective if used in combination with tetrasodium (Na_4) EDTA.

Aluminum acetate. One valuable technique is the Ponder-Vail technique, which uses aluminum acetate (AlAc) in fruit and vegetable extractions, including spices, flours and breads, to enhance the saturation of product during digestion and soaking in the aqueous phase.[8-10] A small amount (1–2 g) of AlAc added to the digestion or saturation reagent can greatly reduce the amount of interfering product material that reaches the extraction filter paper. The use of AlAc does not appear to improve the cleanliness of filter papers prepared from matrices containing animal products. AlAc is added before mixing paraffin or mineral oil in methods using a percolator. For trap flask methods, add AlAc to the analyte that has been transferred to the flask after the wet sieving but before boiling.

III. MACROSCOPIC FILTH ANALYSIS

The most basic method for detecting filth in food products is simply to examine the food items with the naked eye, often after spreading the food on a white glazed paper, assisted by a portable 4–6 in. diameter lens with a circular fluorescent light. Suspect mold on seeds and fruits, rodent and insect excreta pellets, and/or whole insects are picked out and placed in a petri dish, lined with filter paper, for examination, identification and confirmation under a dissecting (widefield) microscope. Such a method is known as a macroscopic or "macro" exam. This suffices for foodstuffs such as beans, rice, chili peppers, large fish and other whole or chipped dry products. Special variations in the precise method of examination and tabulation of results may be required.[2,4]

Next in complexity is the addition of a dry sieving step in the visual examination of filth from foods. The product is placed in a #8 or 10 mesh sieve, 8 or 12 in. diameter, or on a stack of sieves of different meshes and shaken. Large nonfood items such as sticks, stones, glass and cigarette butts may be picked out from the top sieve (termed "pick outs") and examined with or without magnification as necessary. The siftings are usually transferred to a petri dish and then examined for filth and other extraneous materials.

IV. MICROSCOPIC FILTH ANALYSIS

A. TECHNIQUES

Microscopic filth exams involve preparation of food products to extract minute filth elements. The easiest method is to place whole or cut vegetables, fruit (canned or frozen) or dried whole seafood on a #8–10 mesh sieve nested in a #140 or #230 mesh sieve. The food materials are washed well with a forced stream of water, the temperature depending on the analytical method, using a glass or metal rod to turn the solids. It is helpful to rest the sieve on two #10 cans with the lids removed from each to keep the sieve at a convenient height and allow proper drainage of liquids through it. The cans will also attract any stray stirring bars when emptying trap flasks after extraction. Residues retained on the #140 or #230 mesh sieves ("overs") are transferred by washing them out of the sieve with a wash bottle containing distilled water (DW) or 40% isopropyl alcohol (IPA) into a beaker and then directly (aided by washing with DW or IPA) into a Hirsch or Buchner funnel containing ruled filter paper mounted on a vacuum filtering flask (Figure 2). The sieve should be held at a 45–85° angle from the vertical and be rotated manually during the washing. If the residue is thick, it may be washed to one corner of the sieve with water. After the residues (overs) are poured from the beaker into the filter apparatus, a "rubber policeman," which consists of a glass rod with an inserted rubber tip, is useful for transferring possible filth elements adhering to the inside of the beaker. The filtrate will be more evenly distributed over the filter paper if a disk of #16–30 mesh screen, cut from a broken sieve or a clean window screen, is placed between the paper and the funnel underneath it. It also may help to more evenly distribute the filtrate over the filter paper if the paper is wetted and pulled to the flat surface in the funnel by vacuum suction. Then the contents of the beaker are added to the funnel with the vacuum on. Next the beaker is thoroughly rinsed onto the filter paper and any material that may be present on the upturned sides of the filter paper is rinsed toward the flat center area. The filter paper is then transferred to a plastic or glass petri dish of the proper size, usually 100 × 15 mm for 9-cm paper, suitable for microscopic examination under a dissecting microscope. The filter paper is placed in the bottom (smaller diameter) half of the petri dish and the top (larger diameter) half is used for its intended function as a lid to keep airborne contaminants from settling onto an untended dish. A few drops of glycerin alcohol, consisting of 1:1 (v/v) glycerin and ethyl alcohol, are added as needed to the petri dish before the filter paper is placed in the dish to keep the sample material moist and free of mold. Filth and other material on the papers may then be easily examined. Additional small amounts of this solution can be added as necessary while examining the plates under a microscope, to reduce glare and maintain optimum viewing conditions. If material is very thick on the filter paper, the solution will make the material softer and more easily examined. The amount of liquid in the petri dish should always be enough to thoroughly wet all material in the dish but never so much that it allows material to float about.

In those cases where excessive materials may collect on the sieve, heptane or mineral oil flotation procedures are a necessary pretreatment. The principle is basically that of the separatory funnel used in organic chemistry. The aqueous solution containing the product is mixed with a flotation liquid. Insect and other arthropod material, hairs and other filth collect in the flotation layer, whereas most extraneous plant and animal materials remain in the aqueous phase below. This involves placing the material into a 1- or 2-L Erlenmeyer flask, then adding water to bring the volume up to 600–700 ml, as directed. The mixture is then briefly boiled while it is gently stirred, with a magnetic stirring bar on a magnetic stirrer, so that the vortex formed reaches down to expose the top of the stirring bar. The purpose of this step is to remove the air from product materials so they will not float. Some foods, such as frozen seafood, may not require this boiling step. A trapping rod with a rubber disk or stopper at the base (Figure 1) is then placed in the flask. The disk (or stopper) should be of a diameter that is slightly larger than the flask neck so that it needs to be compressed in order to fit through the neck of the flask. Thirty milliliters (for 1-L trap flasks) to 50 ml (for 2-L flasks) of heptane or mineral oil is added by pouring it gently down the rod so as to minimally disturb the material on the bottom of the trap flask. The solution is then magnetically stirred 3–5 min (after clamping the rod so it does not interfere with the stir bar) at a speed that creates a vortex that exposes the stir bar without audible or visible splashing. This thoroughly mixes the aqueous phase and flotation liquid. The trap rod may be used to manually stir and mix the system in lieu of magnetic stirring. If the trap rod is used to mix the system, the flask is held at an angle in one hand and the rod rotated at a rate of 200 strokes per minute for 2 or 3 min. After stirring, the system is usually allowed to sit for a short time to allow the flotation liquid to separate from the aqueous phase. The lipophilic filth elements float up with the layer of flotation liquid that forms above the aqueous phase. Additional aqueous solution is then poured down the stirring rod, to bring the level of the bottom of the aqueous layer even with the bottom of the lower portion of the neck of the flask (Figure 3). The system is allowed to equilibrate for an additional period of time. Filth materials visible to the naked eye often adhere to the inside

FIGURE 3. Interface of aqueous phase and flotation liquid in neck of trap flask. (Photo taken by W.E. Savary.)

surface of the flask, down in the aqueous layer. A stir bar or other magnetic object can be applied to the outside of the flask next to the stir bar inside the flask and used to pull the inside bar up to the filth elements and dislodge them. When dislodged, they usually rise up into the flotation layer. The rubber disk or stopper of the trapping rod may also be used to dislodge such filth elements. Two gentle stirs with the rod may be given at 5–10-min intervals. The contents of the flask should not be disturbed for 10 min after the final stirring. The next step in trapping is to take the trapping rod and gently pull it up so that the disk or stopper fits into the neck, trapping a small amount of aqueous solution and all of the flotation liquid in the neck of the flask (Figure 4). While the pulling pressure is maintained, the flotation layer in the flask neck is poured directly into a beaker or other collection vessel.

Most AOAC methods using the trap flask require two trappings. The neck of the trap flask and trapping rod should be washed with the aqueous phase solvent before the second trapping. This is accomplished by maintaining the pulling pressure on the trapping rod so as to keep the contents of the flask separated from the outside while squirting the aqueous solvent into the neck and onto the rod. The aqueous phase rinse should then be poured off into the collection

FIGURE 4. Interface with trapping rod wafer pulled into position to pour off flotation liquid.

vessel and the trapping rod gently placed down into the lower portion of the trapping flask. Next, fresh flotation liquid is added for the second trap, usually 30 ml (rather than 50 ml as for the first trapping) when using 2-L flasks and 25 ml when using 1-L flasks. The trapping rod is then manually stirred with a circular motion to churn up the sediment on the bottom of the flask and to thoroughly mix both aqueous and flotation phases together. Additional aqueous medium is added to the flask, bringing the liquid volume to the original level in the neck of the flask as in the first trapping procedure. After 10 min, one half minute of circular stirring with the trapping rod assists in filth recovery. After 20 min, trap off the flotation layer into the same collection vessel as the first trapping. The combined trappings are then poured onto a filter paper fitted into a suction funnel.

For certain materials, after boiling or autoclaving, the 2-L glass percolator (Figure 5) or the Kilborn funnel methods are preferred for cleaning up the solvent phase. Filth in finely ground and comminuted products (e.g., flours, cornmeal, cinnamon, ground leafy spices and crabmeat) may be more successfully floated off using these methods. It is best to use hot water initially and then incrementally cool the water in successive drains. High bran content products require

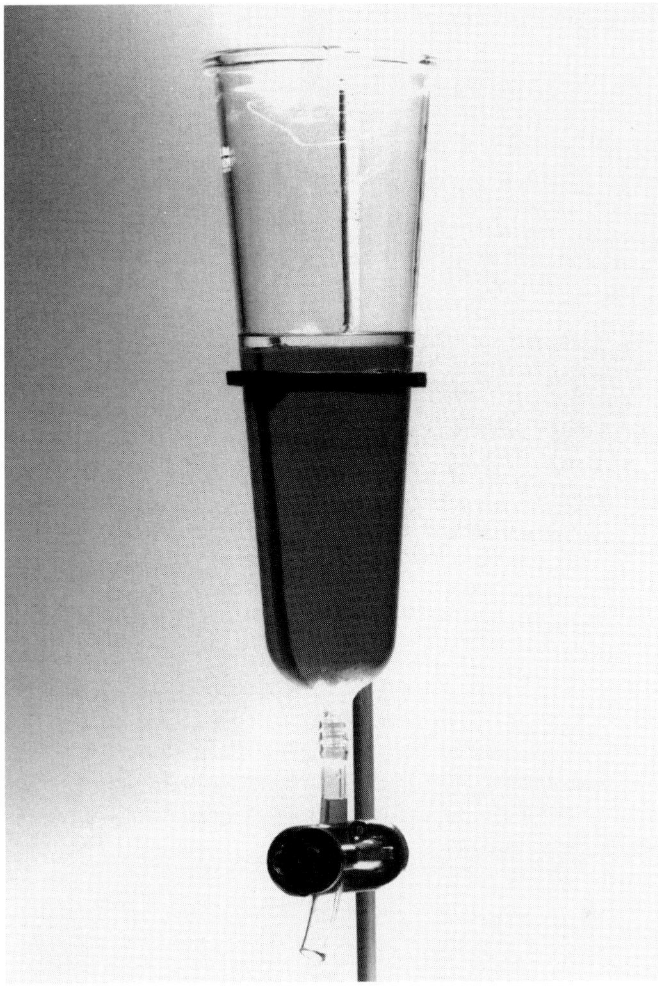

FIGURE 5. Percolator. (Photo taken by W.E. Savary.)

decreasing the last two drains to 1 min. If water is added to the system, use a long hose to supply water to the funnel or percolator, being certain not to aerate the solution. After draining, to the 250 ml level (150 ml level in the Kilborn), the hose should be squeezed tightly (to increase the pressure of the water) and the flow aimed at the top 4 in. of the upper surface of the vessel and rotating the spray so that adhering materials and oil are removed from the surface. If 40% IPA or 60% ethyl alcohol is the aqueous phase, the percolator will be filled directly from the 2-L beaker. Bring volumes of water or alcohol to 1700 ml in the percolator and 700 ml in the Kilborn. The final drain should have 50 ml of 95% ethyl alcohol or of 100% IPA added by using a plastic squeeze bottle with a tight lid. Adding the alcohol aids in cleaning the vessels before the final bolt (drain) of the remaining materials into the original clean beaker. High fat content products may require that the inner surface of the vessel be cleaned with a filtered detergent solution, filtered tap water, and finally alcohol. The glass vessels, beakers and glass coffer dam (if using a Buchner funnel) should be cleaned to a spotless state immediately to avoid adhesion of material. The accuracy of filth recoveries is increased by using good extraction, cleaning and filtration techniques. Filter slowly, if using a Hirsch funnel, to prevent flotation of filth elements over the paper. Metal screens cut from surplus

#10–30 sieves or window screens are necessary to spread the materials evenly on the filter paper.

Kilborn funnels require a special sized rack for support. A wooden rack with two longitudinal boards, the bottom board grooved into the frame, will provide support after circular holes of the proper size are cut into both boards. The holes in the top board will be larger than those in the bottom board. The rack should hold six funnels and should be constructed so that it may be placed across the laboratory sink. The water hose required need not be as long as the water hose for the percolator. The hoses at the base of both vessels should be 7–9 in. long.

Percolator racks may be produced easily by using ring stands weighted for support or secured to the bench. Ring support stand apparatus parts are listed in most scientific equipment and supply company catalogs. Three longitudinally cut pieces of tubing placed over the metal ring provide cushioning between the percolator and the ring. If talent and funds are available, a metal rack similar to the Kilborn rack may be built for the percolators. The rack has large holes in the upper shelf to receive the percolator vessels, smaller holes in the lower shelf, and a large V-shaped (in cross-sectional view) trough or gutter below the vessels resting near the base of the rack. A short length of pipe is placed into a hole drilled through the end of the trough and the side support of the rack so that a 1-1/2 to 2 ft long hose can be attached to its outer end to drain water and acceptable waste into the sink. The flow of materials from the vessels may be completely stopped at will by placing a plastic or metal tubing clamp on the hose at the base of the vessel.

Microscopic filth analyses often require additional extraction processes that break down plant and animal tissues so filth elements can be recovered. Ground grain products, dried, chopped or ground spices, and baked goods are examples of materials that require chemical and boiling treatment followed by separation with the trapping or percolator methods to collect filth. Acid hydrolysis using 30% hydrochloric acid (30 ml in approximately 1000 ml of aqueous product with the addition of a few drops of silicate antifoam) at 121°C for 30 min in the autoclave is one of the most commonly used procedures. The final step (filtration) should produce materials relatively free of plant and animal debris so that filth elements may be easily detected and identified using the widefield microscope. Depending on the results, it may be possible to omit the trapping or percolator step by proper wet sieving on a #230 sieve.

B. QUANTITATIVE TRANSFERS

Each time an analytical portion is transferred from one vessel to another during the course of a filth analysis, there is a risk that small amounts of filth that are not visible to the unaided eye may be inadvertently left behind in the old vessel. Although some small loss appears inevitable, there are procedures that analysts can follow to keep losses to an absolute minimum and improve the overall reliability and reproducibility of their results. These procedures are grouped into the conceptual category of quantitative transfer procedures.

For analytical methods involving wet chemistry digestion or defatting steps, the wash bottle is the analyst's most versatile tool for quantitative transfer of the wet analytical portion from one vessel to another. Many methods specify the formulation of wash bottle liquids and care must be taken to use the correct wash liquids to avoid interference with other chemical reactions that are necessary for a successful analysis.[(3)] The stream of liquid from the wash bottle is used to rinse microscopic filth from the interior of the old vessel into the next vessel. Because the filth is not visible to the unaided eye, the wash bottle technique is critical to ensuring complete transfer. Beginning the transfer wash, the stream of water from the wash bottle must be directed at the inside surface of the old vessel that is at the highest elevation above the bench top. This is not necessarily the bottom of the inverted vessel because the vessel is usually held at an angle. Nor is it the area farthest from the wash bottle nozzle. Because liquids flow downward, the analyst must begin the wash transfer at the highest point and direct the stream progressively downward from there in a pattern that covers the entire inside surface of the vessel, arriving at the lowest point last. This procedure ensures the microscopic filth is moved progressively closer to the transfer point and not simply moved from one side of the vessel to another. The procedure applies equally to beakers, trap flasks, percolators, sieves, product containers or any other vessel where washing is part of the transfer operation. The procedure is repeated three or more times, rotating the vessel to ensure that all residues are transferred to the next vessel.

Losses of minute amounts of material from dripping, leaking or splashing must also be prevented. The usual causes for this type of loss are cracked or chipped vessels (at the rim or lip where poured liquid exits), excessive heating, boiling or stirring on a hot plate, and overzealous sieving with a forced stream of hot water. Sieves, glassware and other vessels must be inspected before and after each use for cracks, rips or deformities that could result in losses.

Proper sieving technique uses the least amount of water pressure that is necessary to get the job done. Excessive force from water pressure causes splashing and may deform the delicate weave of fine mesh sieves, allowing microscopic filth to fall through. Pouring material onto a sieve down a clean glass rod avoids splashing as the liquid strikes the sieve fabric. Alternatively, the sieve may be tilted at a slight angle to avoid splashing. Sieves should be thoroughly cleaned and backwashed after each use, including uses between consecutive portions of a sample.

Other techniques include the use of a rubber policeman to dislodge sticky material from the side of a vessel, well-constructed artist's brushes to transfer dry residues from sieve pans and faucet aerators, which can be used as a wash bottle when large amounts of material are involved. In the latter case, a final transfer using a regular wash bottle is recommended. Once the analyst is convinced that all material has been transferred from one vessel to another, the quantitative transfer procedure is to repeat the last transfer operation one more time to be absolutely confident that nothing remains in the old vessel.

C. SAMPLE CONTAMINATION

Every microanalytical entomology laboratory must have procedures in place to guard against contamination from microscopic filth. Two types of contamination can occur: extrinsic contamination and cross-contamination. The potential sources of extrinsic contamination are glassware, equipment, reagents, clothing and the general environment. In order to prevent extrinsic contamination, glassware, equipment and reagents must be stored in clean, enclosed cabinets when not in use. Glassware and other equipment must be scrupulously clean. Immediately prior to use, these items should be given an extra cleaning by rinsing with clean water or wiping with fresh, clean towels, regardless of the general standard of glassware cleanliness in the laboratory. All reagents should be filtered. Tap water should be treated as if it were a reagent and be periodically examined for absence of filth contaminants. Where problems exist, inline filtration can be installed to remove contaminants from the water supply. Typical water contaminants include hairs, feather barbules and aquatic invertebrates, usually from unprotected local reservoirs.

Procedures must also be established to prevent the analyst's person or clothing from contributing extrinsic contaminants to an analysis. A clean laboratory coat or frock is the best preventive measure in this regard. Each analyst should have a generous number of garments for exclusive personal use and the laboratory should provide professional laundering on a frequent basis. Other types of containment clothing such as headgear, snoods and gloves are optional, depending on the conditions in a particular laboratory. Rather than relying solely on containment garments, each analyst should develop personal techniques for handling samples and equipment that avoid bodily contact or close proximity to the sample or sample contact surfaces. Spoons, scoops, tongs, wash bottles and, occasionally, rubber gloves are useful for this purpose. Hair (both human and pet) from clothing and lint are the primary contaminants that the analyst must guard against. These procedures combined with reasonable personal hygiene will preclude the possibility of direct extrinsic contamination by the analyst.

During the analysis, equipment and exposed food products must be protected from the remote possibility of airborne contamination. Even though most laboratories have air-handling systems that filter the air in the laboratory, it is advisable to take extra prophylactic measures to guard against invisible airborne contaminants in order to ensure the overall integrity of the analysis. Containers such as trap flasks, beakers and percolators that are to be left standing for any period of time during an analysis should be protected with a cover or aluminum foil over the open top. Bench tops and other furniture surfaces must be cleaned regularly to avoid accumulations of dust and residues. Bench tops should also be cleaned thoroughly before and after each analysis. Finally, petri dishes that contain filth extraction filter papers should be used in the conventional manner, with the larger diameter being the top or lid, to prevent airborne contaminants from settling on exposed edges of the paper.

It is also a good practice to restrict access to the laboratory area to those scientists and technicians who are directly involved in microanalytical entomology analyses. One of the greatest potential sources of airborne contaminants is the coworker or casual visitor who is not sensitive to the special procedures that are required to prevent extrinsic contamination.

Cross-contamination is the inadvertent transfer of microscopic filth from one sample, or sample subunit, to another. A few common sense rules are sufficient to prevent cross-contamination.

1. Never use the same piece of glassware or equipment (including sieves) twice without thoroughly washing it between uses, including between uses for consecutive portions of the same sample.
2. Implements such as stirring rods, rubber policemen, etc., should be rinsed into the container where they were used and then thoroughly cleaned prior to subsequent use.
3. Never return unused reagents to their original containers.
4. Dispose of any reagent whose purity or integrity is in doubt.
5. Never allow aerators or wash bottle nozzles to contact another piece of equipment or the analyte.
6. Never allow any part of the analyst's body, covered or not, to contact the inside surfaces of vessels. "Analyst's thumb" is a common source of cross-contamination, if not extrinsic contamination.

D. DISPOSAL

Many of the chemicals used in these analytical procedures must be captured for proper disposal or treatment after their use by the analyst. Each laboratory should consult their chemical hygiene officer or consultant about the proper management of the chemical byproducts of these or any other analyses performed by the laboratory. Maintaining the quality of the environment and protecting the public from hazardous wastes are the individual responsibility of every person who uses these chemicals and compounds.

V. EXAMINATION OF FILTH ELEMENTS

A. GENERAL TECHNIQUES

Once the filth has been isolated, it must be measured, tabulated and identified by visual means. Large, or macroscopic, filth items are examined under low magnification to observe minute micromorphological structures that are taxonomically important. Low magnifications (10–30×), are also required to locate microscopic filth, such as the filth isolated onto filter papers using the trap flask or percolator techniques. Dissecting probes and tweezers are used to manipulate the filth elements on filter papers and microscope slides. Jeweler's forceps are ideal for very fine work. A useful microprobe can also be fashioned by inserting a minuten pin (see chapter 14) into one end of a thin wooden dowel. The wooden stem of a common cotton swab, with the swab clipped off, is suitable. Probes, tweezers and forceps are also used to move debris on a filter paper to reveal any filth that may lie underneath.

When examining filter papers for microscopic filth, it is important to be methodical. The examiner should adjust the microscope so that the field of view allows one to keep track of the ruled lines on the filter paper and not lose one's place or wander aimlessly as the paper is moved under the microscope lens. The entire surface of the paper must be thoroughly examined and every filth item on the paper must be noted and its presence, dimensions and identity recorded at the time the item (filth element) is discovered. Care must be taken to look underneath debris for hidden filth elements and to tease apart aggregations of material to find any embedded filth elements. Filth elements that are likely to bear minute morphological structures that aid identification should be segregated and separately examined at higher magnifications of 50–400× or more. If necessary, these elements can be removed from the filter paper and separately mounted on glass microscope slides for examination at the higher magnifications.

B. MICROSCOPE SPECIFICATIONS

A laboratory requires two basic types of microscopes to conduct microanalytical analyses: widefield, or dissecting, microscopes and compound microscopes. Most manufacturers of microscopes produce instruments suitable for microanalytical work, but care must be taken to select instruments that are configured to the following general specifications.

Widefield microscopes require a wide, flat base that allows the microscopist ample surface to position petri dishes, products and exhibits. The working distance between the microscope's objective lens and specimen should be at least 3 in., to allow the use of probes, forceps and other tools. The body height of the microscope should be adjustable to allow the examination of large exhibits such as product containers. The widefield microscope must be equipped with a lens system that allows examinations at 15× and 30× magnifications. A higher magnification option proves useful for confirming the identity of suspicious particles and for examining minute structures on insect specimens. The field of view at 15× magnification must be at least 5 mm diameter. The light source must be adjustable to provide a 70° incidental angle of illumination.

Compound microscopes should provide an array of magnifications between 100× and 200× with options up to 400× if possible. Standard instruments that are designed to examine specimens on 1 × 3 in. glass microscope slides are acceptable. It is wise to choose instruments, both widefield and compound, with the highest quality optics to assure optimal resolution and to minimize the eye fatigue of the microscopist.

C. CARE AND USE OF THE MICROSCOPE[11]

1. Caveats

1. Never dust a lens by blowing on it. Saliva will inevitably be deposited on lenses and is harmful even in minute amounts.
2. Filter pressurized air for dusting to trap oil and other contaminants.
3. Do not use dry lens tissue on a lens.
4. Never use facial tissues to clean lenses. They may contain glass filaments which can scratch lenses. Linen or chamois may be used for cleaning but may not be as convenient as lens tissue. Do not confuse lens tissue with bibulous paper, which should never be used to clean lenses. Follow the manufacturer's recommendation for using cleaning

solvents other than water. Lens mounting glues are often soluble in alcohol. Xylene used sparingly is generally acceptable for serious stains, such as residual oils.

5. Never leave microscope tubes open. Always keep them closed with dust plug, eyepiece, or objective, as appropriate.
6. Avoid touching lenses. Even light fingerprints, especially on objectives, can seriously degrade image quality.
7. Do not attempt to take optics apart for cleaning. Internal optics should not need routine cleaning and should be professionally serviced if cleaning is needed.
8. Use proper immersion liquid on immersion objectives as specified by the manufacturer. Avoid getting immersion liquid on nonimmersion objectives; it can damage the lens mounting glue.
9. Keep microscopes covered when not in use. Avoid extremes of temperature and high humidity. In work areas with consistently more than 60% relative humidity, store microscopes in circulating air if possible. In very high humidity, optical parts should be stored in tightly covered containers with desiccant and kept very clean to prevent mold growth on the optic coating.
10. Do not use the substage diaphragm to control brightness.

2. Cleaning

1. Body of the microscope
    a. Use alcohol or soapy water on cloth to wipe body.
    b. Lubricate sliding parts with a petroleum jelly, such as Vaseline, or use a manufacturer-recommended lubricant.
2. Lenses and optics
    a. Grit and dust can scratch lenses and coatings.
    b. Blow away dust with rubber bulb.
    c. For light cleaning, breathe (do not blow) on lens to fog it, and then use lens tissue as described below. The fogging is basically water and is not harmful.
    d. For dirtier lenses use lens cleaner solution, such as that available from any camera supply store. Remove stubborn stains by using xylene sparingly.
    e. Use the following procedure to properly clean lenses of microscopes or other optical equipment. Crumple a piece of lens tissue to create many folds to trap dirt without grinding it into the lens. Do not touch the part of the tissue that will be applied to the lens; excessive touching transfers natural oils from the fingers to the lens tissue. Apply a small amount of lens cleaning solution to the lens tissue and blot the tissue against absorbent material to prevent fluid from entering the lens mount. Wipe the lens very lightly to remove gross dirt that was not blown away by the rubber bulb. If necessary, repeat the cleaning process with a new piece of lens tissue and with more pressure to remove oily or greasy residue. Complete the process by using the light cleaning procedure (breath and lens tissue) described above.

D. MICROSCOPES AND ACCESSORIES

1. Types of Microscopes

A microanalytical laboratory must have the following basic microscopes available for the examination of the filth that is extracted from food products and of exhibits collected from sites where food is prepared under insanitary conditions.

a. Widefield Stereoscope

This instrument has a wide variety of applications and is the most frequently used instrument in filth analysis. The main application is for reading (examining) filth extraction papers from analyses for microscopic filth. Extraction papers are usually read at approximately 30× magnification. Higher magnifications are used to confirm the identity of very small objects. The effective upper range of this type of instrument is about 100×.

Eye fatigue is usually the most important limiting factor when working with microscopes of any kind. Proper instrument adjustment is the best way to reduce eye fatigue. Adjustment begins with the light source, which should be daylight blue or white (not frosted). Light strength should be adjusted with the transformer so that details can be seen clearly with a minimum of glare. The light source should be positioned slightly to one side and at about a 70° angle from horizontal. The light field should be centered so that both eyes receive the same intensity.

The eyepieces (oculars) must be adjusted to the individual analyst's eyesight while the eyes are in their most relaxed state. To do this, adjust the interpupillary distance of the oculars so that the oculars are centered on the respective pupils. Determine which ocular is independently focusable. Place a specimen on the stage of the instrument. Center it in the field of view. Hold an index card between the focusable ocular and the eye, blocking vision on that side without closing the eye. Using the main focusing adjustment knob, clearly focus the image of the specimen for the eye that is not blocked. Remove the index card from the first position and use it to block the vision of the other eye. Using the ocular focusing mechanism for the focusable ocular (not the main adjustment knob), clearly focus the image of the specimen. The specimen image should now be clearly in focus for both eyes in their most relaxed state.

This procedure should be followed for each magnification. It is a mistake to adjust the oculars by closing one eye and adjusting for the other. This places unnatural strain on the open eye.

The wearing of eyeglasses is a matter of individual preference. People with an astigmatism may find that wearing their eyeglasses will reduce eye fatigue. If eyeglasses are to be worn, the ocular adjustments should also be made with eyeglasses on.

b. Compound Microscope

The microscopist should read and use the manuals supplied by the instrument manufacturer to obtain familiarity with a laboratory's compound microscopes. Oculars should be adjusted in the same manner as for widefield stereoscopes. The compound microscope is used primarily to examine individual filth elements at high magnifications. Usually, the specimens are mounted on a clear glass microscope slide. Mounting techniques and media are discussed in chapter 14.

The microscopist should learn how to achieve optimal or Koehler illumination. The Koehler Principle calls for focusing the field iris in the same plane as the specimen in order to obtain maximum resolution. Step-by-step instructions for achieving Koehler illumination are found in instrument manuals or in the FDA technical Bulletin No. 2.[(6)] Additionally, the trainee should learn how to tighten loose focusing mechanisms, properly clean lenses, and change bulbs for all the instruments the trainee will be using. These and other maintenance operations are described in the instruction manuals accompanying each instrument.

c. Phase-Contrast Microscope

The phase-contrast microscope is a compound microscope that has special substage accessories and objective lenses designed to produce optical contrast between the specimen and the mounting medium. Using phase-contrast optics, the analyst can observe many details that are obscure or indiscernible under a conventional compound microscope. This type of microscope is widely used for examining mites and small, somewhat transparent insects.

Success with the phase-contrast microscope is highly dependent on the refractive indices of the specimen and the mounting medium. For good contrast, the mounting medium should have the greatest possible difference between the refractive indices used to create variations in light intensity. To the observer, it appears that a halo of intense light is surrounding the very dark edge and surface structures of the specimen, starkly silhouetting otherwise vague details.

Adjustment of the instrument proceeds in the same manner as a conventional compound microscope. Once optimal illumination is achieved, the phase-contrast optics must be aligned. The phase objective lenses each have a metal ring of a specific size imbedded between two of the lens casings. In the condenser are annular diaphragms that, when positioned by rotating the condenser to different positions, are usually marked "1," "2," "3" and "clear" (the latter position corresponding to the condenser of a conventional microscope). Each objective has a corresponding condenser position. Thus, a Ph 2 objective must be matched with the number 2 position on the condenser, for example.

The instrument, with objective lens and condenser position matched, is then focused on a specimen using the regular oculars. With the specimen in focus, a telescoping ocular is used to simultaneously observe the ring in the objective lens (dark) and the ring of light from the condenser (bright). Using the knobs on the condenser, the bright ring is carefully centered within the dark ring. Once this is accomplished, the instrument will give optimal illumination with standard oculars. This alignment must be checked each time the objective lens is changed or the condenser is rotated.

d. Polarizing Microscope

A polarizing microscope is essentially a compound microscope fitted with polarizing prisms, called "Nicols," below and above a rotating circular stage.

When the two Nicols are placed in the optical train, the first acts as a polarizer and the second as an analyzer. The vibration direction of the plane polarized light produced by the polarizer should be known (conventionally the north–south direction). If the analyzer, which can be rotated in most instruments, is set in the same relative position as the polarizer (parallel Nicols), light is transmitted; but, if the analyzer is rotated through 90° so that its plane of vibration is at right angles to the polarizer (crossed Nicols), no light will pass.

Optically active substances show interference colors when placed between crossed Nicols. Observations of optical activity can be useful to the analyst for identifying to some extent such diverse things as glass fragments, synthetic fibers, crystals, starches and mites.

2. Accessories

a. Eyepiece Micrometer

This accessory consists of a clear disk with a graduated scale printed on it, which is inserted in the eyepiece of the microscope for use in measuring specimens. To provide meaningful measurements, the micrometer must be calibrated for each magnification. Calibration is performed using a stage micrometer with a graduated scale of known increments (usually 0.1 or 0.01 mm).

To calibrate, place the stage micrometer on the microscope stage and focus on the stage micrometer scale. The eyepiece micrometer scale should always be in focus; if not, the disk is probably improperly inserted. Move the stage micrometer so that the zero end of the eyepiece scale aligns with the zero end of the stage micrometer and the two scales are superimposed on each other over their entire lengths. Reading the eyepiece scale, find the farthest division from zero that coincides with a division on the stage micrometer.

Record the following information:

1. Magnification.
2. The number of eyepiece scale divisions between the zero coincident and the far eyepiece scale coincident. This value is designated EMD (eyepiece micrometer divisions).

3. The number of stage micrometer scale divisions between the zero coincident and the far stage scale coincident. This value is designated SMD (stage micrometer divisions).
4. The millimeter value of one division of the stage near scale. This value is usually found printed on the micrometer. This value is called mm/SMD (millimeters per stage micrometer division).

Calculate the following:

$$(\text{mm/SMD}) \times (\text{SMD/EMD}) = (\text{mm/EMD})$$

The value derived from this equation (mm/EMD) is the number of millimeters per eyepiece micrometer division for that particular magnification. This value is used to convert eyepiece micrometer divisions to millimeters by multiplication.

Example:

At 10× magnification, 12 EMD coincides with 18 SMD on a stage micrometer in which 1 SMD = 0.1 mm.

$$(0.1\ \text{mm/SMD}) \times (18\ \text{SMD}/12\ \text{EMD}) = 0.15\ \text{mm/EMD}$$

An object is observed to be 5 eyepiece divisions (EMD) long at 10× magnification.

$$(0.15\ \text{mm/EMD}) \times (5\ \text{EMD}) = 0.75\ \text{mm}$$

The object's length is calculated as 0.75 mm, the last digit being only an approximation. The significance of calculated numbers should be carefully considered in light of the mathematical rules concerning significant figures.

b. Mechanical Stage Micrometer

This accessory consists of two graduated scales that are inscribed on the mechanical stage. One scale is on the moving portion and the other scale is directly parallel to the first on the stationary portion of the stage. As the length of the specimen's image is moved through a fixed point in the field of view, the number of divisions on the stationary stage scale can be counted by observing the starting and finishing positions of the sliding scale's end point. By substituting this number for the EMD in the equation found in the previous section, the microscopist can calculate a conversion factor using a stage micrometer with a graduated scale of known increments. Quite often, mechanical stage micrometers will have provision for interpolating the final digit of a reading, which adds to their accuracy.

The eyepiece micrometer has more versatility; it can position over the image at exactly the angle desired for determining longest and shortest dimensions.

VI. LABORATORY INFORMATION BULLETINS

LIB is an abbreviation for laboratory information bulletin. The LIB provides a noncollaborated analytical method or technique that has been developed and tested by FDA laboratories for a product that lacks an official AOAC method or for other situations not covered by manuals and compendia. In testing an interim method such as the LIB for recoveries of filth elements, it is sometimes necessary to spike practice samples with known amounts and types of filth to test recovery levels. The usual spike material consists of beetle elytra and rodent hairs.[12,13] The address for requesting a specific LIB is U.S. Food and Drug Administration, Office of Regulatory Affairs (ORA), Office of Regional Operations (ORO), Division of Field Science, Room 12-39, Rockville, MD 20857.

REFERENCES

1. Food and Drug Administration, *Laboratory Procedures Manual*, chapter 18 (Entomologist Training), U.S. Food and Drug Administration, Rockville, MD, 1-79, 1994.
2. FDA, *Macroanalytical Procedures Manual*, FDA Tech. Bull. No. 5, Assoc. Off. Anal. Chem., Arlington, VA, 1-210, 1984.
3. Boese, J.L. and Bandler, R., Extraneous Materials: Isolation, in *Official Methods of Analysis of the Association of Official Analytical Chemists*, 15th ed., Vol. 1, Helrich, K., Ed., Assoc. Off. Anal. Chem., Arlington, VA, 368-406, 1990.
4. FDA, *Microanalytical Methods*, U.S. Dept. Health, Educ., Welfare, Washington, DC, 1-180, 1959.
5. Gorham, J.R., Ed., *Principles of Food Analysis for Filth, Decomposition, and Foreign Matter*, 2nd ed., U.S. Food and Drug Admin. Tech. Bull. No. 1., Washington, DC, 1-286, 1981.
6. Gorham, J.R., Ed., *Training Manual for Analytical Entomology in the Food Industry*, Food and Drug Admin. Tech. Bull. No. 2, Assoc. Off. Anal. Chem., Arlington, VA, 1-145, 1977.
7. Smith, F.R., The effect of natural and laboratory digestion on the microscopic appearance of certain animal hairs, *J. Assoc. Off. Agric. Chem.*, 41, 479-480, 1958.
8. Washbon, E.C., *The Ponder-Vail Method for Extraction of Light Filth in Corn Meal and Whole Flour*, FDA Lab. Information Bull. 3604, 1-3, 1991.
9. Zimmerman, M.L., Madenjian, J.J., and Roberson, M.E., *Improved Method for Extraction of Light Filth in Mole Paste (Vail's Method)*, FDA Lab. Information Bull. 2865, 1-3, 1985.
10. Zimmerman, M.L. and Madenjian, J.J., *Improved Method for Extraction of Light Filth in Ground Capsicums Excluding Paprika (Vail's Method)*, FDA Lab. Information Bull. 2866, 1-3, 1985.
11. Poelma, P.L. and Bryce, J.R., Microscopic examination of foods, in *FDA Bacteriological Analytical Manual*, 7th ed., Assoc. Off. Anal. Chem., Arlington, Va, 11-16, 1992.
12. Brickey, P.M., Gecan, J.S., Thrasher, J.J., and Vazquez, A.W., Notes on microanalytical techniques in the analysis of foods for extraneous materials, *J. Assoc. Off. Anal. Chem.*, 51, 872-876, 1968.
13. Brickey, P.M., Gecan, J.S., Thrasher, J.J., and Vazquez, A.W., Notes on microanalytical techniques in the analysis of foods for extraneous materials. Part II, *J. Assoc. Off. Anal. Chem.*, 55, 51-56, 1972.

Chapter 16

FORENSIC EVIDENCE DEVELOPMENT

Michael L. Zimmerman and Paris M. Brickey, Jr.

CONTENTS

I. Introduction ... 261

II. Inspectional Evidence ... 262
A. Rodent Filth ... 262
1. Rodent Droppings ... 262
2. Rodent Gnawing ... 262
3. Rodent Urine ... 262
4. Rodent Sightings ... 263
B. Insect Filth ... 263
1. Identifying Insects in the Field ... 263
2. Life Cycle/Stages of Insects ... 263
3. Location of Insects ... 263

III. Analytical Evidence ... 264
A. Procedural Issues ... 264
B. Documentation of Analytical Results ... 264
C. Performing the Analysis ... 265

IV. Compliance ... 267

I. INTRODUCTION

The U.S. Food and Drug Administration (FDA) and similar agencies that regulate food sanitation operate like other enforcement agencies in gathering and developing good forensic evidence to prove violations of the laws that an agency is mandated to enforce. For example, one can compare the FDA with the Federal Bureau of Investigation (FBI) and find many similarities. Both agencies deploy field investigators to search for evidence. Both have expert laboratory personnel who are highly trained in scientifically identifying and developing the evidence found by the investigators and relating the evidence to a specific regulatory or legal issue. Finally, both have paralegal personnel who are responsible for interpreting the evidence developed by investigators and laboratory scientists to determine the best and proper charges or legal actions to be filed in court.

There are two basic types of forensic testimony. Most investigators and laboratory scientists provide testimony of fact, relating what they saw or did and the circumstances surrounding their forensic activities. The expert witness is called to provide testimony to the court that interprets the significance of the scientific facts of the case. Attaining expert status is beyond the scope of this chapter; however, the basic procedures of evidence development apply equally to fact and expert witnesses. In this chapter, each of the basic areas of evidence gathering, or development, will be discussed with emphasis placed on the inspectional and analytical evidence development relating to microanalytical entomology and the detection of filth in foods.

II. INSPECTIONAL EVIDENCE

An inspection by a government agency is usually initiated with an official notification (or court warrant in some cases). Once the investigator has issued the necessary legal documents, a walk-through inspection is performed to become familiar with the establishment and become alert to any potential sanitation problems. Then a more thorough probing of specific infestations, adulterations or insanitary conditions is begun. All of the investigator's observations and any exhibits or items of evidence that the investigator collects are carefully documented in written investigative reports. There are specific types of evidence that the investigator should keep in mind when making an inspection.

A. RODENT FILTH

The investigator should have a good knowledge of the habits of rats and mice. Given the furtive nature of rodents, it is rare that actual specimens of rodents are encountered or collected. More frequently, it is the evidence of rodent activity that is detected and collected. This situation requires the investigator to be familiar with the ecology of these pests, including such things as where they nest, their home ranges, food preferences and activity patterns. The most common types of physical evidence of rodent activity are discussed below.

1. Rodent Droppings

Rodent droppings are characteristically shiny black or dark brown with one or both ends tapered. Fresh pellets are covered with a clear to milky surface coating of mucus that can persist for days. Confirmation of the origin of suspect rodent pellets is performed in the laboratory by examining the hairs that are typically embedded in rodent pellets (see chapter 12). The distribution and numbers of pellets throughout the establishment should be recorded along with notations as to their freshness (appearance). Approximate size can also be important because it will indicate whether a rat or mouse infestation is present. The distribution, relative appearances and size ranges of pellets will indicate the extent of an infestation.

2. Rodent Gnawing

Mice and rats differ in the type of gnawing they produce, either to reach food from a food container or to collect nesting material. Mice are "nibblers" and will move from place to place and gnaw small holes. Rats, on the other hand, generally gnaw large holes in a selected, limited number of packages. The collection of examples of gnawing must be selective and careful. Bag cuttings around a "gnawed" hole should be done with care to not disturb any attached rodent hairs around the hole. Any gnawed hole showing fluorescence under ultraviolet light (see below) around the circumference should be collected and the ultraviolet phenomenon should be noted on the accompanying report of evidence collection.

3. Rodent Urine

Mammalian urine will usually fluoresce a characteristic greenish yellow under shortwave ultraviolet (UV) light. Many other materials also fluoresce, including some packaging glues. Therefore, examples of fluorescent material must be collected and tested in the laboratory to confirm the origin as urine. Because rats and mice dribble urine as they scurry around and feed, a good description of the urine pattern as well as photographs under UV light should be made at the time that the fluorescent stain or material is discovered. If multilayered bags are encountered, a swatch of all layers should be collected and the layers kept in their original order (i.e., outside, second, third, etc.). One way to document relative positions of layers of packaging is to mark the outer layer with an "O" and the innermost layer with an "I." Descriptions of urine patterns have proven invaluable in several cases in which the amount and pattern proved (along with lab analysis) urine to be of human origin rather than rodent.

4. Rodent Sightings

An eyewitness sighting is one of the best pieces of evidence of rodent infestation. Because rodents are generally nocturnal in nature and very shy, the investigator must be alert to any signs of rodents. Daylight sightings of live rodents generally indicate a large population. Observations of dead rodents should be recorded along with descriptions of the appearances of the rodent carcasses.

B. INSECT FILTH

There are a number of critical pieces of insect evidence that can be pieced together so that the court has a clear picture as to the insanitary condition under which a food product was held, prepared or packed.

1. Identifying Insects in the Field

Although it is not usually possible to specifically identify insects at the site of discovery in a manufacturing or storage facility, it is necessary for the investigator to be able to characterize most food-infesting insects to their taxonomic order. Descriptive terms such as "beetlelike" or "mothlike" are excellent adjectives and can help tremendously in developing inspectional evidence. In determining what kind of insect is involved, the investigator can form a strategy for further investigation. For example, sightings of moths flying around will generally indicate that there may be infested materials in the area, but not necessarily very near the sighted insects because the flight ranges of moths can be relatively large. Crawling or flying beetles, on the other hand, may indicate an infestation of foods in close proximity to the insects. Once the source of the insect is located, the investigator can collect as many different types and life stages of the insects as are needed to document the infestation.

2. Life Cycle/Stages of Insects

One of the most critical pieces of evidence is the description and collection of as many life stages of the insects as possible. The expert witness can use this evidence to show when, where and/or how the infestation occurred. It will also indicate the duration of the infestation. Such evidence in many cases will be the determining factor in proving the duration of the infestation.

3. Location of Insects

The expert witness must know not only the identity and life cycle of the insects found, but also the precise location where the insects were discovered and collected. Without this information, a 402(a)(3) charge (see chapter 2) may not be possible. Using a food warehouse inspection as an example, bag seams should be examined and signs of insect boring through cardboard boxes and cellophane wrappers should be noted. If insects are found on or in shipping cartons, the individual packages inside the cartons should be examined for infestation.

Some stored-product pests, such as dermestids, can frequently be found in areas such as fuse boxes, gear housings, etc. Other insects, such as confused flour beetles, tend to stay in the food they are infesting, and others, such as the sawtoothed grain beetle, can frequently be seen rapidly crawling on the outside of boxes of cookies. The location of the insects along with the types of insects found are evidence of the extent of infestation. For example, infestations of insects that rarely fly or cannot fly will probably be somewhat localized, whereas infestations of insects that are good flyers can be expected to be more widespread.

The few examples listed above illustrate how important the ecology of stored-product insects can be in the development of evidence.

III. ANALYTICAL EVIDENCE

The microanalytical entomology laboratory is probably one of the most important links in evidence development. Laboratory scientists provide scientific support by testing and examining the evidence collected by the investigator during an inspection. Other types of services provided by the food sanitation section within the laboratory include the analysis of consumer complaints, suspected tampering, surveillance samples and research ranging from methods development to insect fragment identification papers. On occasion, the lab is called on to identify unknowns, foreign objects, solvents, etc. In this section, the focus will be primarily on entomology and biological sciences as they relate to food contaminants, both avoidable and unavoidable elements.

A. PROCEDURAL ISSUES

The first and probably most important steps taken by any analyst prior to beginning the analyses center around the sample controls, quality assurance, environmental and reagent checks. These checks within the food sanitation section would involve maintaining an in-house reference collection (see chapter 14) of analysts' body hairs, including that of their pets (used to rule out sample contaminations by the analyst), running consecutive reagent blanks, placing filters on water sources, sample controls such as backwashing sieves, pre-rinsing glassware, and maintaining a 702(b) sample reserve. Most of these checks and balances can be taken care of through properly documented laboratory standard operating procedures (SOPs) and quality assurance programs (QAPs). Notes/comments on the backs of worksheets attesting to these steps will enable the analyst when questioned to verify that yes, in fact, checks and controls including SOPs and QAPs were followed during the analyses to ensure sample integrity.

B. DOCUMENTATION OF ANALYTICAL RESULTS

Now that the analytical integrity has been established, the next most important step is the worksheet write-up—documentation. People who possess great analytical expertise but cannot document their work and findings onto worksheets are of no use to the FDA. Documentation in the laboratory involves the completion of analytical worksheets. These forms are used to show the continuation of sample integrity through to analytical completion and compliance. The analyst records the condition of the sample when received, the amount received, from whom, the sample number and, in more detail, exactly what they received. The methods used are stated, the results are calculated and tabulated and, finally, the reserve sample is identified and its disposition stated. These steps are a must to show the chain of custody, the ability to reproduce the scientific work in cases where it may be challenged and to put into writing the condition of the product and description of the exhibits identified in order to show our unbiased approach in obtaining the results.

The back of the worksheet is used to show calculations, initial results, and to make notations on observations. This information may be included in the final results write-up. Quality controls that are incorporated into the analyses are also noted including the results of these checks. On completion of the work, the analyst completes the front of the worksheet and transfers the final results from the back to the front. The worksheet, like the collection report prepared by the investigator, is a written testimony as to the work performed and the written results from observations. It is critical that the work performed, including the condition of the sample and the results obtained, are properly and correctly identified and completed. When and if an analyst is contacted to verify his or her work on a particular sample, more often than not, several years have passed since the work was done. So, the more detailed and accurately written the worksheets, the easier it will be for the analyst when explaining the steps taken and possibly testifying in court.

One final comment under worksheet documentation—be consistent! When preparing worksheets, wording should be consistent with the stated methodologies, texts, and results of fellow analysts. For example, evidence presented in a recent FDA court case involving a $15 million seizure of cocoa beans held in five warehouses showed that the buildings were insect/ rodent infested with insects in various life stages and that the product showed evidence of feeding, tunnelling, etc. The three analysts working on the samples stated that the commodity was "insect defiled," "insect infested" and "insect damaged." The defendants did not challenge the science nor the results of the FDA's case but instead focused on the inconsistency of the terminology used. Needless to say, the lawyers for the defendants had a field day with the government witnesses in trying to show inconsistencies with the case when, in fact, all three statements were finally shown to mean the same thing. Be consistent!

C. PERFORMING THE ANALYSIS

Entomology, as it applies to the FDA is the study of insects found contaminating foods. Although insects are considered a food staple in many parts of the world, sometimes even imported as delicacies (e.g., canned silk worm larvae, steamed ant eggs and frozen water bugs), the FDA's concern is with the insects found contaminating foods and their relationship to insanitation of food. Unlike with drugs, devices and a few other areas that the FDA regulates, the FDA does not have Memoranda of Understanding (MOUs) which allow investigators overseas to inspect and document the condition of food firms prior to their trip to the United States. Thus, the FDA relies on insects as one way of indicating sanitary conditions or the lack thereof—forensics.

In most cases of insanitary conditions, insects are the primary filth elements involved. Other contaminants that analysts may encounter include rodents, birds, mammals (body parts, hairs, excreta, for example), other arthropods, foreign objects, decomposition/molds, and unknowns. The insects encountered could be from the natural growing habitats, intercepted during harvesting, during transportation or storage, or even collected from store shelves or from the consumer's possession. They may be classified as either filth/storage or field/ incidental types. The insects known as filth/storage types are those that thrive and reproduce under manmade conditions such as those found in warehouses, bakeries and canneries. Some inhabit filthy, dirty environments including excrement and carrion. The insects (or fragments) for which identities are unknown, or from field environments, may be classified as incidentals because of the nature of product contamination. Filth/storage types may be encountered outside buildings but are not classified as field insects because of their ability to thrive and reproduce inside establishments.

The FDA entomologists and other analysts who perform sanitation analyses are unique forensic scientists by definition of the work that they perform. They are called on to identify insects by their feeding, damage or physical presence including their fragments, and by combining the corresponding written studies on their biology, habits, etc., to provide testimony as to the conditions that lead to the infestation for those species. Additionally, these analysts may interpret the time of infestation, types of processing performed, and origins of contamination, by using the scientific literature available and the observations during the inspections, and data from the worksheets regarding the insect filth size ranges, life stages, etc. This information is then relayed to the compliance branch for regulatory consideration.

The approach to take before beginning the analysis for suspected insect contamination is as follows: once the sample is received and the proper documentation, controls and paperwork are started, organize the lab bench with the required glassware and reagents for the analyses. Be sure that the microscope work station is equipped with an adjustable chair and properly adjusted and aligned compound and stereomicroscopes and that micro-tools, microscope slides/cover slips and a hot plate are available. Then ask yourself these questions: What is the commodity? Why was it collected? What would adulterate the product? What is the current condition of the goods and what was it when it was collected? These questions lay the groundwork prior to beginning the analyses. Use these questions to formulate an approach to the sample.

If the food is preserved by air drying, then live insects may be encountered, as is most often the case with dried fish. If the food is packaged in a glass jar or a can or has received additional treatment such as cooking or freezing, then dead whole insects or just fragments may be found. For instance, if a canned fish product were analyzed and fly maggots discovered, by identifying the maggots to the species level (if they are third instar), a determination can be made as to the time the infestation occurred prior to the packaging process. If you discovered live insects, then the food item was adulterated after the treatment, probably in transit or subsequent storage, and you should look for entry/exit holes into the containers.

To select the correct filth extraction procedure, the analyst will need to know what is being collected and for what reason? Did the investigator observe objectionable conditions during the inspection (compliance samples) or was the product sampled as a result of a routine inspection (surveillance samples)? When an investigator does collect physical evidence to document existing filth conditions at the firm, these types of samples are called filth exhibits. These exhibits may be confirmed in the lab using available authentic specimens from the lab's vial collection, comparisons with reference books or, in some cases, consultation with outside experts or even institution collections or museums (see chapter 3). If a food product was collected and there is no existing methodology for extracting filth, then consider the product and country of origin as to what types of insects might be found. What is the physical state of the product itself? Coarse ground versus finely ground? For example, if the product was milled and the suspect was grain insects, then be on the lookout for ground insect parts, and the method would involve some type of chemical extraction procedure. If the product was sun dried and contained a fruit base with sugars, one may find insects that would be feeding on the outside of the product and on the sugar coating and not boring types, so a simple washing procedure may be all that is needed as a filth extraction method. Be sure to record each step of the analytical method that is performed, including the sample size or counts, glassware, reagents, steps, and time elements to allow for duplication of recovery procedures. Remember, with proper methodology, it is the responsibility of the analyst to ensure that the validity of the results will stand up in the court of law unchallenged. For example, when the defendants are presented with the FDA's evidence during discovery, most often they agree to accept the findings and begin to negotiate a settlement without going to trial.

To document the results clearly, accurately and chronologically, identify the filth by first examining the integrity of the package, noting entry/exit holes, tears, and gnawing marks. Be sure to state number of openings seen and the dimensions. Identify those whole specimens collected as exhibits or encountered during the analyses, then the extracted filth on the filter papers. Record the identifications using the current Latin and/or common names if available, state which life stage, whether alive or dead, if intact, and be sure to give accurate numbers and size ranges. For example, the sawtoothed grain beetle, *Oryzaephilus surinamensis* L., has a small strain variety only found in certain parts of the Mediterranean and Far East. Thus, when the size ranges are determined, the lower limits may help identify this pest as the small strain variety. Many times, the whole insects encountered are also the fragments extracted from the product, so identification of the whole insects makes identifying the fragments that much easier.

When the work is completed, the analyst will have recorded his or her observations and findings from the moment the sample was signed out through to sample descriptions, method extractions and recording of the results. Then the work is verified and confirmed by a second analyst prior to leaving the laboratory branch. One of the benefits of this type of lab work is the fact that the findings are physical, making the confirmations easier and quicker. It is extremely important that the same counts, sizes, and other results are confirmed. When verifying approximations, be sure to ask the original analyst what formula was used and duplicate it. Remember, be consistent.

IV. COMPLIANCE

As with every forensic science organization, the FDA has a responsibility to be fair and unbiased when rendering its decisions to the defendant and to protect the consumer when pursuing a legal action. The last stage of review of a sample package within a field office is the compliance branch. They have the assignment of interpreting the findings submitted from the investigations and laboratory branches and determining if there is enough evidence on which to build a case and, if so, what type of regulatory action is to be taken.

They are charged with reviewing the submitted data for accuracy and sample integrity (including chain of custody) and ensuring that proper techniques were employed during sample collection and analyses, and that results were verified, confirmed and accurate. The final recommendation is then forwarded to headquarters for their concurrence and routing to the agency's attorneys at the Office of the General Counsel.

Chapter 17

REFLECTIONS ON FOOD-BORNE FILTH IN RELATION TO HUMAN DISEASE

J. Richard Gorham

CONTENTS

I. Introduction 269

II. What Filth Means 270

III. The Pure Food Law 270
- A. Filth and Disease 270
- B. Defect Action Levels 270

IV. Reactions to the Food, Drug and Cosmetic Act 270
- A. Indicators of Insanitation 271
- B. Hazards Tolerated 271
- C. Evidence of Hazards 271
- D. Toxins 272
- E. Allergens 272
- F. Pathogens 272

V. Integrated Pest Management 272
- A. Pesticides and Biocontrols 272
- B. Enhanced Expertise 273
- C. Nutritional Losses Due to Pests 273
- D. Management of Infestible Foods 273

VI. High-Risk Situations 273

References 274

I. INTRODUCTION

Filth in food comes from the pests that contaminate food—insects, mites, birds and rodents. Although the presence of minute quantities of filth in food is inevitable, the federal Food, Drug and Cosmetic Act (FD&C Act) prohibits food contamination by filth. The Food and Drug Administration (FDA) has taken a pragmatic approach in establishing defect action levels (DALs), which represent current industry standards for food purity. Although much of the food industry conforms to high standards, incidents of gross adulteration continue to occur. Some contaminants may not be altogether harmless to the consumer. The presence of high levels of filth in food indicates that at some point the food has been subjected to insanitary handling. Depending on whether the contamination occurred before or after the last heat-processing (sterilizing, biocidal) step in manufacture or preparation, the significance of filth in food ranges from merely aesthetically unappealing to allergenic, toxic, pathogenic or directly traumatic. Much remains to be learned about the significance for human health of filth in food.

II. WHAT FILTH MEANS

Since the health significance of filth in food has been reviewed previously, only the more salient points will be discussed in this chapter.[1-3] The term *filth* as used here is defined in the *Official Methods of Analysis* of AOAC International as extraneous matter (things that obviously do not belong in food) derived from mites, insects, birds, rodents and other animals.[4] This filth (more specifically, "light filth") may be in the form of recognizable particles (such as whole insects and mites, sufficiently complete as to warrant a name), fragments of insects and mites, feather barbules, hairs and feces, or in a form so finely dispersed, such as urine, that chemical rather than mechanical or visual means are required to separate it from the food and identify it.

III. THE PURE FOOD LAW

Around the turn of the century, disregard for the public's welfare by certain persons in the food and drug trade was so blatant, food and drug adulteration so common and the public outcry against these practices so plainly heard in Washington that Congress reacted decisively to create a new, progressive and strict standard for food and drug sanitation, a standard that has protected public health for more than 88 years.

A. FILTH AND DISEASE

Although the Food and Drug Act of 1906 and its subsequent revisions are ostensibly public health measures, these laws do not require the government to prove a causal relationship between filth in food and illness.[5] That relationship, assumed in the law, has been upheld as valid with remarkable uniformity by the courts.

B. DEFECT ACTION LEVELS

It is difficult, without unconscionable waste and exorbitant expense, to produce filth-free food. The "pure food law," however, permits no filth in food [402(a)(3)] and no filth around food [402(a)(4)] that enters interstate commerce or is imported from other countries.[5] This administrative dilemma has been resolved by allowing low levels of filth in certain foods, thus sparing the consumer unnecessary expense, while adhering at least to the spirit, if not precisely to the letter, of the law. Any filth in food present at this allowable level, the DAL, or lower is declared to be unavoidable with regard to production and innocuous with regard to health.[6] The DALs, directed principally at pests that originate in farm fields, show no leniency toward storage pests.

Many things over which we may have no control are much more life-threatening than filth in food. Automobile accidents, for example, are a far more serious public health problem than filth in food. But food intake is a universal requirement. Since most of us have little chance to exercise control over the specific ingredients (especially the unstated ones) in the food we buy, we consumers should be able to assume the purity of food we purchase from commercial sources.

IV. REACTIONS TO THE FOOD, DRUG AND COSMETIC ACT

Criticisms of parts 402(a)(3) & (4) of the FD&C Act come from several directions.[5] Some critics claim that there is no need for the government to inspect foods because modern food processing facilities are so technically advanced and so expertly managed that it is highly improbable that filthy food would emerge from them, and that today's food processors are so image-conscious they would not risk tarnishing their corporate reputations by producing filthy food. There is much truth in that claim. Food sanitation has never been better in this country than it is today. But one need only look at the list of enforcement actions reported in each issue of the *FDA Consumer* or in the FDA's *Monthly Import Detention List* to realize that food adulteration is still a common occurrence.

Other critics allege that no harm can come from eating many kinds of filth in food; eating flour beetles is comparable, they might suggest, to eating shrimp. Again, there is truth in that claim. It is well known that many people around the world eat certain insects by choice whenever they can, and that certain insects are rich in calories and various nutrients important in human nutrition.[7,8]

It has been suggested that the dietary use of insects could be greatly expanded, both by eating insects directly and by processing them (in much the same way that fish protein concentrate is prepared) and incorporating them into other foods (see chapter 13).[7] It is necessary here to make a careful distinction between defective sanitation on the one hand and good manufacturing practice (GMP)-style food processing on the other. A random cockroach accidentally embedded in today's lunchtime yogurt would indicate insanitation and would be a violation of 402(a)(3). Greater wax moth larvae (*Galleria mellonella*) (see chapter 7) reared under sanitary conditions, heat-sterilized, dried, pulverized and blended with pancake mix for commercial distribution, all done according to an approved GMP protocol (to cite a hypothetical example), would represent good sanitation.

A. INDICATORS OF INSANITATION

Certain mites, insects, birds, bats, rodents and other animals are indicators of insanitation. Sparrows in a food warehouse, rats in a grocery store, sawtoothed grain beetles in a flour mill and roaches in a restaurant—all these and many others are indicators of insanitation. The implication is (and this is often borne out during regulatory inspections) that if the management of a food establishment cannot keep pests under control, it may also permit other lapses in its quality control program.

Critics of 402(a)(3) often point out that people in other cultures avidly eat insect-infested and pest-damaged foods and do so with no apparent ill effects. Furthermore, the revulsion experienced by many people in the United States when they find their food to be insect infested is, the critics suggest, purely a culturally biased, aesthetic reaction, a sort of taboo that has no scientific basis.[9] Again, there is some truth in these criticisms, but the arguments are not as definitive as they might at first seem.

B. HAZARDS TOLERATED

The threat to health associated with eating insect-infested food in a person who has grown up eating such foods is minimal, or at least much less than it would be for a citizen of an affluent society where ingestion of high levels of insect filth would be unusual. But it must be remembered that the person accustomed to eating insects has developed physiological defenses to a degree that he or she can successfully cope with most of the allergenic, toxigenic and pathogenic hazards attendant to eating such foods. It is also likely that our "healthy" eater of insect-infested foods is but a remnant of a larger age-class, some of whom did not survive such environmental perversities as consuming insect-infested foods in infancy and childhood.

C. EVIDENCE OF HAZARDS

If we had to judge the health significance of insect-infested or pest-damaged food in the United States based on the number of reported illnesses resulting from eating such food, we might easily conclude that there is no health significance. But three qualifying thoughts color that conclusion: (1) There is no reliable information about case rates of illness associated with eating insect-infested or pest-damaged foods; (2) if the case rate actually is low in the United

States, that could easily be due to the widespread use of GMPs in the food industry and to the vigorous enforcement of parts 402(a)(3) & (4) of the FD&C Act, as well as comparable local and state statutes; and (3) although the evidence is meager, there is reason to believe that toxic, allergenic and pathogenic hazards do attend the consumption of certain pest-infested or pest-damaged foods or certain foods that have been in contact with food pests.[10-30]

D. TOXINS

Biologically active compounds, either directly toxic or tumorigenic, are produced by certain cockroaches, beetles and other insects and mites associated with food.[11,16-19,31] Attacks by pests on crops in the field may cause some food plants to produce phytoalexins and may predispose other food plants to fungal invasion, proliferation and subsequent production of mycotoxins.[10,13,14,32]

E. ALLERGENS

Bernton and Brown discovered an allergen from cockroaches that remains allergenic after heating for one hour at 100°C.[21] They also showed that some people have antibodies to certain stored-food insects.[20]

F. PATHOGENS

Numerous pathogens, recovered from foods and food-contact surfaces, have been traced to various kinds of food pests: ants, roaches, flies, birds, rodents and many others.[24-26,33] Although the number of documented cases of human illness caused by pathogens carried by food pests is very small, the habits of food pests, principally their indiscriminate travels between filth and food or food-contact surfaces, are so well known that the presence of these pests in food establishments is inconsistent with acceptable sanitary practice.

V. INTEGRATED PEST MANAGEMENT

A. PESTICIDES AND BIOCONTROLS

Pesticides are still the most effective practical tool for reducing contamination and damage due to pests in food crops in the field, in storage and during processing.[34] Although pest infestation and contamination of food are likely to persist for many years to come, some success in pest control might eventually be achieved by integrating chemical and biological controls. Although integrated pest management might be advantageous to the environment by reducing chemical pesticide usage, consumers might have to accept food with higher levels of pest infestation (filth) and higher levels of pest damage (levels of control achieved using biocontrol agents are typically much lower than those commonly achieved by the use of "hard" pesticides). Under some circumstances, there could also be an added arthropod component in the food in the form of insects and mites that had been used as biocontrol agents.

The fact that there is no experimental evidence that biocontrol agents are safe with regard to human health set the stage for a regulatory dilemma. One approach would have been to require an FDA Food Additive Petition, which would have allowed predatory and parasitic arthropods to be added to human food for the purpose of controlling specific stored-food pests (see chapter 8).[35] That alternative, however, was abandoned in favor of a proposal from the Environmental Protection Agency (EPA) that defines arthropod biocontrol agents as "pesticides" within the meaning of the Federal Insecticide, Fungicide, and Rodenticide Act, and then exempts them from having to meet any particular tolerance levels in bulk commodities. Any foods derived from bulk commodities "treated" with these biological "pesticides" would still have to satisfy the standards of the food DALs.[36] Experiments are underway to determine the effectiveness of some of the biocontrol agents exempted by the EPA.[37]

B. ENHANCED EXPERTISE

If integrated pest management, at least with regard to employing animate biocontrol agents as "pesticides," were ever to be broadly applied to the food-production industry, it would be necessary for food-purity enforcement personnel to acquire the expertise that would enable them to more clearly distinguish between pests (and their fragments and metabolic products) of sanitary importance and those extraneous elements (the biocontrol agents) in foods that are now presumed to be innocuous. The situation could become even more complex, and therefore demand greater expertise to resolve, if traditional foods were deliberately, safely and legally enriched with insect tissues.

C. NUTRITIONAL LOSSES DUE TO PESTS

Regrettably, many people around the world have, on occasion, been happy to have even insect-infested food to eat. (Bird- and rodent-contaminated foods seem to have few champions anywhere.) However, it must be recognized that any pest infestation or pest damage of food is invariably accompanied by a decline in the nutritional quality of that food.(12,15-17,31)

D. MANAGEMENT OF INFESTIBLE FOODS

There should be four goals in dealing with infestible foods. First, every effort should be expended to prevent damage, infestation and contamination of foods. Second, any infestation should be stopped at the earliest possible moment, preferably before the food becomes inedible. Third, food that is declared unfit for human consumption because of pest damage, infestation or contamination should, whenever possible, be converted to animal feed or to compost. Any live pests in the food should be killed as soon as possible to prevent their spread to other commodities. Fourth, if farmers have in the past used pesticides merely to reduce infestations of crop pests such as aphids or thrips (which are presumably safe for people to eat) just to stay within the DAL guidelines, as Pimentel and others suggest has happened, then the benefits and risks of that practice will have to be carefully re-evaluated.(38)

VI. HIGH-RISK SITUATIONS

Inspectors or other quality control officials must be watchful for those circumstances in which pathogens carried by food pests or the toxins produced by them may survive or persist long enough to affect humans. Although filth in food that goes through commercial sterilization will be pathogen-free, this kind of processing does not guarantee destruction of toxins or allergens (some allergens and toxins are heat stable).

The most dangerous situations are those in which a pathogen gains access to a food, either by direct carriage by a pest or indirectly from a food-contact surface contaminated by a pest, after the last biocidal step in food processing, or when there is no processing step that is fatal to pathogenic microorganisms or denaturing to toxins or allergens.

Klowden and Greenberg(27) developed experimental evidence that is highly pertinent to the question of the health significance of filth in food.

> The post-mortem persistence (up to 60 days) of *Salmonella* in the carcasses of cockroaches infected before their death is important in view of the minimal levels of "filth" composed of insect parts and rodent hairs which food processors are allowed, by federal law, to include in their foods. Cockroach remains are undoubtedly major contributions to this tolerable filth. Unpasteurized foods, such as spices and packaged cereals, might be "seasoned" with a portion of infected cockroach carcass, and become a route of *Salmonella* dissemination. The incorporation of contaminated cockroach debris into the environment of the home, restaurant, and food processing plant is another avenue which has not been explored.

A classic case of human disease associated with food-borne filth occurred in Indiana in 1964.[28] Specimens of *Trogoderma* (dermestid beetles) were recovered from the feces of a 4-month-old child who had eaten a high-protein baby food (a packaged, dry cereal). The child's illness, ulcerative colitis (an unusual disease in infants), was tentatively attributed to some pathogenic effect, perhaps mucosal invasion by the hastisetae, of the larval beetles.

REFERENCES

1. Gorham, J.R., Filth and Extraneous Matter in Food, in *Encyclopedia of Food Science and Technology*, Hui, Y.H., Ed., Wiley-Interscience, NY, 847-868, 1991.
2. Gorham, J.R., Food Pests as Disease Vectors, in *Ecology and Management of Food-Industry Pests*, Gorham, J.R., Ed., AOAC International, Arlington, VA, 477-482, 1991.
3. Gorham, J.R., Food, Filth, and Disease: A Review, in *Foodborne Disease Handbook*, Vol. 3, Hui, Y.H., Gorham, J.R., Murrell, K.D., and Cliver, D.O., Eds., Marcel Dekker, NY, 627-638, 1994.
4. Helrich, K., Ed., *Official Methods of Analysis of the Association of Official Analytical Chemists*, 15th ed., Vol. 1, Assoc. Off. Anal. Chem., Arlington, VA, 1-684, 1984.
5. Food and Drug Administration, *Federal Food, Drug and Cosmetic Act, as Amended*, DHHS Publication No. (FDA) 93-1051. U.S. Dept. Health and Human Services, Rockville, MD, 1-223, 1993.
6. Food and Drug Administration, *The Food Defect Action Levels*, Food and Drug Administration, Washington, DC, 1-14, 1989.
7. Tippett, R.C., The need and practicality of using insects as a source of food, *Maryland Entomol.*, 1, 2-6, 1978.
8. DeFoliart, G.R., Insects as a source of protein, *Bull. Entomol. Soc. Am.*, 21, 161-163, 1975.
9. Olkowski, H. and Olkowski, W., Entomophobia in the urban ecosystem, *Bull. Entomol. Soc. Am.*, 22, 313-317, 1976.
10. Bailey, J.A. and Skipp, R.A., Toxicity of phytoalexins, *Ann. Appl. Biol.*, 89, 354-358, 1978.
11. Bignell, D.E. and Mullins, D.E., A preliminary investigation of the effects of diets on lesion formation in the hindgut of adult female American cockroaches, *Can. J. Zool.*, 55, 1100-1109, 1977.
12. Cusak, P.D., Evans, G.O., and Brennen, P.A., The origin and sources of mite infestation of stored grain and related products in the Republic of Ireland, *Ann. Appl. Biol.*, 82, 178-179, 1976.
13. Doherty, J. and Buescher, R., Occurrence of the phytoalexin phaseollin in pods of *Phaseolus vulgaris*, *J. Sci. Food Agric.*, 29, 853-856, 1978.
14. Mislevic, P.B., Toxigenic Fungi in Foods, in *Mycotoxins in Human and Animal Health*, Rodricks, J.V., Hesselline, C.W., and Mehlman, M.A., Eds., Pathotox, Park Forest, IL, 469-477, 1977.
15. Pant, K.C. and Susheela, T.P., Effect of storage and insect infestation on the chemical composition and nutritive value of grain sorghums, *J. Sci. Food Agric.*, 28, 963-970, 1977.
16. Rajan, P., Effect of feeding diet containing insect infested maize and cowpea on the growth and histological changes in livers of albino rats, *Nutr. Rep. Int.*, 13, 347-354, 1976.
17. Warner, C.E. and Bohane, S.J., Preliminary study of a diet infested with *Acarus siro* (L.) on adult food consumption and foetal status in mice, *J. Stored Prod. Res.*, 14, 13-17, 1978.
18. Wirtz, R.A., Taylor, S.L., and Semey, H.G., Concentrations of substituted *p*-benzoquinones and l-pentadecene in the flour beetles *Tribolium confusum* J. du Val and *Tribolium castaneum* (Herbst), *Comp. Biochem. Physiol.*, 61B(1), 25-28, 1978.
19. Wirtz, R.A., Taylor, S.L., and Semey, H.G., Concentrations of substituted *p*-benzoquinones and l-pentadecene in the flour beetles *Tribolium madens* (Charp.) and *Tribolium brevicornis* (LeC.), *Comp. Biochem. Physiol.*, 61C(2), 287-290, 1978.
20. Bernton, H.S. and Brown, H., Insects as potential sources of ingestant allergens, *Ann. Allergy*, 25, 381-387, 1967.
21. Bernton, H.S. and Brown, H., Insect allergy: The allergenic potentials of the cockroach, *Southern Med. J.*, 62, 1207-1210, 1969.
22. Herranz, G. and Herranz, M.P., Hypersensitivity reaction to the ingestion of mites (*Tyrophagus farinae*): Pathogenic study of a fatal case, *Rev. Med. Univ. Navarra*, 7, 137-147, 1963 (in Spanish, English summary).
23. Zschunke, E., Contact urticaria, contact dermatitis, and asthma from cockroaches, *Arch. Dermatol.*, 114, 1715-1716, 1978.
24. Berg, R.W. and Anderson, A.W., Salmonellae and *Edwardsiella tarda* in gull feces: A source of contamination in fish processing plants, *Appl. Microbiol.*, 24, 501-503, 1972.

25. Ebeling, W., The Cockroach as Implicated in Public Health, in *Proceedings of the 1977 Seminar on Cockroach Control*, Westphal, K., Ed., Bureau of Rodent Control, New York State Department of Health, Albany, NY, 15-21, 1977.
26. Frishman, A.M. and Alcamo, I.E., Domestic cockroaches and human bacterial disease, *Pest Control*, 45(6), 16, 18, 20, 46, 1977.
27. Klowden, M.J. and Greenberg, B., Effects of antibiotics on the survival of *Salmonella* in the American cockroach, *J. Hyg.*, 79, 339-345, 1977.
28. Okumura, G.T., A report of canthariasis and allergy caused by *Trogoderma*, *Calif. Vector Views*, 14(3), 19-22, 1967.
29. Stek, M., Peterson, R.V., and Alexander, R.L., Retention of bacteria in the alimentary tract of the cockroach, *Blattella germanica*, *J. Environ. Health*, 41, 212-213, 1979.
30. Wallace, G.D., Intermediate and transport hosts in the natural history of *Toxoplasma gondii*, *Am. J. Trop. Med. Hyg.*, 22, 456-464, 1973.
31. Swaminathan, M., Effect of insect infestation on weight loss, hygienic condition, acceptability, and nutritive value of foodgrains, *Indian J. Nutr. Diet.*, 14, 205-216, 1977.
32. Widstrom, N.W., Aflatoxin contamination of maize as it relates to insect damage, *J. Environ. Qual.*, 8, 5-11, 1979.
33. Kopanic, R.J., Jr., Sheldon, B.W., and Wright, C.G., Cockroaches as vectors of *Salmonella*: Laboratory and field trials, *J. Food Prot.* 57(2), 125-132, 1994.
34. Furtick, W.R., Uncontrolled Pests or Adequate Food? in *Pesticides and Human Welfare*, Gunn, D.L. and Stevens, J.G.R., Eds., Oxford Univ. Press, NY, 3-13, 1977.
35. Anon., Food Additive Petition needed for adding beneficial insects to food, *Food Chem. News*, 30(21), 7-9, 1988.
36. Anon., Predaceous insects in grain EPA exemption OK-ed by FDA, *Food Chem. News*, 32(45), 48-50, 1991.
37. Flinn, P.W., Hagstrum, D.W., and McGaughey, W.H., Suppression of Insects in Stored Wheat by Augmentation with Parasitoid Wasps, in *Proceedings of the 6th International Working Conference on Stored-Product Protection*, Vol. 2, (17-23 April, Canberra), Highley, E., Ed., CAB International, Wallingford, 1103-1105, 1994.
38. Pimentel, D., Pesticides, insects in foods, and cosmetic standards, *BioScience*, 27, 178-185, 1977.

SUBJECT INDEX

A

Acacia ant, 115
Acalyptrate flies, 84, 86, see also specific types
Acaroid mites, 124-129, see also specific types
Acid alcohol solutions, 246
Adulteration, defined, 2
Aerators, 243
Aerial web spiders, 143, see also specific types
Alcohol solutions, 246
Allergens, 41, 271–273
Almond moth, 94, 98
Aluminum acetate, 247
Ambrosia beetle, 63–64
American cockroach, 33–34
American house dust mite, 124
American house spider, 137, 148
American spider beetle, 63
Ametabolous insects, 29–37, see also specific types
Analytical documentation, 17
Analytical evidence, 264–267
Angora rabbit, 190, 195
Angoumois grain moth, 57, 94–95, 99
Anthomyiid flies, 74, see also specific types
Antiforam A, 246
Antlike flower beetles, 49, 52–53, see also specific types
Ants, 8, 105–115, see also specific types
- acacia, 115
- Argentine, 110–111
- army, 111
- bigheaded, 114
- cornfield, 112
- crazy, 109, 113
- disease and, 272
- families of, 109–115
- fire, 108
- fragments of, 111
- habitats of, 110–115
- house, 110–111
- human consumption of, 113, 222
- identification of, 108–109
- morphology of, 106–107
- New World army, 111
- odorous house, 110–111
- pavement, 109, 115
- pharaoh's, 107–108, 114
- pyramid, 110
- significance of, 107–108
- subfamilies of, 109–115
- synanthropic, 106, 110–111, 113
- thief, 114
- tree, 110
- velvety tree, 110
- weaver, 222–223

Argentine ant, 110–111
Army ant, 111
Asian cockroach, 33
Asian musk shrew, 7–8, 161–162, 171, 184–185, 192
Astigmatid mites, 124
Asynanthropic flies, 87, see also specific types
Australian cockroach, 33–34
Autoclaves, 243

B

Bacteria, 33, 72, 177, see also specific types
Bamboo powderpost beetle, 53
Banana spider, 137
Bandicoot rat, 161, 183
Bark beetles, 49–50, 54, 56–58, 63–64, see also specific types
Bat parasites, 88, see also specific types
Bats, 7, 8, 88, 124, 159, see also specific types
- brown, 162, 186, 201
- disease and, 271
- free–tailed, 32
- habitats of, 88, 159
- hair from, 159, 162, 169–170, 186, 201, 237
- little brown, 162, 186, 201
- pipistrelle, 186

Bean weevil, 41, 53
Bed bugs, 162
Bee louse, 88
Bees, 22, 106, 121, 125, 218–219, see also specific types
Beetles, 25, 39–66, see also Weevils; specific types
- allergens and, 41
- ambrosia, 63–64
- American spider, 63
- antennae of, 44
- antlike flower, 49, 52–53
- bamboo powderpost, 53
- bark, 49–50, 54, 56–58, 63–64
- carpet, 59, 86, 152
- categories of, 41
- checkered, 49, 55–56
- cigarette, 43, 49, 52
- clerid, 8
- common names for, 49–52
- confused flour, 263
- confused sap, 62
- corn sap, 62
- cosmopolitan species of, 48
- countries of origin of, 40
- darkling, 51–52, 64–65
- dermestid, 44, 50, 59–60
- disease and, 271–272
- driedfruit, 44, 62
- drugstore, 49, 52
- equipment for study of, 41–42
- excreta from, 60
- false powderpost, 49, 53
- families of, 48–66
- featherwinged, 62–63
- flat bark, 49–50, 56–58

flat grain, 57–58
flour, 43–44, 65, 263, 271
flower, 49, 52–53
fragments of, 42–45, 53, 59
fungus, 43, 55–56, 60, 62, 64–65
genitalia of, 42, 45, 58–59
grain, 43, 49–50, 54, 57–58, 65, 94–95, 266
ground, 45, 49, 54
habitats of, 59, 61, 63, 65–66
hairy cellar, 65
hairy fungus, 62
handsome fungus, 65
hide, 44
hister, 60
hump, 63
identification of, 45–48, 58, 60–65
khapra, 59
lizard, 60–61
minute bark, 54
minute brown scavenger, 61
minute fungus, 56
minute tree fungus, 55
morphology of, 42–45
oval grain, 54
parasitoids of, 120
pleasing fungus, 60
ports of entry of, 40
powderpost, 49, 53, 61–62
red flour, 43, 65
redlegged ham, 44
reference collections of, 230
rove, 8, 64
rusty grain, 57
sap, 51, 62
sawtoothed grain, 43, 57, 266
scavenger, 61
seed, 49, 53–54
Siamese grain, 58, 65
sightings of, 263
significance of, 41
silken fungus, 56
spider, 51, 63
squarenecked grain, 57
squarenosed fungus, 43
techniques for study of, 41–42
thorax of, 44–45
tropicopolitan species of, 48
Turkish grain, 57
twohanded fungus, 64
varied carpet, 59
whitemarked spider, 63
yellowbrown sap, 62
Benzoquinones, 41, see also specific types
Bigheaded ant, 114
Biocontrols, 272
Birds, 88, 124–125, 177, 237, 265, 270–271, see also Feathers; specific types
Biting midges, 74, 81, see also specific types
Black blow fly, 81
Black Death plague, 160
Black earwig, 32
Black rat, 160
Black scavenger flies, 74, 86, see also specific types
Black striped filth fly, 82
Black widow spider, 137, 148
Blenders, 243
Blow flies, 74, 80–81, 85, see also specific types
Bluebottle flies, 81
Booklice, 29, 35–37, 152
Book pseudoscorpion, 152
Boreal combfoot spider, 137
Borers, 44, 53, 63, see also specific types
Bot fly, 72
Broadnosed grain weevil, 58
Brownbanded cockroach, 33
Brown bats, 162, 186, 201, see also specific types
Brown house moth, 94, 99
Brown rat, 160
Brown spiders, 137, 147, see also specific types
Brown widow spider, 137
Bruchids, 53–54, see also specific types
Bubonic plague, 161
Bugong worm, 218

C

Calcium chloride, 246
Canthariasis, 6
Carcinogens, 41, see also specific types
Cardinal spider, 137
Carnid flies, 74, see also specific types
Carob moth, 98
Carpet beetles, 59, 86, 152, see also specific types
Casemaking clothes moth, 94, 98
Caterpillars, 94, 97, 99, 219–220, 222, see also Moths; specific types
Cat hair, 158, 162, 173, 175, 237
pictures of, 187–189, 192–193, 201, 203
Cave spiders, 137, 145, see also specific types
Ceiling spider, 137
Cellar spiders, 137, 146, see also specific types
Centipedes, 136, 154, see also specific types
Checkered beetles, 49, 55–56, see also specific types
Cheese skipper fly, 84
Chemical disposal, 254
Chicken mite, 124
Chortoglyphid mites, 131
Cigarette beetle, 43, 49, 52
Classification vs. identification, 22–23
Clearing reagents, 235
Clerid beetles, 8, see also specific types
Clothes moths, 94, 98–99, 152, see also specific types
Clover mite, 124, 132
Cluster fly, 81
Cobweb spiders, 137, 148, see also specific types
Cobweb weaver spiders, see also specific types
Cocker spaniel, 193
Cockroaches, 8, 25, 29, 33–35
allergens and, 33
American, 33–34
Asian, 33
Australian, 33–34

brownbanded, 33
cats and, 162
disease and, 33, 271–273
dogs and, 162
eggs of, 119
excreta from, 33
families of, 33–34
food–borne disease and, 33
fragments of, 34–35
German, 33
honey production and, 219
identification of, 33
morphology of, 33
oriental, 34
parasitoids of, 119
reference collections of, 230
as scorpion food, 153
smokeybrown, 34
Codling moth, 99
Coffee bean weevil, 46–47, 49, 53
Coffee berry borer, 63
Combfoot spider, 137
Common silverfish, 30
Compound microscopes, 255, 257
Confused flour beetle, 263
Confused sap beetle, 62
Cornfield ant, 112
Corn sap beetle, 62
Cosmopolitan species of beetles, 48, see also specific types
Cosmotropical spotted isometrus, 153
Cottontail rabbit, 180, 190–191
Cow hair, 163, 175, 197, 237
Cowpea weevil, 53
Crab spiders, 137, 145–148, see also specific types
Crazy ant, 109, 113
Credibility, 12–19
Crickets, 223
Cryptostigmatid mites, 125, see also specific types

D

Daddy longlegs spider, 146, 154
DALs, see Defect action levels
Darkling beetles, 51–52, 64–65, see also specific types
Dark–winged fungus gnats, 74, 86, see also specific types
Deer mice, 163
Defect action levels (DALs), 269–270, 272–273
Dermestid beetles, 44, 50, 59–60, see also specific types
Desert locust, 217
Diagnostic fragments, 26–27, 34–35
Direct comparison, 24
Disease, 7, 269–274, see also Pathogens; Toxins; specific types
ants and, 107–108, 114, 272
bats and, 162, 271
beetles and, 271–272
birds and, 177, 270–271
cats and, 162
cockroaches and, 33, 271–273
dogs and, 162
flies and, 72–73, 83, 85, 87, 272
food–borne, 7, 33
hair and, 159, 160, 162, 270
insects and, 270–272, see also under specific insects
mice and, 161, 163
mites and, 270–272
moths and, 271
rabbits and, 162
rodents and, 270–272
shrews and, 162
squirrels and, 162
vectors for, see Vectors
Dissecting microscopes, 41
Documentation, 17, 264–265
Dog hair, 162, 170, 173, 175, 189, 193–194, 237
Dragonfly, 119
Driedfish mites, 124, 129, 131, see also specific types
Driedfruit beetles, 44, 62, see also specific types
Driedfruit mites, 124, 129, 131, see also specific types
Driedfruit moths, 94, 98
Drugstore beetle, 49, 52
Dump flies, 84, see also specific types
Dung flies, 74, 85, 87, see also specific types
Dust mites, 124, 129–132, see also specific types
Dwarf spiders, 145, see also specific types

E

Earwigs, 7–8, 29, 31–33
East African mouse, 32
Ecology, 6
Economic entomology, 6
Ectoparasites, 7–8, 88, 161, see also specific types
Emperor moth, 220
Emulsifiers, 246
Endoparasites, 80, see also specific types
English sparrow, 177, 198–199
Entomology, defined, 1–3, 6
Entomophagy, see Human consumption of insects
Environmental control, 15–16, 19
Equipment capabilities, 15
Equipment quality, 14–15
Equivalent insects, 25–26
Ethnic items from insects, 219–222
Ethnoentomology, 217–224
European earwig, 32
European grain moth, 94, 98
European house spider, 137, 142
Evidence development, 261–267
Excreta, 3–4, 7–8, 158–161, 174, 265, 270
from beetles, 60
from booklice, 36
from cockroaches, 33
disease and, 270
from flies, 81–82
honey production and, 219

from moths, 93
from rodents, 237, 262
Extraction methodology, 21, 242–247
Eye gnat, 74, 82
Eyepiece micrometers, 258–259

F

False powderpost beetle, 49, 53
False stable fly, 84
Families
of ants, 109–115
of beetles, 48–66
of booklice, 36–37
of cockroaches, 33–34
of earwigs, 32–33
of firebrats, 30–31
of flies, 77–88
of mites, 128–132
of moths, 97–99
of parasitoids, 119–120
of silverfish, 30–31
of spiders, 142–148
FD&C, see Food, Drug and Cosmetic Act
Featherlegged spiders, 137, 148, see also specific types
Feathers, 4, 7–8, 177, 270
food insects and, 224
fragments of, 177, 233, 237
identification of, 177–179, 198–199
morphology of, 177–179
pictures of, 198–199
sources of, 177
Featherwinged beetle, 62–63
Feeding damage, 5–6
Feral pigeons, 177
Filbertworm, 99
Filter paper, 243
Filth, defined, 2–3, 270
Filth flies, 72–73, 76, 80–84, 88, see also specific types
acalyptrate, 86
black striped, 82
cats and, 162
dogs and, 162
mites and, 132
parasitoids of, 121
synanthropic, 87
yellow striped, 82
Fire ants, 108, see also specific types
Firebrats, 29–31, see also specific types
Flat bark beetles, 49–50, 56–58
Flat grain beetle, 57–58
Fleas, 161
Flesh flies, 74, 85, see also specific types
Flies, 25, 32, 71–88, 266, see also Gnats; Maggots; Midges; specific types
acalyptrate, 84, 86
anthomyiid, 74
asynanthropic, 87
bats as hosts to, 162
black blow, 81
black scavenger, 74, 86
black striped, 82
blow, 74, 80–81, 85
bluebottle, 81
bot, 72
carnid, 74
as centipede food, 154
cheese skipper, 84
cluster, 81
disease and, 72–73, 272
dump, 84
dung, 74, 85, 87
excreta from, 81–82
false stable, 84
families of, 77–88
filth, see Filth flies
flesh, 74, 85
fragments of, 73–76, 80, 82–88
fruit, 82, 88
greenbottle, 81
habitats of, 72, 80–88
heleomyzid, 74
house, 75, 83–85
humpbacked, 74, 84
identification of, 77–80, 88
latter blow, 81
little house, 84
march, 87
Mediterranean fruit, 88
milichiid, 74
minute black scavenger, 74, 85–86
morphology of, 73–76
moth, see Sewer flies
muscid, 74
mycteribiid, 162
Oriental latrine, 81, 132
parasitic, 81
parasitoids of, 120–121
picture–winged, 74, 85
pomace, 74, 82
red–tailed flesh, 85
reference collections of, 230
scavenger, 74, 85–86
screw–worm, 72, 80–81
sewer, 74, 84–85
shore, 74, 82, 88
significance of, 80
skipper, 74, 84
small dung, 74, 87
soldier, 74, 87
stable, 84
streblid, 162
synanthropic, 72, 81, 83–85, 87–88
warble, 72
window, 74, 86
wings of, 75–76
yellow dung, 85
yellow striped, 82
Flotation liquids, 246, 250
Flour beetles, 43–44, 65, 263, 271, see also specific types
Flour moths, 94–95, 98, see also specific types

Flower beetles, see also specific types
Food and Drug Act of 1906, 3, 270
Food, Drug and Cosmetic Act of 1938, 3, 12, 224, 269
Food poisoning, 161
Forensic entomology, 2, 6, 30, 88
Forensic evidence development, 261–267
Fragments, 4, 7, 237, 265, 270
of ants, 111
of bees, 121
of beetles, 42–45, 53, 59
of booklice, 36
of cockroaches, 34–35
diagnostic, 26–27, 34–35
disease and, 273
earwig, 33
elytral, 26
of feathers, 177, 233, 237
of flies, 73–76, 80, 82–88
food insects and, 219, 224
of hair, 159, 174–175, 233
of honeybees, 121
honey production and, 219
identification of, 2, 15, 25–27, 264, 266
of mites, 270
of moths, 93, 95–100
mounting of, 233–236
of parasitoids, 121
of wasps, 121
Frass, 3
Free–tailed bats, 32
French mini lop–eared rabbit, 194–195
Fruit flies, 82, 88, see also specific types
Fungus beetles, 43, 55–56, 60, 62, 64–65, see also specific types
Fungus gnats, 74, 86, see also specific types
Fungus weevils, 49–51, 53, see also specific types
Funnels, 243–244, 252, see also specific types
Funnel weaver spider, 137
Funnel–web spiders, 142, see also specific types

G

Gamasid mites, 132, see also specific types
Garden spiders, 137, see also specific types
Genitalia, 25–26
of beetles, 42, 45, 58–59
of flies, 82
of mites, 126
of moths, 94, 96–98
reference collections of, 236
of spiders, 140
German cockroach, 33
Giant crab spiders, 145, see also specific types
Giant rats, 32, see also specific types
Giant water bugs, 221–222, see also specific types
Giant wood spider, 137, 218
Glue boards, 232
Glycerin jelly, 27, 42, 169, 174, 176, 234, 236
Gnats, 74, 82, 86, see also Flies; specific types
Goat hair, 163, 237
Golden silk spider, 137, 147
Grain beetles, 43, 49–50, 54, 57–58, 65, 94–95, 266, see also specific types
Grain borers, 44, 53, see also specific types
Grain mites, 124, 128–129, 132, see also specific types
Grain moths, 94–95, 98–99, see also specific types
Grain weevils, 58, see also specific types
Granary mite, 124
Granary pseudoscorpion, 152
Granary weevil, 58
Grass spiders, 142, see also specific types
Gray silverfish, 30
Greater wax moth, 94, 97, 271
Greenbottle fly, 81
Ground beetles, 45, 49, 54, see also specific types
Groundnut bruchid, 53

H

Habitats
of ants, 110–115
of bats, 88, 159
of beetles, 59, 61, 63, 65–66
of booklice, 36–37
categories of, 5
of centipedes, 154
of daddy longlegs, 154
of firebrats, 30
of flies, 72, 80–88
of harvestmen, 154
of insects, 5
of mites, 124–125
of parasitoids, 121
of scorpions, 153
of silverfish, 30
of spiders, 147–148
of sunspiders, 153
of whipscorpions, 154
Hair, 4, 7–8, 16, 157–158, 237, 265
from Asian musk shrew, 161–162, 171, 184–185, 192, 237
from bats, 159, 162, 169–170, 186, 201, 237
from bees, 121
from cats, 158, 162, 173, 175, 237
pictures of, 187–189, 192–193, 201, 203
characteristics of, 168–176
diagnostic, 168–176, 200, 202, 204, 206, 208, 210, 212–214
tables of, 168–169
clothing as sources of, 159
from cows, 163, 175, 197, 237
development of, 163
diagnostic characteristics of, 168–176, 200, 202, 204, 206, 208, 210, 212–214
disease and, 159–160, 162, 270
from dogs, 162, 170, 173, 175, 189, 193–194, 237
from flies, 81
fragments of, 159, 174–175, 233
from goats, 163, 237
growth of, 163
from horses, 163, 175, 237
human, 7, 163–164, 166–168, 175, 195–196

identification of, 157–177
as adulterant, 157–158
diagnostic characteristics and, 168–176, 200, 202, 204, 206, 208, 210, 212–214
figures for, 180–197, 200–214
key for, 169–173
modes of contamination and, 158–159
morphology and, 163–168
problems with, 173–175
reference collection for, 170, 176
regulatory significance and, 159–163
sources and, 158
from kolinskies, 237
from livestock, 163, 166, 175, 237
from mice, 158, 172–173, 181–183, 191, 201, 203, 237
modes of contamination from, 158–159
from monkeys, 237
morphology of, 158, 163–168
mounting of, 176–177
from oxen, 237
from pets, 7, see also specific types
from rabbits, 159, 162, 170, 172, 175, 180, 190–191, 194–195, 201, 203, 207, 237
from rats, 158–160, 172–173, 175, 182–183, 191, 201, 203, 237
reference collections of, 170, 176, 264
from rodents, 160–163, 169–170, 172, 175, 202, 204, 219, 224
from sheep, 163, 175, 196, 237
from shrews, 161–162, 171, 175, 184–185, 192, 237
significance of, 157–163
sources of, 158–159
from spiders, 138–139, 148
from squirrels, 162, 172, 187, 192, 209, 237
structure of, 163–165
from swine, 163, 175, 197, 237
types of, 165–168
from voles, 163, 172
from weasels, 173, 190
Hairy cellar beetle, 65
Hairy fungus beetle, 62
Hammock spider, 137
Handsome fungus beetles, 65
Harvestmen, 136, 151–152, 154, see also specific types
Hazards, 7, see also specific types
ants and, 107
evidence of, 271–272
flies and, 72
nonpathogen, 7–8
pathogenic, 271–272
potential, 7
tolerated, 271
Heleomyzid flies, 74, see also specific types
Helminths, 162
Hemimetabolous insects, 29–37, see also specific types
Hide beetle, 44
High–magnification microscopes, 41
High–risk situations, 273–274
Hister beetle, 60
Histoplasmosis, 162
History of entomology, 3–4
Hobo spider, 137, 140, 142
Honey, 218–219
Honeybees, 22, 106, 121, 218–219
Horse hair, 163, 175, 237
House ants, 110–111, see also specific types
House centipedes, 136, 154, see also specific types
House dust mites, 124, 129–131, see also specific types
House flies, 75, 83–85, see also specific types
House mice, 160–161, 181–182, 191, see also specific types
House mites, 124, see also specific types
House moths, 94, 99, see also specific types
House pseudoscorpion, 152
House spiders, 137–138, 142, 145, 148, see also specific types
Hoyer's solution, 234–236
Human consumption of insects, 113, 217–224
future trends in, 222–224
history of, 217–218
regulation of, 224
Human hair, 7, 163–164, 166–168, 175, 195–196
Humpbacked flies, 74, 84
Hump beetle, 63
Hunting spiders, 144–145, see also specific types
Huntsman spider, 137–138, 145

I

Identification, 21–27, 266
of ants, 108–109
of beetles, 45–48, 58, 60–65
of booklice, 35–36
classification vs., 22–23
of cockroaches, 33
defined, 23
by direct comparison, 24
of earwigs, 31–32
of evidence, 261
of feathers, 177–179, 198–199
field, 263
of firebrats, 30
of flies, 77–80, 88
of food insects, 222–224
of fragments, 2, 15, 25–27, 264, 266
of hair, 157–177
as adulterant, 157–158
diagnostic characteristics and, 168–176, 200, 202, 204, 206, 208, 210, 212–214
figures for, 180–197, 200–214
key for, 169–173
modes of contamination and, 158–159
morphology and, 163–168
problems with, 173–175
reference collection for, 170, 176
regulatory significance and, 159–163
sources and, 158
labels in, 230–231, 233, 235, 237
of maggots, 266
methods of, 23–25

of mites, 123, 126–128
of moths, 93–97, 100–102
of parasitoids, 116–118
procedures in, 22–25
reference collections and, see Reference collections
resources for, 24–25
of silverfish, 30
of spiders, 141–142
of wasps, 116–118
of whole insects, 266
Incidental filth, 7–8
Indiameal moth, 94, 96, 98
Indian lac, 218
Insects, 19, 269, see also Beetles; Cockroaches; Flies; Mites; specific types
ametabolous, 29–37
categories of, 5–6
disease and, 270–272, see also under specific insects
equivalent, 25–26
ethnic items from, 219–222
habitats of, 5
hemimetabolous, 29–37
human consumption of, 113, 217–224
future trends in, 222–224
history of, 217–218
regulation of, 224
life cycles of, 263
location of, 263
preservation of in liquids, 232–233
preservation of whole, 229–233
in processed foods, 4–9
scale, 218
stored–product, see Stored–product insects
whole, 25–26, 229–233, 266
Inspectional evidence, 262–263
Integrated pest management, 272–273
Invertebrate vectors, 7–8

J

Jumping spiders, 137, 146–147, see also specific types

K

Khapra beetle, 59
Kilborn funnels, 243, 252
Kolinsky hair, 237

L

Laboratory administrative systems, 18
Laboratory information bulletins (LIBs), 259
Laboratory procedures, 241–259, see also specific types
apparatus for, 242–247, see also specific types
extraction methodology and, 242–247
general, 254–255
microscopic analysis and, 247–255, see also Microscopes
quantitative transfers and, 252–253
reproducibility and, 241
sample contamination and, 253–254
Laboratory standard operating procedures, 264
Large pale clothes moth, 94, 98
Larger grain borer, 53
Latter blow fly, 81
Least weasel, 190
Leptospirosis, 161
Lesser bandicoot rat, 161, 183
Lesser grain borer, 44, 53
Lesser mealworm, 64
Lesser wax moth, 94, 97
LIBs, see Laboratory information bulletins
Lice, 29, 35–37, 88, 152, see also specific types
Life cycles, 263, 265–266
Light filth, 4, 270
Little brown bat, 162, 186, 201, see also specific types
Little earwig, 32
Little house flies, 84, see also specific types
Livestock hair, 163, 166, 175, 237
Lizard beetle, 60–61
Locusts, 217, 223–224, see also specific types
Longbodied cellar spider, 137
Longlegged spider, 146
Lop–eared rabbit, 194–195

M

Maggots, 74, 87, 266, see also Flies; specific types
Maguey worm, 219
Maine Coone cat, 188
Maize weevil, 58
Managerial responsibilities, 16–17
March flies, 87, see also specific types
Meadow mouse, 163
Meal caterpillars, 94, 97, see also specific types
Meal moths, 94, 97–98, see also specific types
Mealworms, 64, see also specific types
Medical entomology, 6
Mediterranean flour moth, 94–95, 98
Mediterranean fruit fly, 88
Mexican bean weevil, 53
Mexican jumping–bean moth, 99
Mice, see also specific types
deer, 163
disease and, 161
East African, 32
hair from, 158, 172–173, 181–183, 191, 201, 203, 237
house, 160–161, 181–182, 191
meadow, 163
Webster, 180
white–footed, 163, 183, 191
Microanalytical entomology, defined, 1–3, 6
Micrometers, 258–259
Microscopes, 170, 174, 176–177, 233–236, 256–259, see also specific types
accessories for, 258–259
care of, 235–236, 255–256
cleaning of, 256
compound, 255, 257

dissecting, 41
phase–contrast, 126, 257–258
polarizing, 258
quantitative transfers and, 252–253
specifications for, 255
stereoscopic, 174–175, 178, 233, 265
storage of, 235–236
techniques in use of, 247–252
types of, 256–258
widefield, 255, 257
Midges, 74, 81–82, 218, see also Flies; specific types
Migratory locusts, 217, see also specific types
Milichiid fly, 74
Minute bark beetle, 54
Minute black scavenger fly, 74, 85–86
Minute brown scavenger beetle, 61
Minute fungus beetle, 56
Minute jumping spider, 146
Minute tree fungus beetle, 55
Mites, 4, 7–8, 119, 123–132, 152, 154, 237, 269, see also specific types
acaroid, 124–129
allergens and, 129
American house dust, 124
astigmatid, 124
bats as host to, 162
chicken, 124
chortoglyphid, 131
clover, 124, 132
cryptostigmatid, 125
disease and, 270–272
driedfish, 124, 129, 131
driedfruit, 124, 129, 131
dust, 124, 129–132
families of, 128–132
filth flies and, 132
food insects and, 224
fragments of, 270
gamasid, 132
grain, 124, 128–129, 132
granary, 124
habitats of, 124–125
house, 124
house dust, 124, 129–132
identification of, 123, 126–128
mold, 124–125, 128
morphology of, 125–126
prostignatid, 125
scaly grain, 124, 129
slime, 124, 129–132
stored–product, 128, 129–132
straw itch, 124
tortoise, 124
warty grain, 124, 132
Mold mites, 124–125, 128, see also specific types
Molds, 41, 65, 72
Money spider, 137
Monkey hair, 237
Mopane caterpillar, 220, 222
Morphology
of ants, 106–107
of beetles, 42–45
of booklice, 35
of centipedes, 154
of cockroaches, 33
of earwigs, 31
of feathers, 177–179
of firebrats, 30
of flies, 73–76
of hair, 158, 163–168
of harvestmen, 154
of mites, 125–126
of moths, 93–97, 100
of parasitoids, 106–107
of pseudoscorpions, 151–152
of scorpions, 153
of silverfish, 30
of spiders, 138–140
Mosquitoes, 88, see also specific types
Moth (sewer) flies, 74, 85–86, see also specific types
Moths, 25, 93–102, see also Caterpillars; specific types
almond, 94, 98
angoumois grain, 57, 94–95, 99
brown house, 94, 99
carob, 98
casemaking clothes, 94, 98
clothes, 94, 98–99, 152
codling, 99
disease and, 271
driedfruit, 94, 98
emperor, 220
european grain, 94, 98
excreta from, 93
families of, 97–99
flour, 94–95, 98
fragments of, 93, 95–100
grain, 94–95, 98–99
greater wax, 94, 97, 271
house, 94, 99
identification of, 93–97, 100–102
immature, 99–102
indiameal, 94, 96, 98
large pale clothes, 94, 98
lesser wax, 94, 97
meal, 94, 97–98
Mediterranean flour, 94–95, 98
Mexican jumping–bean, 99
morphology of, 93–97, 100
pallid emperor, 220
raisin, 94, 98
reference collections of, 230
rice, 94, 97
sightings of, 263
stored nut, 94, 98
tapestry, 94
tobacco, 94, 98–99
tropical tobacco, 94, 99
tropical warehouse, 98
warehouse, 94, 98
wax, 94, 97, 271
webmaking clothes, 94, 98
whiteshouldered house, 94, 99
Mouse spider, 137, 145

Murky meal caterpillar, 94, 97
Muscid flies, 74
Mycotoxins, 41, see also specific types
Mycteribiid flies, 162, see also specific types
Myiasis, 6

N

New World army ant, 111
Nonpathogen hazards, 7–8
Norway rat, 160–161, 180, 191
Nutritional losses from pests, 273

O

Odorous house ant, 110–111
Orb weaver spider, 142–143
Oriental cockroach, 34
Oriental latrine fly, 81, 132
Outrageous adulteration, 7–8
Oval grain beetle, 54
Oxen hair, 237

P

Pallid emperor moth, 220
Palm weevil, 223
Pancreatin, 247
Parasites, see also specific types
  bat, 88
  bats as hosts to, 162
  birds and, 177
  ecto–, 7–8, 88, 161
  endo–, 80
  life cycles of, 6
Parasitic flies, 81, see also specific types
Parasitoids, 105, 115–120, see also specific types
  families of, 119–120
  of filth flies, 121
  of flies, 121
  fragments of, 121
  habitats of, 121
  identification of, 116–118
  morphology of, 106–107
  significance of, 115
Parson spider, 137, 145
Pathogenic hazards, 271–272
Pathogens, 7, 15, 27–274, see also Disease; specific types
  ants and, 114
  bats and, 162
  flies and, 72, 83, 87
  food insects and, 224
  hair and, 159–160
  harvestmen and, 154
Pavement ant, 109, 115
Pea weevil, 41, 53
Pecan weevil, 41
Percolators, 243, 251
Pesticides, 224, 272
Pest management, 272–273
Pet hair, 7, see also specific types
Phantom midge, 218
Pharaoh's ant, 107–108, 114
Phase–contrast microscopes, 126, 257–258
Picture–winged flies, 74, 85, see also specific types
Pigeon louse, 88
Pigeons, 7, 177, 198–199
Pig hair, 163, 175, 197, 237
Pigs, see Swine
Pink scavenger caterpillar, 94, 99
Pin mounts, 230
Pipistrelle bat, 186
Pirate spider, 137, 145
Plagues, 160–163, see also specific types
Pleasing fungus beetle, 60
Point mounts, 231–232
Polarizing microscopes, 258
Polynesian rat, 160
Pomace flies, 74, 82, see also specific types
Potential hazards, 7
Powderpost beetle, 49, 53, 61–62
Prostignatid mites, 125, see also specific types
Protozoa, 162, 177
Pseudoscorpions, 151–152, see also specific types
Public health categories, 6
Public health significance, 15, 270, see also Disease
Pyramid ant, 110

Q

QA, see Quality assurance
QC, see Quality control
Qualitative tests, 24
Quality assessment, 19
Quality assurance, 11–20
  activities in, 18–19
  credibility and, 12–19
  defined, 11
  elements of, 13–14
  regulatory role of, 11–14
  training in, 13
Quality control, 13
Quality reviews, 19
Quantitative transfers, 19, 252–253
Quinones, 41, 65, see also specific types

R

Rabbit hair, 159, 162, 170, 172, 175, 237, see also specific types of rabbits
  pictures of, 180, 190–191, 194–195, 201, 203, 207
Rabies, 162
Raisin moth, 94, 98
Rat–bite fever, 160–161
Rat fleas, 161
Rats, see also specific types
  bandicoot, 161, 183
  black, 160
  brown, 160
  giant, 32
  hair from, 158–160, 172–173, 175, 182–183, 191, 201, 203, 237
  lesser bandicoot, 161, 183

Norway, 160–161, 180, 191
Polynesian, 160
roof, 160–161, 183
Rat–tailed maggots, 74, 87, see also specific types
Reagents, 15–16, 242–247, see also specific types
Recluse spiders, 137, 140, 144, 147, see also specific types
Red flour beetle, 43, 65
Red jumping spider, 137, 147
Redlegged ham beetle, 44
Red–tailed flesh fly, 85
Reference collections, 15, 27, 170, 176, 227–238, 264
equipment needed for, 228–229
of hair, 170, 176, 264
management of, 236–237
sources of, 237–238
techniques used for, 229–236
Reproducibility, 241
Research reports, 19
Rice moth, 94, 97
Rice weevil, 58
Rickettsial pox, 161
Ring–legged earwig, 32
Risk assessment, 8–9
Risk categories, 6–9
Roaches, see Cockroaches
Rodents, 32, 124, 265, 269, see also specific types
disease and, 270–272
excreta from, 237, 262
gnawing of, 262
hair from, 160–163, 169–170, 172, 175, 202, 204, 219, 224
honey production and, 219
inspection of filth from, 262–263
sightings of, 263
Roof rat, 160–161, 183
Rove beetle, 8, 64
Rusty grain beetle, 57

S

Sac spiders, 137, 140, 143, see also specific types
Salmonellosis, 160–161
Sampling, 17–19, 253–254, 267
Sap beetles, 51, 62, see also specific types
Sawtoothed grain beetle, 43, 57, 266
Scale insects, 218, see also specific types
Scaly grain mite, 124, 129
Scavenger beetles, 61, see also specific types
Scavenger caterpillars, 94, 99, see also specific types
Scavenger flies, 74, 85–86, see also specific types
Scorpions, 136, 151–154, see also specific types
Screw–worm flies, 72, 80–81, see also specific types
Seaside earwig, 32
Seed beetles, 49, 53–54, see also specific types
Sewer flies, 74, 84–85, see also specific types
Sheep hair, 163, 175, 196, 237
Sheet–web weaver spider, 145
Shih tzu dog, 194
Shore earwig, 32
Shore flies, 74, 82, 88, see also specific types
Shortbodied cellar spider, 137
Shrews, 7–8, 161–162, 171, 184–185, 192
Siamese cat, 187–188, 193
Siamese grain beetle, 58, 65
Siberian weasel, 190
Sieves, 244–245
Significance
of ants, 107–108
of beetles, 41
of flies, 80
of hair, 157–163
of parasitoids, 115
of pseudoscorpions, 152
public health, 15, 270
of wasps, 115
Silken fungus beetle, 56
Silk spiders, 137, 147, 223, see also specific types
Silkworm, 219–220
Silverfish, 29–31, 153, see also specific types
Skins, 3, 26, 33, 36
Skipper flies, 74, 84, see also specific types
Slime mites, 124, 129–132, see also specific types
Small dung flies, 74, 87, see also specific types
Smokeybrown cockroach, 34
Soldier flies, 74, 87, see also specific types
SOP, see Standard operating procedures
South American brown spider, 137
Southern cowpea weevil, 53
Sparrows, 7, 177, 198–199
Spider beetles, 51, 63, see also specific types
Spiders, 7, 123, 135–148, see also specific types
aerial web, 143
American house, 137, 148
banana, 137
black widow, 137, 148
boreal combfoot, 137
brown, 137, 147
brown widow, 137
cardinal, 137
cave, 137, 145
ceiling, 137
cellar, 137, 146
as centipede food, 154
cobweb, 137, 148
cobweb weaver, 137
combfoot, 137
crab, 137, 145–148
daddy longlegs, 146, 154
dwarf, 145
eggs of, 119
European house, 137, 142
examples of, 137
families of, 142–148
featherlegged, 137, 148
funnel weaver, 137
funnel–web, 142
garden, 137
giant crab, 145
giant wood, 137, 218
golden silk, 137, 147
grass, 142
habitats of, 147–148
hammock, 137

harvestmen, 136, 151–152, 154
hobo, 137, 140, 142
house, 137, 142, 145, 148
human consumption of, 218, 223
hunting, 144–145
huntsman, 137–138, 145
identification of, 141–142
jumping, 137, 146–147
longbodied cellar, 137
longlegged, 146
money, 137
morphology of, 138–140
mouse, 137, 145
orb weaver, 142–143
parasitoids of, 119–120
parson, 137, 145
pirate, 137, 145
recluse, 137, 140, 144, 147
red jumping, 137, 147
sac, 137, 140, 143
sheet–web weaver, 145
shortbodied cellar, 137
silk, 137, 147, 223
South American brown, 137
spitting, 147
sun–, 136, 151–153
synanthropic, see Synanthropic spiders
tropical cellar, 137
tropical crab, 147
true jumping, 146
venomous, 140
wandering, 137, 140, 143–144
weaver, 137, 142–143, 145
web, 137, 143, 148
web weaver, 145
western black widow, 137
white–eyed, 142
widow, 137, 140, 148
wood, 137, 218
zebra, 137, 146
Spitting spiders, 147, see also specific types
Spitz dog, 194
Squarenecked grain beetle, 57
Squarenosed fungus beetle, 43
Squirrel hair, 162, 172, 187, 192, 209, 237
Stable flies, 84, see also specific types
Staining reagents, 235, see also specific types
Standard operating procedures (SOP), 264
Standards, 15–16
Staphylococci, 162
Starlings, 177, 198–199
Stereoscopic microscopes, 174–175, 178, 233, 265
Stingless bees, 121, 219, see also specific types
Stored nut moth, 94, 98
Stored–product insects, 5–8, 124
ants as, 114–115
beetles as largest group of, 40–41
defined, 3
flies as, 72–73, 80, 86
moths as, 93, 100
parasitoids as, 116–118
wasps as, 116–118
Straw itch mite, 124
Streblid flies, 162, see also specific types
Striped earwig, 32
Sunspiders, 136, 151–153, see also specific types
Swine hair, 163, 175, 197, 237
Synanthropic ants, 106, 110–111, 113, see also specific types
Synanthropic birds, 88, see also specific types
Synanthropic flies, 72, 81, 83–85, 87–88, 120, see also specific types
Synanthropic spiders, 136–148, see also specific types
examples of, 137
families of, 142–148
general, 136–138
identification of, 141–142
morphology of, 138–140
Systematics, 22

T

Tailless whipscorpions, 136, 151–152, 154, see also specific types
Tamarind weevil, 58
Tapestry moth, 94
Tarantulas, 137, 148, see also specific types
Taxonomy, 22, 24
Tequila worm, 219
Thai giant water bug, 221
Thief ant, 114
Thrips, 119, see also specific types
Ticks, 162
Tobacco moths, 94, 98–99, see also specific types
Tortoise mite, 124
Toxins, 41, 272–273, see also Disease; specific types
Toxoplasmosis, 162
Training, 13
Trap flasks, 243–244, 249
Tree ants, 110, see also specific types
Trichinosis, 160–161
Trophic levels, 5
Tropical cellar spider, 137
Tropical crab spiders, 147, see also specific types
Tropical tobacco moth, 94, 99
Tropical warehouse moth, 98
Tropicopolitan species of beetles, 48, see also specific types
True bugs, 230, see also specific types
True jumping spiders, 146, see also specific types
Tularemia, 161
Turkish grain beetle, 57
Tween, 80, 247
Twohanded fungus beetle, 64
Typhus fever, 160–161

V

Varied carpet beetle, 59
Vectors, 7–8, see also Disease
ants and, 107–108, 114
Asian musk shrew and, 161
beetles and, 41

flies and, 72–73, 80, 83
hair and, 160
invertebrate, 7–8
vertebrate, 7–8
Velvety tree ant, 110
Venomous spiders, 140, see also specific types
Vertebrate specimens, 236–237
Vertebrate vectors, 7–8
Viruses, 177
Vole hair, 163, 172

W

Wandering spiders, 137, 140, 143–144, see also specific types
Warble fly, 72
Warehouse moths, 94, 98, see also specific types
Warty grain mite, 124, 132
Wasps, 106, 115–118, 121, see also specific types
Water bugs, 221–222, see also specific types
Waterstrider, 119
Wax moths, 94, 97, 271, see also specific types
Weasel hair, 173, 190
Weaver ants, 222–223, see also specific types
Weaver spiders, 137, 142–143, 145, see also specific types
Webbing, 3
Webmaking clothes moth, 94, 98
Web spiders, 143, see also specific types
Webster mouse, 180
Web weaver spiders, 145, see also specific types
Weevils, 43, 50, 58–59, see also Beetles; specific types
bean, 41, 53
broadnosed grain, 58
coffee bean, 46–47, 49, 53
cowpea, 53
fungus, 49–51, 53
grain, 58
granary, 58
maize, 58
Mexican bean, 53
palm, 223
pea, 41, 53
pecan, 41
rice, 58
southern cowpea, 53
tamarind, 58
Western black widow spider, 137
Westhighland terrier, 194
Whipscorpions, 136, 151–152, 154, see also specific types
White Angora rabbit, 190, 195
White–eyed spider, 142
White–footed mouse, 163, 183, 191
Whitemarked spider beetle, 63
Whiteshouldered house moth, 94, 99
Whole insects, 25–26, 229–233, 266
Widefield microscopes, 255, 257
Widow spiders, 137, 140, 148, see also specific types
Window flies, 74, 86, see also specific types
Windscorpions, 136, 153, see also specific types
Wings, 25
booklice, 36
of earwigs, 32
of flies, 75–76, 82–85, 87–88
of moths, 94–96, 99
of parasitoids, 116, 119, 121
Wood gnat, 74
Wood spiders, 137, 218, see also specific types
Wool, 7
Worksheets, 19
Worms, see also specific types
bugong, 218
filbert–, 99
maguey, 219
meal–, 64
screw–, 72, 80–81
silk–, 219–220
tequila, 219

Y

Yeasts, 72
Yellowbrown sap beetle, 62
Yellow dung fly, 85
Yellowjacket, 121
Yellow striped filth fly, 82

Z

Zebra spider, 137, 146

TAXONOMIC INDEX

A

Acanthomyops spp., 112
Acanthoscelides obtectus, 41, 49, 53–54
Acari, 123–132, 152, see also specific genera and species
  families of, 128–132
  identification of, 126–128
  morphology of, 125–126
Acaridae, 124, 126, 128, see also specific genera and species
Acarus
  siro, 124, 128
  spp., 128
Achaearanea
  spp., 145
  tepidariorum, 137, 148
Achroia grisella, 94, 97, 101
Acrididae, 217, see also specific genera and species
Acrotelsia collaris, 31
Actidium spp., 63
Adistemia watsoni, 50
Aegiale hesperianus, 219
Aeroglyphidae, 124, 128, 132, see also specific genera and species
Aeroglyphus
  robustus, 124, 132
  spp., 128
Agaonidae, 117, 120, see also specific genera and species
Agelenidae, 137, 140, 142, see also specific genera and species
Aglenus brunneus, 66
Aglossa caprealis, 94, 97
Agrotis infusa, 218
Ahasverus
  advena, 49, 56–58
  spp., 47, 57
Alaptus spp., 117, 119
Aleuroglyphus spp., 127–128
Allochernes powelli, 152
Allowithius spp., 152
Alphitobius
  diaperinus, 51, 64
  laevigatus, 51
  spp., 65
Alphitophagus bifasciatus, 51, 64
Amaurobiidae, 137, 140–142, 144, see also specific genera and species
Amaurobius ferox, 137, 142
Amblypygi, 136, 152, 154, see also specific genera and species
Amblyseius spp., 124
Ameroseiidae, 124, 132, see also specific genera and species
Ammotrechidae, 153, see also specific genera and species
Ammotrechula
  peninsulana, 153
  spp., 153
Anagasta kaelmeilla, 94–96, 98, 102
Andricus foliatus, 120
Androlaelaps casalis, 124
Aneuretinae, 108, 110, see also specific genera and species
Anisolabis maritima, 32
Anisopodidae, 73–74, 78, see also specific genera and species
Anisopteromalus calandrae, 118, 120
Anobiidae, 43, 47, 49, 52, see also specific genera and species
Anobium punctatum, 52
Ansopodidae, 88, see also specific genera and species
Anthicidae, 46, 49, 52–53, see also specific genera and species
Anthicus
  floralis, 49
  heroicus, 52
  spp., 52
Anthomyia spp., 80
Anthomyiidae, 74, 79–80, see also specific genera and species
Anthrenocerus australis, 50
Anthrenus
  fasciatus, 50
  flavipes, 50
  oceanicus, 50
  pellio, 50
  scrophulariae, 50, 59
  spp., 59–60
  unicolor, 50
  verbasci, 50
  vorax, 59
Anthribidae, 46, 53, see also specific genera and species
Anthropoda, 136, see also specific genera and species
Aphaenogaster spp., 113
Aphidiinae, 116, see also specific genera and species
Aphonopelma spp., 137
Apidae, 121, 218, see also specific genera and species
Apis
  mellifera, 22–23, 218–219
  spp., 23
Apotropina spp., 82
Arachnida, 123–132, 135–148, see also specific genera and species
  families of, 128–132, 142–148
  identification of, 126–128, 141–142
  morphology of, 125–126, 138–140
Araecerus fasciculatus, 53
Araneae, 135–148, 152, 218, 223, see also specific genera and species
  families of, 142–148

identification of, 141–142
morphology of, 138–140
Araneidae, 137, 142–143, 147, see also specific genera and species
Araneus spp., 143
Argiope spp., 137, 142
Aridius nodifer, 50
Arixeniidae, 32, see also specific genera and species
Aruecerus fasciculatus, 49
Ascidae, 124, 132, see also specific genera and species
Atherigona
orientalis, 84
spp., 84
Atomaria
linearis, 49, 56
spp., 56
Attagenus
spp., 59–60
unicolor, 59
Atta spp., 113
Auricularia spp., 55

B

Badumna
longiquus, 137, 144
spp., 144
Baeus spp., 116, 120
Bandicota
bengalensis, 161, 169, 175, 183
spp., 172–173, 209
Basilia spp., 88
Bdellidae, 124, 132, see also specific genera and species
Beamys spp., 32
Belaphotroctes spp., 36
Belostomatidae, 221, see also specific genera and species
Bercaea haemorrhoidalis, 85
Berginus
pumilus, 51, 62
tamarisci, 62
Bethylidae, 116–117, 120, see also specific genera and species
Bibionidae, 87, see also specific genera and species
Blaps
macronata, 51
spp., 47
Blatta orientalis, 34
Blattaria, 29, 33–35, see also specific genera and species
Blattella
asahinae, 33
germanica, 33
Blattellidae, 33, see also specific genera and species
Blattidae, 33, see also specific genera and species
Blattisocius tarsalis, 124
Blomia
freemani, 124, 132
spp., 128
Bombycidae, 219, see also specific genera and species
Bombyx mori, 219–220
Bostrichidae, 44, 47, 49, 53, see also specific genera and species
Bovidae, 163, see also specific genera and species
Bracon hebetor, 115
Braconidae, 116, 119, see also specific genera and species
Braula coeca, 88
Braulidae, 88, see also specific genera and species
Bruchidae, 47, 49, 53–54, see also specific genera and species
Bruchus
ervi, 49
pisorum, 41, 49, 53–54
rufimanus, 49
spp., 54
Bryobia praetiosa, 124, 132
Buthidae, 153

C

Cadra
calidella, 98
cautella, 94, 98, 102, 115
figulilella, 94, 98, 102
Cadrema
pallida, 82
spp., 77
Calisoga longitarsis, 137, 145
Calliphora
spp., 81, 120
vicina, 73
Calliphoridae, 72–74, 77, 79–81, 121, 132, see also specific genera and species
Callosobruchus
analis, 49, 54
chinensis, 49, 53, 54
maculatus, 49, 53, 54
spp., 54
Caloglyphus spp., 128
Camponotus spp., 109, 112
Campylobacter jejuni, 72
Canaceidae, 88, see also specific genera and species
Canis familiaris, 162
Capiennius salei, 137
Carabidae, 47, 49, 54, see also specific genera and species
Carcinophoridae, 32, see also specific genera and species
Carcinops
pumilio, 50, 60
troglodytes, 50
Carnidae, 74, 78, 80–81, see also specific genera and species
Carnus hemapterus, 81
Carpoglyphidae, 124–125, 128–129, see also specific genera and species
Carpoglyphus
lactis, 124, 129, 131
spp., 128–129
Carpophilus
dimidiatus, 51, 62

hemipterus, 51, 62
lugubris, 51
marginellus, 51
mutilatus, 51, 62
obsoletus, 51, 62
spp., 44, 51, 62
Cartodere
constricta, 50
punctulata, 50
Caryedon serratus, 49, 53–54
Cataglyphus spp., 109
Cathartus quadricollis, 49, 57–58
Caulophilus oryzae, 50, 58
Cecidomyiidae, 87, see also specific genera and species
Centruroides spp., 153
Cephalonomia
gallicola, 116–117, 120
spp., 121
tarsalis, 115–116, 120
Cerapachinae, 108, 110, see also specific genera and species
Ceraphronidae, 117, 119, see also specific genera and species
Ceratitis capitata, 88
Ceratopogonidae, 74, 78, 81, see also specific genera and species
Cerophagus spp., 128
Cerylonidae, 46, 54, see also specific genera and species
Chalcididae, 118, 120, see also specific genera and species
Chalcidoidea, 116, see also specific genera and species
Chaoboridae, 218, see also specific genera and species
Chaoborus edulis, 218
Cheiracanthium
inclusum, 137, 143
mildei, 137, 143
spp., 140, 143
Cheiridiidae, 152
Cheiridium museorum, 152
Chelicerae, 142, 144, 146, see also specific genera and species
Chelifer cancroides, 152
Cheliferidae, 152, see also specific genera and species
Chelisoches morio, 32
Chelisochidae, 32, see also specific genera and species
Cheyleltus eruditus, 124
Cheyletidae, 124, 132, see also specific genera and species
Chilopoda, 154, see also specific genera and species
Chironomidae, 74, 78, 81–82, see also specific genera and species
Chiroptera, 162, see also specific genera and species
Chloropidae, 74, 77, 80, 82, see also specific genera and species
Choetospila spp., 118, 120
Chortoglyphidae, 124, 128, 131, see also specific genera and species
Chortoglyphus
arcuatus, 124, 131
spp., 128
Chrysomya
megacephala, 81, 121, 132
spp., 77, 81
Ciidae, 47, 49, 55, see also specific genera and species
Cirina forda, 220
Cis
asiaticus, 49, 55
chinensis, 49, 55
Clathonius terrachelatus, 152
Cleridae, 44, 47–49, 55–56, 129, see also specific genera and species
Clostridium spp., 107
Clubionidae, 137, 140–141, 143, see also specific genera and species
Cochliomyia hominivorax, 80–81
Coleoptera, 39–66, 152, 223, see also specific genera and species
categorization of, 41
equipment for study of, 41–42
families of, 48–66
identification of, 45–48
morphology of, 42–45
parasitoids of, 115, 119
significance of, 41
techniques for study of, 41–42
Colydiidae, 66, see also specific genera and species
Conomyrma
insana, 110
spp., 110
Coproica spp., 87
Corcyra cephalonica, 94, 97, 101
Corus
medicinalis, 137, 142
spp., 142
Corylophidae, 46, 49, 56, see also specific genera and species
Cosmopterigidae, 99, see also specific genera and species
Crematogaster spp., 108, 113–114
Cricetomys spp., 32
Cryotphagidae, 47, see also specific genera and species
Cryptolestes
agandae, 50
capensis, 49, 57
cornutus, 49, 57
divaricatus, 57
ferrugineus, 49, 57
klapperichi, 49, 57
pusilloides, 49, 57
pusillus, 49, 57
spp., 42, 47, 57–58
turcicus, 50, 57
ugandae, 57
Cryptophagidae, 47, 49, 56, see also specific genera and species
Cryptophagus
acutangulus, 49
cellaris, 49
pilosus, 49
spp., 47, 56

Cryptophilus
integer, 60–61
spp., 50, 60–61
Ctenidae, 137, 140–141, 143–144, see also specific genera and species
Ctenolepisma
lineata, 31
longicaudata, 30–31
spp., 31
urbana, 31
Cucujidae, 42–43, 45, 47, 49–50, 56–58, see also specific genera and species
Culicidae, 88, see also specific genera and species
Cunaxa setirostris, 124
Cunaxidae, 124, 132, see also specific genera and species
Cupiennius salei, 144
Curculio caryae, 41
Curculionidae, 42–43, 46, 50, 58–59, 223, see also specific genera and species
Cydia
latiferreana, 99
pomonella, 99
Cynaeus angustus, 51

D

Dacne
japonica, 50, 60
picta, 50, 60
spp., 60–61
Dactylopidae, 218, see also specific genera and species
Dactylopius coccus, 218
Damon spp., 154
Dasyhelea spp., 81
Delia spp., see *Paregle* spp.
Dermanyssidae, 124, 132, see also specific genera and species
Dermanyssus gallinae, 124
Dermaptera, 29, 31–33, see also specific genera and species
Dermatophagoides
pteronyssinus, 124
spp., 128
Dermestes
ater, 50
carnivorus, 50
frischii, 50
haemorrhoidalis, 50
lardarius, 50
masculatus, 50
peruvianus, 50
spp., 44, 48, 59–60, 129
undulatus, 50
Dermestidae, 7, 44, 48, 50, 59, 86, 115, 129, 152, see also specific genera and species
Desidae, 137, 140–141, 144, see also specific genera and species
Desmometopa spp., 83
Diapriidae, 117, 120, see also specific genera and species
Diapriinae, 117, see also specific genera and species
Dictyna
spp., 144
sublata, 137, 144
Dictynidae, 137, 140–142, 144, see also specific genera and species
Dienerella spp., 51
Digamaellidae, 124, 132, see also specific genera and species
Digamasellus spp., 124
Dinocheirus panzeri, 152
Dinoderus
bifoveolatus, 53
minutus, 49, 53
spp., 53
Diptera, 71–88, 132, 218, see also specific genera and species
families of, 77–88
identification of, 77–80, 88
morphology of, 73–76
parasitoids of, 119–121
reference collections of, 230
Dirhininae, 118, 120, see also specific genera and species
Discomyza spp., 82
Dolichoderinae, 108–111, see also specific genera and species
Dorylinae, 108, 111, see also specific genera and species
Dorypteryx spp., 37
Drosophila
melanogaster, 82
spp., 77, 82
Drosophilidae, 74, 77, 80, 82, see also specific genera and species
Dryomyzidae, 88, see also specific genera and species
Dysdera crocata, 137, 144
Dysderidae, 137, 141, 144, see also specific genera and species

E

Echimyopodidae, 124, 128, 132, see also specific genera and species
Ecitoninae, 108, 111, see also specific genera and species
Eciton spp., 111
Eidmannella pallida, 137, 145
Embidopsocus spp., 36
Embioptera, 120, see also specific genera and species
Encyrtidae, 118–119, see also specific genera and species
Endomychidae, 47, 50, 61, 65, see also specific genera and species
Endrosis sarcitrella, 94, 99, 102
Enicmus consimilis, 51
Entoamoeba hisstolytica, 72
Ephestia
elutella, 94, 98, 102
kaehniella, 98
Ephydridae, 74, 80, 82, see also specific genera and species

Equidae, 163, see also specific genera and species
Eremobates spp., 153
Eremobatidae, 153, see also specific genera and species
Eremorhax spp., 153
Erigoninae, 145, see also specific genera and species
Eriococcidae, 217, see also specific genera and species
Eristalinus aeneus, 87
Eristalis
  spp., 77
  tenax, 73, 87
Ero furcata, 137, 145
Erotylidae, 48, 50, 60, see also specific genera and species
Erythmelus spp., 119
Escherichia coli, 72
Euborellia annalipes, 32
Eucolidae, 116, 120, see also specific genera and species
Eufallia seminvenus, 51
Eugamasus butleri, 124
Eulophidae, 118–119, see also specific genera and species
Euroglyphus spp., 128
Eurytomidae, 118, 120, see also specific genera and species

F

Fannia
  canicularis, 73, 84
  scalaris, 73, 84
  spp., 77, 82, 84
Fanniinae, 79, 82, 84, see also specific genera and species
Felis catus, 162
Ficus spp., 120
Filistatidae, 137, 140–141, 144, see also specific genera and species
Forficula auricularis, 32
Forficulidae, 32, see also specific genera and species
Formica spp., 109, 112
Formicidae, 106–115, 222, see also specific genera and species
Formicinae, 108–110, 112–113, see also specific genera and species

G

Galeodes spp., 153
Galeodidae, 153, see also specific genera and species
Galleria mellonella, 94, 97, 101, 271
Galleriinae, 97–98, 101, see also specific genera and species
Ganoderma lucidum, 55
Gelechidae, 99, see also specific genera and species
Gibbium psylloides, 51, 63
Glavia spp., 153
Glycarcaridae, 125, see also specific genera and species
Glycyphagidae, 124, 128, 130–131, see also specific genera and species
Glycyphagus
  domesticus, 124, 130
  spp., 128
Gnaphosidae, 137, 141, 144–145, see also specific genera and species
Gnatocerus
  cornutus, 51
  maxillosus, 51
  spp., 65
Gohieria spp., 128, 130
Gomimbrasia belina, 220
Gonatocerus spp., 117, 119

H

Habrobracon spp., 116, 119
Hadraule blaisdelli, 55
Haplotinea
  ditella, 99
  insectella, 94, 99
  spp., 102
Haptoncus luteolus, 62
Heleomyzidae, 74, 79, 88, see also specific genera and species
Hemeromyia spp., 81
Hemimeridae, 32, see also specific genera and species
Hemiptera, 119–120, 221, 230, see also specific genera and species
Henoticus
  californicus, 49, 56
  spp., 47, 56
Hermetia
  illucens, 73, 87
  spp., 78
Herpyllus
  ecclesiasticus, 137, 145
  spp., 144
Heteropoda venatoria, 137, 145
Heteropodiadae, 137–138, 141, 145, see also specific genera and species
Heteroptera, 120, see also specific genera and species
Hippelates spp., 82
Hippoboscidae, 88, see also specific genera and species
Histeridae, 46, 50, 60, see also specific genera and species
Histiogaster spp., 127–128
Histiosoma spp., 124, 128, 131
Histiostomatidae, 124, 128, 131, see also specific genera and species
Histoplasma capsulatum, 162
Hloparamecus depressus, 51
Hofmanophila
  laeolus, 94, 99
  pseudopretella, 94, 99
  spp., 102
Holepyrus spp., 115
Holocnemis plachei, 137, 146
Holoparamecus spp., 61
Homoptera, 119, 217–218, see also specific genera and species

Hydrotaea spp., 84
Hylemya spp., 80
Hymenoptera, 105–121, 218, 222, see also specific genera and species
 families of, 109–115, 119–120
 identification of, 108–109, 116–118
 morphology of, 106–107
 significance of, 107–108, 115
Hypoclinea spp., 109–110
Hypothenemus
 areccae, 51, 63–64
 columbi, 63
 crudiae, 63–64
 eruditus, 51, 63–64
 hampei, 51, 63–64
 obscrusu, 63–64
 seriatus, 63–64
 spp., 63–64

I

Ichneumonidae, 116, see also specific genera and species
Insecta, see Coleoptera; Diptera; Hymenoptera; Lepidoptera
Iridomyrmex
 humilis, 107, 111
 spp., 109–111
Isometrus maculatus, 153

K

Kleemania plumosus, 124
Kukalcania hibernalis, 137, 144

L

Labidura riparia, 32
Labiduridae, 32, see also specific genera and species
Labiidae, 32, see also specific genera and species
Lacciferidae, 218, see also specific genera and species
Laccifer lacca, 218
Lachesilla pedicularia, 37
Lachesillidae, 36–37, see also specific genera and species
Laelius trogodermatis, 115
Laemophloeinae, 45, 56, see also specific genera and species
Lamprochernes minor, 152
Languridae, 48, 50, 60–61, see also specific genera and species
Larca granulata, 152
Lardoglyphidae, 124–126, 128, 131, see also specific genera and species
Lardoglyphus
 angelinae, 129
 konoi, 124, 129
 spp., 127–129, 131
 zacheri, 129
Lasioderma serricorne, 43, 49, 52, 120
Lasiopleura spp., 82
Lasius
 alienus, 112
 spp., 109, 112
Laspeyresia saltitans, 99
Latheticus oryzae, 51, 65
Lathridiidae, 43, 48, 50–51, 61, see also specific genera and species
Lathridius
 anthracinus, 51
 minutus, 43, 51
 pseudominutus, 51
 spp., 51
Lathromeroidea spp., 118–119
Latrodectus
 gemoetricus, 137
 hesperus, 137
 mactans, 137, 148
 spp., 140, 148
Leiodinychus krameri, 124
Lentinus
 edodes, 55, 60
 polychrous, 55
Lepidoglyphus spp., 128, 131
Lepidoptera, 93–102, 152, 218–220, see also specific genera and species
 families of, 97–99
 identification of, 93–97, 100–102
 immature, 99–102
 morphology of, 93–97, 100
 parasitoids of, 119–120
 reference collections of, 230
Lepidotrichidae, 31, see also specific genera and species
Lepinotus spp., 37
Lepisma saccharina, 30–31
Lepismatidae, 30–31, see also specific genera and species
Leporidae, 162, see also specific genera and species
Leptanillinae, 108, 113, see also specific genera and species
Lepthyphantes spp., 137, 145
Leptocera spp., 73, 77, 87
Leptometopa spp., 83
Leptomyrmex spp., 109
Leptospira spp., 162
Lethocerus indicus, 221
Leucobimatium arundinaceum, 50, 60–61
Limosiniae, 87, see also specific genera and species
Linyphiidae, 137, 142–143, 145, see also specific genera and species
Liometopum
 apiculatum, 222
 spp., 109
Liposcelidae, 36, 152, see also specific genera and species
Liposcelis spp., 35–36
Lipscelidae, 36–37, see also specific genera and species
Listeria monocytogenes, 107, 177
Locusta migratoria, 217
Lophocateres pusillus, 52, 58, 65

Loxosceles
lueeta, 137, 147
reclusa, 137, 147
spp., 140, 147
Loxoscelidae, 144, see also specific genera and species
Lucilia spp., 81
Lyctidae, 48, 51, 61–62, see also specific genera and species
Lyctoxylon
japonum, 62
spp., 62
Lyctus
brunneus, 51, 61–62
spp., 62

M

Macrocheles muscaedomesticae, 124, 132
Macrochelidae, 124, 132, see also specific genera and species
Madiza spp., 81, 83
Marava arachidis, 32
Masicodamon spp., 154
Megaphragma spp., 118–119
Megaselia spp., 73, 77, 84
Megaspilidae, 117, 119, 201, see also specific genera and species
Megathymidae, 219, see also specific genera and species
Meoneura spp., 78, 81
Merophysiidae, 61, see also specific genera and species
Mezium
affine, 51
americanum, 51, 63
Michaelopus spp., 128
Microtus spp., 162–163, 171–172, 209
Milichia spp., 83
Milichidae, 74, 78, 80, 82–83, see also specific genera and species
Milichiella spp., 78, 83
Mimetidae, 137, 142, 145, see also specific genera and species
Minthea rugicollis, 62
Mitopus morio, 154
Molosidae, 32, see also specific genera and species
Monanus concinulus, 50, 57–58
Monomorium
destructor, 108
gracillimum, 109
latinode, 107
minimum, 109, 114
pharaonis, 107–108, 114
spp., 108, 113–114
Monotoma spp., 46, 65–66
Muridae, 163, see also specific genera and species
Murmidius
ovalis, 49, 54
segregatus, 49
spp., 46, 54
Mus
musculus, 160, 163, 169, 180–182, 191
spp., 173, 205, 209
Musca
domestica, 73, 83–84, 120
sorbens, 83
spp., 77
Muscidae, 73–74, 77, 79, 83–84, see also specific genera and species
Muscina
spp., 84
stabulans, 73, 84
Muscinae, 79, see also specific genera and species
Mustela
nivalis, 190
siberica, 190
Mutillidae, 106, see also specific genera and species
Mycetaea
hirta, 65
subterranea, 50, 65
Mycetoglyphus spp., 128
Mycetophagidae, 6, 48, 51, 62, see also specific genera and species
Mycetophagus quadrigutatus, 51
Mymaridae, 117, 119, see also specific genera and species
Mymar spp., 117, 119
Myotis
lucifugus, 186
spp., 162, 201
Myrmeciinae, 108, 113, see also specific genera and species
Myrmicinae, 108–110, 113–115, see also specific genera and species

N

Nanosella spp., 63
Necrobia
ruficollis, 49, 55
rufipes, 44, 49, 55
spp., 55–56, 129
violacea, 49, 55
Nemapogon granella, 94, 98, 102
Nemesiidae, 137, 141, 145, see also specific genera and species
Nephila
clavipes, 137
maculata, 137, 147, 218
spp., 147–148, 223
Nephilengys cruentata, 137, 147
Nesticidae, 137, 142–143, 145, 148, see also specific genera and species
Nesticodes rufipes, 137, 148
Nesticus cellulanus, 137, 145
Neuroptera, 119–120, see also specific genera and species
Neviamyrmex spp., 111
Nicoletidae, 31, see also specific genera and species
Niptus
holoeucus, 51
hololeuca, 120
Nitidula
bipunctata, 51
spp., 62

Nitidulidae, 44, 46, 51, 62, see also specific genera and species
Noctuidae, 218, see also specific genera and species
Nothomyrma spp., 115
Nothomyrmeciinae, 115, see also specific genera and species
Notoxus monodon, 52
Nuctenea umbratica, 137, 143
Nycteribidae, 88, see also specific genera and species

O

Oecobiidae, 137, 140–141, 146, see also specific genera and species
Oecobius
  annulipes, 137
  parietalis, 137, 146
Oecophoridae, 99, see also specific genera and species
Oecophylla
  smaragdina, 113, 222–223
  spp., 112–113
Oestridae, 72, see also specific genera and species
Oligota granaria, 51
Oonipidae, 137, 141, 146, 148, see also specific genera and species
Oonops domesticus, 137
Ophyra spp., 84
Opiliones, 136, 152, see also specific genera and species
Orchestina saltitans, 137, 146
Orthocis auriculariae, 49, 55
Orthoperus spp., 56
Orthoptera, 120, 217, see also specific genera and species
Oryzaephilus
  mercator, 50, 57–58
  spp., 47, 57–58, 86
  surinamensis, 43, 50, 57–58, 115, 120, 266

P

Paisochelifer callus, 152
Palorus
  ratzburgi, 51
  spp., 64, 65
  sybdeoressys, 52
Paracharon spp., 154
Paralimna spp., 82
Paralipsa galaris, 94, 98, 101
Paraphrynus spp., 154
Parasitidae, 124, 132, see also specific genera and species
Paratillus carus, 55
Paratrechina
  longicornis, 109, 113
  spp., 109, 112–113
Paregle spp., 80
Passandrinae, 45, 56, see also specific genera and species
Periplaneta
  americana, 33–34
  australasiae, 34
  fulignosa, 34
Peromyscus
  leucopus, 163, 183, 191
  spp., 162–163, 172–173, 205, 209
Phaenicia spp., 73, 81
Pharaxonotha kirschi, 50, 60–61
Pheidole
  megacephala, 114
  spp., 108–109, 113–114
Pheidologeton spp., 108, 113
Phidippus formosus, 137, 147
Philodromidae, 137, 141, 146, see also specific genera and species
Philodromus
  rufus, 137
  spp., 146
Pholcidae, 137–138, 141, 146, see also specific genera and species
Pholcus phalangioides, 137, 146
Phoneutria fera, 137, 144
Phoridae, 73–74, 77, 79, 84, see also specific genera and species
Phormia regina, 81
Phrynus spp., 154
Phthiraptera, 36, see also specific genera and species
Phycitinae, 95–98, see also specific genera and species
Physocyclus globosus, 137
Phytoseiidae, 124, 132, see also specific genera and species
Piophila casei, 73, 84
Piophilidae, 73–74, 79, 84, see also specific genera and species
Pipistrellus hesperus, 186
Pityohyphantes costatus, 137, 145
Platydema
  americanum, 52, 64
  ruficorne, 52
Platygasteridae, 117, 120, see also specific genera and species
Platystromatidae, 74, 78–79, 85, see also specific genera and species
Plodia interpunctella, 94, 96, 98, 101, 115
Poecilosomella spp., 87
Pollenia rudis, 81
Polyphagidae, 33, see also specific genera and species
Polyporaceae, 55, see also specific genera and species
Ponerinae, 108, 110, 115, see also specific genera and species
Prenolepis spp., 109
Prostephanus truncanus, 49, 53
Pselaphochernes spp., 152
Pseudolynchia canariensis, 88
Pseudomonas aeruginosa, 107
Pseudomyrmecinae, 108, 115, see also specific genera and species
Pseudomyrmex spp., 115
Pseudoscorpiones, 136, 151–152, see also specific genera and species

Psilochorus simoni, 137
Psocatropus microps, 37
Psocoptera, 29, 35–37, 152, see also specific genera and species
Psoquilla spp., 37
Psoquillidae, 36–37, see also specific genera and species
Psychoda spp., 77, 85
Psychodidae, 74, 77–78, 84–85, see also specific genera and species
Psyllipsocidae, 36–37, see also specific genera and species
Psyllipsocus spp., 37
Pteromalidae, 118, 120, 201, see also specific genera and species
Ptiliidae, 46, 51, 62–63, see also specific genera and species
Ptiliola spp., 63
Ptinidae, 47, 51, 63, 120, see also specific genera and species
Ptinus
  fur, 120
  ocellus, 51
  spp., 63
  tectus, 120
  villiger, 51
Pullimosina spp., 87
Pycnosoma spp., 120
Pyemotes tritici, 124
Pyemotidae, 124, 132, see also specific genera and species
Pygmephoridae, 124, 132, see also specific genera and species
Pygmephorus spp., 124
Pyralidae, 97–98, 104, see also specific genera and species
Pyralinae, 97, see also specific genera and species
Pyralis farinalis, 94, 97, 101
Pyroderces rileyi, 94, 99
Pyroglyphidae, 7, 124, 128–129, see also specific genera and species
Pyroglyphus spp., 128

R

Rattus
  exulans, 160, 182
  norvegicus, 160, 163, 180, 182, 191
  rattus, 160, 183
  spp., 169, 173, 175, 205, 209
Rhizoglyphus spp., 128
Rhizophagidae, 46, 65, see also specific genera and species
Rhodesiella spp., 82
Rhynchophorus phoenicis, 223
Rhypoglyphus
  indicus, 128
  spp., 127–128
  trigona, 128
Rhypsocus spp., 37
Rhyzopertha dominica, 44, 49, 53

S

Salmonella
  enteritidus, 160
  newport, 160
  spp., 72, 107–108, 160–163, 177, 218, 273
  typhimurium, 160
Salticidae, 137, 141, 146–147, see also specific genera and species
Salticus scenicus, 137, 146, 147
Sarcophaga
  haemorrhoidalis, 73, 85
  spp., 78, 120
Sarcophagidae, 73–74, 78–79, 84–85, see also specific genera and species
Saturniidae, 220, see also specific genera and species
Scathophaga stercoraria, 85
Scathophagidae, 74, 78–80, 85, see also specific genera and species
Scathophagus spp., 78
Scatopse
  notata, 85
  spp., 85
Scatopsidae, 73–74, 78, 85–86, see also specific genera and species
Scelionidae, 116–117, 120, see also specific genera and species
Scelioninae, 116–117, 120, see also specific genera and species
Scenopenidae, 74, 78–79, 86, see also specific genera and species
Scenopinus
  fenestralis, 86
  glabrifrons, 86
  spp., 78
Schistocerca gregaria, 217
Scholastes spp., 78, 85
Sciaridae, 74, 78, 86, see also specific genera and species
Sciuridae, 162, see also specific genera and species
Sciurus spp., 187, 192
Scolytidae, 47, 51, 63–64, see also specific genera and species
Scorpiones, 136, 152–153, see also specific genera and species
Scotophaeus
  blackwalli, 137, 145
  spp., 144
Scutigera coleoptrata, 154
Scutigeridae, 154, see also specific genera and species
Scutigeromorpha, 136, 154, see also specific genera and species
Scytodes thoracica, 137, 147
Scytodidae, 137, 141, 147, see also specific genera and species
Selenopidae, 137, 141, 147, see also specific genera and species
Selenops spp., 137, 147
Sepsidae, 74, 77, 79, 86, see also specific genera and species

Sepsis spp., 77, 86
Sericoderus spp., 56
Setomorpha rutella, 94, 99, 102
Shigella
flexner, 108
spp., 72
Sicariidae, 137, 140–141, 147, see also specific genera and species
Silvaninae, 45, 56, see also specific genera and species
Silvanoprus
cephalotes, 57
longicollis, 57
Silvanus
lewisi, 57
proximus, 57
spp., 50
Sitophilus
granarius, 50, 58–59, 120
linearis, 50, 58–59
oryzae, 42, 50, 58–59, 115, 120
spp., 42–43, 59
zeamais, 42, 50, 58
Sitotroga cerealella, 57, 94–95, 99–100
Solenopsis
geninata, 107–108
molesta, 114
spp., 108, 113–114
Solifugae, 136, 152, see also specific genera and species
Spalangia spp., 118, 120
Spermophora
senoculata, 137
spp., 146
Sphaeroceridae, 73–74, 77, 79, 87, see also specific genera and species
Spinibdella spp., 124
Staphylinidae, 31, 46, 51, 64, see also specific genera and species
Staphylococcus
aureus, 72
spp., 107
Steatoda
bipunctata, 137, 148
grossa, 137, 148
spp., 148
Stegobium paniceum, 49, 52, 120
Stomoxys
calcitrans, 73, 84, 120
spp., 84
Stratiomyidae, 73–74, 78, 87, see also specific genera and species
Streblidae, 88, see also specific genera and species
Streptococcus spp., 107
Suidae, 163, see also specific genera and species
Suidasia
nesbitti, 129
pontifica, 124, 129–130
spp., 127–129
Suidasiidae, 124–126, 128–129, see also specific genera and species
Suncus
murinus, 161–162, 169, 171, 184–185, 192
spp., 171
Supella longipalpa, 33
Sylvicola
fenestralis, 73, 88
spp., 78
Symphyta, 119, see also specific genera and species
Synthesiomyza nudiseta, 84
Syrphidae, 73–74, 77, 79, 82, 87, see also specific genera and species
Systole albipennis, 118, 120

T

Tachinidae, 32, see also specific genera and species
Tamarindus indica, 48
Tapinoma
melanocephalum, 111
sessile, 111
spp., 109, 111
Tarsonemidae, 124, 132, see also specific genera and species
Tarsonemus granarius, 124
Tegenaria
domestica, 137, 142
gigantea, 137, 142
parietina, 137
saeva, 142
spp., 140, 142
Teichomyza fusca, 82
Telenominae, 117, 120, see also specific genera and species
Tenebrio
molitor, 52
obscurus, 52
spp., 47, 65
Tenebrionidae, 43–45, 47–48, 51–52, 64–65, see also specific genera and species
Tenebroides mauritanicus, 52, 65
Tephritidae, 88, see also specific genera and species
Tetragnathidae, 137, 147–148, 218, 223, see also specific genera and species
Tetramorium
caespitum, 109, 115
spp., 109, 114–115
Tetranychidae, 124, see also specific genera and species
Tetraponera spp., 115
Tetrastichinae, 118, see also specific genera and species
Thaneroclerus buqueti, 48–49, 55–56
Thaumatomyia spp., 82
Theraphosidae, 137, 141, 145, 148, see also specific genera and species
Theridiidae, 137, 140, 142–143, 145, 148, see also specific genera and species
Theridion spp., 148
Thermobia domestica, 30–31
Thes bergrothi, 51
Thomisidae, 137, 141, 146, 148, see also specific genera and species
Thorictodes heydeni, 50, 60

Thylodrias
contractus, 50
spp., 59
Thyreophagus spp., 127–128
Thysanoptera, 119, see also specific genera and species
Thysanura, 29–31, see also specific genera and species
Tinea
cloacella, 98–99
infimella, 99
pallescentella, 94, 98
pellionella, 94, 98
ruricolella, 99
salella, 99
spp., 98, 102
Tineidae, 98–99, 152, see also specific genera and species
Tineola bisellliella, 94, 98, 102
Toracochaeta spp., 87
Tortricidae, 99, see also specific genera and species
Toxoplasma gondii, 162
Trabutina manipara, 217
Triaarthria spp., 32
Tribolium
castaneum, 41, 43, 52, 65
confusum, 52, 65, 115
destructor, 52
madens, 52
spp., 43–44, 64–65
Trichogramma spp., 118–119
Trichogrammatidae, 117, 119, see also specific genera and species
Trichophaga tapetzella, 94, 99, 102
Trigona spp., 121, 219
Trigonogenius globulum, 51, 63
Troctomorpha, 35, see also specific genera and species
Trogiidae, 36–37, see also specific genera and species
Trogium spp., 37
Trogoderma
glabrum, 50
granarium, 50, 59
inclusum, 50
spp., 44, 59–60, 274
stermale, 50
variabile, 50
Trogossitidae, 48, 52, 65, see also specific genera and species
Tydeidae, 124, 132, see also specific genera and species
Tydeus interruptus, 124, 132
Typhaea stercorea, 51, 62
Tyroborus spp., 127–128
Tyrolichus spp., 127–128
Tyrophagus
putrescentiae, 7, 124–125, 128
spp., 8, 127–128

U

Uloboridae, 137, 141–142, 144, 148, see also specific genera and species
Uloborus
glomosus, 137, 148
spp., 148
Urophorus humeralis, 51
Uropodidae, 124, 132, see also specific genera and species

V

Venturia canescens, 115
Vespertiliontidae, 162, 170, 201, see also specific genera and species
Vespidae, 121, see also specific genera and species
Vespula
germanica, 121
spp., 121
Vibrio cholera, 72, 177
Vitula
edmandsii, 98
serratilineella, 94, 98
Volvariella volvacea, 60

W

Winterschmidtiidae, 124, 132, see also specific genera and species
Withiidae, 152, see also specific genera and species
Withius niger, 152

X

Xenopysylla cheopis, 161
Xylocoris flavipes, 115
Xysticus spp., 137, 148

Y

Yersinia pestis, 161

Z

Zabrotes subfusciatus, 53–54